Non-oxide and Heteroanionic Materials

Inorganic Materials Series

For a list of titles in this series see:
rsc.li/inorg-mater-series

How to obtain future titles on publication:
A standing order plan is available for this series. A standing order will bring delivery of each new volume immediately on publication.

For further information please contact:
Book Sales Department, Royal Society of Chemistry, Thomas Graham House, Science Park, Milton Road, Cambridge, CB4 0WF, UK
Telephone: +44 (0)1223 420066, Fax: +44 (0)1223 420247
Email: booksales@rsc.org
Visit our website at books.rsc.org

Non-oxide and Heteroanionic Materials

Edited by

Duncan H. Gregory
University of Glasgow, UK
Email: Duncan.Gregory@glasgow.ac.uk

ROYAL SOCIETY OF CHEMISTRY

Inorganic Materials Series No. 16

Hardback ISBN: 978-1-83916-181-0
PDF ISBN: 978-1-83916-592-4
EPUB ISBN: 978-1-83916-593-1
Print ISSN: 2472-3819
Electronic ISSN: 2472-3827

A catalogue record for this book is available from the British Library

The Royal Society of Chemistry is a charity, registered in England and Wales, Number 207890, and a company incorporated in England by Royal Charter (Registered No. RC000524), registered office: Burlington House, Piccadilly, London W1J 0BA, UK, Telephone: +44 (0) 20 7437 8656.

For further information see our website at www.rsc.org

For general enquiries, please contact books@rsc.org

For EU product safety enquiries, please email books@rsc.org or contact Royal Society of Chemistry Worldwide (Germany) GmbH, Römischer Hof, Unter den Linden 10, 10117 Berlin.

Printed in the United Kingdom by CPI Group (UK) Ltd, Croydon, CR0 4YY, UK

Series Preface

Materials chemistry has grown hugely in the past few decades and continues to evolve rapidly. This is especially the case since the subject now finds itself central to many of the major challenges that face global society, such as those concerning energy and the environment; but applications of materials are also found in many everyday situations, with ongoing developments in technology around electronics and the miniaturisation of devices. Although the study of materials is a highly interdisciplinary topic, the role of innovative, fundamental chemistry continues to lie at the heart of the discovery of new materials, the understanding of their atomic-scale structure and the relationship of structure to a useful property.

With the *Inorganic Materials Series*, our aim is to provide a series of themed volumes, each reflecting the diversity of the topic as it continues to develop over the coming years, with many chosen to relate closely to emerging practical applications. The use of the title *Inorganic Materials* is by no means limiting, but instead emphasises the study of the chemistry of all elements in the Periodic Table and how they impact the development of new materials. The chapters in each book will have a pedagogical flavour with a target audience from first-year postgraduate student upwards, but with wide subject coverage ranging from continuous inorganic solids, to molecular and soft matter systems. We will use expert, active researchers as editors and contributors to provide an up-to-date perspective of their field. At the same time our aim is to provide an international perspective, so to reflect the diversity and interdisciplinarity of the now very broad area of the subject.

Duncan W. Bruce
Elizabeth A. Gibson
Dermot O'Hare
Richard I. Walton

Inorganic Materials Series
Non-oxide and Heteroanionic Materials
Edited by Duncan H. Gregory
A catalogue record for this book is available from the British Library

Preface

Inorganic solid-state and materials chemistry has long been dominated by, and is synonymous with, the chemistry of oxides. To a large extent, this dominance is a consequence of the natural conditions on Earth and its thermodynamics. Oxygen forms a large part of the gaseous atmosphere of the planet and as a molecule dioxygen is reactive, with a propensity to form the divalent oxide anion, O^{2-}. Many of the principles of solid-state chemistry have been established by considering the reactions of one or more elements with oxygen to form binary and higher oxides, which can then be combined together in synthesis (often in the solid state). The design of oxide materials has relied on the relationships between synthesis, structure and properties and the modification and optimisation of materials performance is commonly achieved by the substitution, addition or removal of metal or metalloid cations within the constant environment of an anionic oxide sublattice.

The relative ease with which oxides can be synthesised (and handled) and the ubiquity of oxide compounds (as or from minerals) has perpetuated the emphasis on oxides, yet despite this, many of the most fascinating solids and important materials do not contain oxygen. Some of these are naturally-occurring, but over time scientists have developed increasingly reliable and sophisticated ways to prepare *non-oxide* materials. In parallel, the adoption of advanced characterisation techniques has enabled us to learn more and more about non-oxides such that compounds previously considered 'exotic' have become progressively better understood. Families of non-oxides have become well-established with many as well-known as their oxide counterparts. Moreover, materials design strategies now embrace the chemistry of the anion to a greater extent than at any time previously and the tuning of properties by the (partial) replacement of anions (whether oxide or others) is a central theme of *heteroanionic* materials chemistry.

Inorganic Materials Series
Non-oxide and Heteroanionic Materials
Edited by Duncan H. Gregory
A catalogue record for this book is available from the British Library

This book, *Non-oxide and Heteroanionic Materials*, assembles many of the most influential classes of inorganic solids beyond the traditional boundaries of oxide chemistry, providing important examples of state-of-the-art and emerging materials within each category. Chapter 1 begins with the familiar territory of Group 16 but with an emphasis on the elements below oxygen in the Periodic Table. This chapter considers some of the vast expanse of chalcogenides currently known to inorganic chemists, first establishing some of the 'ground rules' of solid-state chalcogenide chemistry before discussing layered dichalcogenides as archetypal examples of fundamentally important compounds that have developed from important bulk materials to cutting edge nanomaterials. The multitude of methods for the synthesis of chalcogenides are reviewed with the suitability of each to certain compound- and structure types analysed. Finally, a selection of some of the most significant contemporary materials is presented, with an emphasis on sustainable energy applications and how these are influenced by the often singular or unusual properties of sulfides, selenides and tellurides. Chapter 2 covers Group 15 and considers the pnictides from phosphorus to bismuth. The contrast to the chalcogenides is striking and the chapter discusses the unusual and intriguing crystal chemistry of this group of compounds while also discussing in some detail how such compounds can be isolated and stabilised through careful and controlled synthesis. The atypical structure and bonding of pnictides also gives rise to a wide range of interesting electronic and magnetic phenomena from superconductors to topological insulators and semi-metals. Useful chemical as well as physical properties exist however; emerging catalytic and electrochemical applications are among those currently being studied.

Chapter 3 discusses the materials chemistry of carbides, a grouping which represents the most widely studied of the tetrelides of Group 14. From recognised beginnings including both molecular and interstitial compounds, much recent interest in carbides has stemmed from the discovery of MXenes. This chapter provides a detailed overview of these fascinating layered nanomaterials and their origins from the well-established MAX phases. In the space of just over a decade, studies of these two-dimensional (2D) materials have uncovered numerous intriguing and useful physical properties which have been put to use in applications including sensors, actuators, light-emitting diodes (LEDs) and membranes. In Chapter 4 the spotlight falls on solid state halides and hydrides (often themselves considered as pseudohalides). This chapter concentrates especially on applications and the enormous impact that both sets of compounds are having in terms of future sustainable energy. Hydrides are inevitably central to the implementation of a hydrogen economy and inherently linked to the generation, purification and storage of hydrogen. With hydrogen as a potential clean fuel for use in both electricity generation and heating, applications as catalysts (for example in fuel cells) and as thermal storage materials are also potentially very important. Coupled with the ascendency of halide perovskites in solar cells plus the use of both sets of materials as components in secondary batteries, hydrides and halides are arguably at the forefront of future energy conversion and storage technologies.

Finally, the book concludes with an in-depth view of the concept of heteroanionic materials (HAMs) in Chapter 5. In principle, there are endless possibilities when one considers the combinations of simple anions in the solid state. This chapter considers one particularly enticing aspect of the area; namely the interface between oxide and non-oxide chemistry. With a focus particularly on combinations of oxide with its nearest neighbours (the anions of nitrogen and sulfur), the chapter collates some of the many intriguing properties—and potential applications—that can arise when these anions are brought together within a single crystalline structure. Just some of the vast array of structures and bonding permutations are considered and the connections between these structures and properties as diverse as superconductivity and photocatalysis are explained.

Although the exploration of non-oxide materials chemistry is a relatively young venture, there is a rich history of these solids and the catalogue of new and beguiling compounds is rapidly expanding. The aim of this book is to capture this burgeoning area in five distinctive and representative chapters. While this volume by no means covers all the many types of non-oxides (some of which might deserve a tome to themselves) or provides a comprehensive explanation of every one of their countless features, I hope that it serves as a valuable and satisfying introduction to the field.

Duncan H. Gregory

Contents

Inorganic Materials Series
Non-oxide and Heteroanionic Materials
Edited by Duncan H. Gregory
A catalogue record for this book is available from the British Library

CHAPTER 1

Contemporary Chalcogenide Materials Chemistry

GUANG HAN[a*], YAO CHEN[a], XIAOYUAN ZHOU[b] AND DUNCAN H. GREGORY[c*]

[a]College of Materials Science and Engineering, Chongqing University, Chongqing 400044, P.R. China; [b]College of Physics, Chongqing University, Chongqing 401331, P.R. China; [c]WestCHEM, School of Chemistry, University of Glasgow, Glasgow, G12 8QQ, UK
*E-mail: guang.han@cqu.edu.cn, Duncan.Gregory@glasgow.ac.uk

1.1 Structures and Properties of Metal Chalcogenides

Chalcogenides, as the name suggests, are chemical compounds containing at least one chalcogenide anion and at least one electropositive element. Generally, oxides are considered separately from chalcogenides—sulfides, selenides, tellurides (and polonides)—because of their quite different chemical properties. In addition, metal chalcogenides, notably containing selenium or tellurium as the chalcogen species, are less known and have not been systematically studied to the same extent as the corresponding oxides.

Here, we focus on compounds consisting of metal cation(s) and three important Group 16 chalcogens (*i.e.*, sulfur, selenium and tellurium), which are defined as metal chalcogenides. Considering binary metal chalcogenides as an example, the relevant chemical formulae can be written as M_aX_b, where

Inorganic Materials Series
Non-oxide and Heteroanionic Materials
Edited by Duncan H. Gregory
A catalogue record for this book is available from the British Library

M is a metallic element and X = S, Se, Te. Metal chalcogenides constitute a tremendous number of chemical compounds with a huge library of component elements drawn from across the periodic table. They form with a vast variety of crystal structures and exhibit an extraordinary array of physical and chemical properties that range from intriguing optical, electrical, thermal, and electrochemical properties to novel magnetic, superconducting, and topological properties. Therefore, they have been investigated and utilised in various fields, such as photovoltaics, thermoelectrics and many aspects of solid-state electrochemistry among others.

1.1.1 Basic Classification of Metal Chalcogenides

Chalcogens possess a rich metal chemistry both in the solid state and in molecular compounds, which can be ascribed to their ability to catenate and bond with multiple metal centres.[1–3] Hence, metal chalcogenides involve numerous compounds that range from straightforward stoichiometric metal-based sulfides, selenides, and tellurides to complex compounds or solid solutions containing diverse metal and/or chalcogen elements with a variety of valence states and proportions.[3–6] Chalcogens tend to form electron-pair compounds where the chalcogens exhibit the divalent state with two lone pairs of electrons.[1]

The chemistry of binary metal chalcogenides is so rich because many of the metal and metalloid elements that bond with a specific chalcogen can form several compounds of different stoichiometry and occasionally the series of such compounds (for a given metal) can be extensive.[1] An example is indium selenide, which exists in the form of InSe, In_2Se_3, In_3Se_4, In_4Se_3, In_6Se_7 (note InSe and In_2Se_3 themselves also form as different phases—polytypes—with different structures).[7] Although complicated structures are not infrequent, generally, the majority of binary metal chalcogenides are derived from basic crystal structure types and can be easily described accordingly.[1,8] Three-dimensional crystal structure types, usually rock salt (NaCl), hexagonal NiAs, zinc blende and wurtzite structures, as well as two-dimensional layered lattice types related to the trigonal CdI_2, constitute the structures of the majority of such chalcogenides.[1,8] Noteworthy is that simple selenides and tellurides can often share the same crystal structure types as the corresponding sulfide analogues, although important variations can exist, especially for tellurides.[1,9] For example, while PbX (X = S, Se, Te) compounds each form with the rock salt (NaCl-type) structure,[10] SnS and SnSe crystallise in an orthorhombic structure whereas SnTe exists with the same rock-salt structure as its lead congener.[11]

Alkali and alkaline earth metal monochalcogenides are colourless, salt-like compounds that are often soluble in water (especially alkali metal monochalcogenides).[12] Alkali metal chalcogenides usually crystallise in the antifluorite structure, while those of the alkaline earth metals crystallise with the cubic NaCl structure.[12] By contrast, transition metal chalcogenides possess various stoichiometries and numerous crystal structures, and the most important and widely studied ones are simple stoichiometries, such as metal : chalcogen atomic ratios of 1 : 1 and 1 : 2.[12] Several metals, principally the early transition metals (*e.g.*, Groups 4–7) can also form corresponding trichalcogenides, such as $NbSe_3$, $ZrSe_3$ and $ZrTe_3$.[1,12,13]

Transition-metal chalcogenides can demonstrate highly covalent and non-ionic behaviours typified by the zinc blend or wurtzite crystal structures in 1 : 1 compounds of the later transition metals (*e.g.*, ZnS).[1] Many transition metals combine with chalcogens to yield MX_2 dichalcogenides with a relatively precise 1 : 2 stoichiometry, which crystallise in either two-dimensional or three-dimensional structures.[1] About two thirds of MX_2 chalcogenides form layered structures, especially for the dichalcogenides of nearly all of the early transition metals in Groups 4–7 (with the exception of Mn).[1] The layered dichalcogenides exhibit a variety of electrical transport properties, which range from insulators (*e.g.*, HfS_2) through semiconductors (*e.g.*, MoS_2) and semi-metals (*e.g.*, WTe_2) to metals (*e.g.*, VSe_2).[1] The M–X units in layered MX_2 typically adopt either a trigonal prismatic coordination similar to that in MoS_2 or an octahedral coordination as in CdI_2.[1,12] The strong chemical bonding between metal atoms and chalcogens is in striking contrast to the weak chalcogen–chalcogen bonding between the layers. Generally, transition metal atoms in MX_2 contribute four electrons to form the bonding state; thus, the metals and chalcogens are endowed with the formal charges of +4 and −2, respectively. For example, the disulfides containing early transition metals such as Ti, V, Mo and W possess valence states of 4, bonding with two chalcogen dianions.[12] By contrast to the above examples, the disulfides containing late transition metals such as Fe, Co and Ni (and including Mn) generally adopt the pyrite (FeS_2, distorted NaCl type) or related marcasite (FeS_2, orthorhombic; space group *Pnnm*) structural motif.[12] Both structure types contain the disulfide S_2^{2-} anion.

Group 13 metal elements (*e.g.*, Al, Ga, In) can combine with chalcogens to form metal chalcogenides with multiple stoichiometries,[7,14–16] which is especially notable for indium selenides.[7,17] For example, with increasing temperature, In_2Se_3 transforms from the room-temperature form, α-In_2Se_3[18,19] to the β-In_2Se_3,[20] γ-In_2Se_3[18,21] and δ-In_2Se_3[18,21] phases consecutively. Specifically, α-In_2Se_3 itself can crystallise in both hexagonal (2H) and rhombohedral (3R) layered structures, with different stacking of Se–In–Se–In–Se layers in which In is both tetrahedrally and octahedrally coordinated by Se.[18,19] In_4Se_3 is also layered but has a very different crystal structure to α-In_2Se_3 in which $(In_3)^{5+}$ clusters are bonded to Se ions along the *bc* plane, while the remainder of the In atoms form quasi one-dimensional chains.[22,23] The relative newcomer, In_3Se_4, (synthesised *via* solvothermal methods) also crystallises with a layered rhombohedral structure, here consisting of Se–In–Se–In–Se–In–Se layers and lattice parameters of a = 3.964(2) Å and c = 39.59(2) Å.[24–26] The chalcogenides of Group 13 metals have found applications in thermoelectric conversion devices,[7,23,27,28] photodetectors,[29–31] Li-ion batteries[32] and in phase-change memory.[33]

Group 14 and 15 elements bond with chalcogens to form MX (M = Ge, Sn, Pb, X = S, Se, Te), MX_2 (M = Ge, Sn, X = S, Se) and M_2X_3 (M = Sb, Bi, X = S, Se, Te) compounds. Specifically, (i) GeS, GeSe, SnS, and SnSe crystallise in orthorhombic layered structures at room temperature,[34] while PbX (X = S, Se, Te) compounds possess a rock salt structure;[8] (ii) Ge, Sn-based MX_2 chalcogenides form layered crystal structures, containing X–M–X covalently bonded layers separated in the third dimension by van der Waals forces;[35–37] (iii) Bi_2Se_3, Bi_2Te_3, and Sb_2Te_3 crystallise in rhombohedral, layered structures, while Sb_2S_3, Sb_2Se_3, and Bi_2S_3 have orthorhombic, non-layered structures.[38–42] The Group 14 and 15 chalcogenides are typically semiconductors with a diverse range of band gaps, which demonstrate potential applications in thermoelectric conversion and photodetectors as well as solar cells and lithium/sodium-ion batteries.[1,43,44]

Ternary metal chalcogenides, containing two different metallic elements or two types of chalcogens, comprise numerous classes of materials. Taking AB_2X_4 (A, B = metallic elements, X = S, Se, Te) as one exemplar stoichiometry, these chalcogenides adopt a variety of structures, exhibit intriguing properties and find widespread applications.[45–47] Particularly, chalcogenide spinels (AB_2X_4), such as chalcochromites MCr_2X_4 (M = Ba, Cd, Co, Zn, Fe, Cu, Hg, Mn, Ni), thiocobaltite $CuCo_2S_4$, thiorhodites $M_xRh_{3-x}S_4$ (M = Cu, Co, Fe), and thioaluminates MAl_2S_4 (M = Zn, Cr) often combine a distinctive tranche of semiconducting, magnetic and optical properties.[48] Among other ternary chalcogenides of note, $CuMX_2$ (M = Ga, In; X = S, Se, Te) chalcopyrites are well-known semiconductors with high absorption coefficients, which have strong potential for utilisation as light absorbing materials in photovoltaic devices.[49–51] Cu_2SnSe_3, a typical ternary diamond-like compound,[52] has been regarded as a good thermoelectric material,[53] while CuAgSe[54] and Cu_5FeS_4 (bornite)[55,56] also behave as promising thermoelectric materials, principally arising from inherently low lattice thermal conductivity. Considering examples of further compositional complexity, quaternary metal chalcogenides, such as Cu_2ABX_4 (A = Zn, Mn, Cd, Co, Fe, Hg; B = Sn, Ge; X = S, Se), have attracted intensive attention due to their promising applications in both photovoltaics and thermoelectrics.[1,44,57–59]

Aside from the many stoichiometric chalcogenides such as those highlighted above, solid solutions are common among structurally-related chalcogenide materials and members of such series (such as Bi_2Se_3–Bi_2Te_3 and Sb_2Te_3–Bi_2Te_3)[60,61] may achieve extraordinary properties that do not exist in either of the individual binary end members.[1] Combinations of different metal chalcogenides can be of significant scientific and technological value since their functional and mechanical properties can be modulated accordingly over a wide range through compositional regulation.[1]

1.1.2 Structural Characteristics of Layered Metal Chalcogenides

One of the important characteristics of metal chalcogenides is that they can exist in various structural configurations, such as amorphous, crystalline and nanocrystalline forms.[12] Obviously, atomic radii, metal-to-chalcogen proportion, and valence electron concentration play decisive roles in determining the structural types and features.[1] Here, we focus on the structural characteristics of typical metal chalcogenides with layered structures.

The development of layered metal chalcogenides has a long-standing history. For example, the crystal structure of the well-known material MoS_2 was first determined by Dickinson and Pauling as early as 1923.[62] Much later, Feldman *et al.* first synthesised MoS_2 as nanotubes (in large quantities with uniform morphology) by utilising the gas-phase reaction between MoO_{3-x} and H_2S in a reducing atmosphere at elevated temperatures (800–950 °C).[63] Various nanostructured layered metal chalcogenides have subsequently appeared within researchers' sights, and the momentum of this research built considerably following the successful exfoliation of two-dimensional monolayer graphene by Novoselov *et al.* in 2004.[64] These two-dimensional chalcogenides, possessing finite band gaps in a departure from graphene's zero band gap, are currently at the forefront of research in the disciplines of physics, chemistry, and materials science.[65] Layered metal chalcogenides are epitomised by stacking stratiform structures composed of covalently bonded two-dimensional planes supported by interplanar van der Waals forces. These materials have rich chemical compositions and exhibit various exotic, anisotropic physicochemical phenomena such as tuneable energy gaps, superconductivity, magnetism, topological properties and charge density waves.[66–68] In addition, layered chalcogenides can be exfoliated relatively easily into materials of N layers ($N \geq 1$) with different physical and chemical characteristics.[69–71] For example, several layered metal chalcogenides (*e.g.*, MoS_2, $MoSe_2$, WS_2 and WSe_2) exhibit a transformation from indirect-band gap semiconductors in the bulk to direct-band gap semiconductors in monolayer form.[65] Such exfoliated chalcogenide layers can be widely applied in the fields of electronics, photoelectronics, energy and catalysis.[69,70]

The electronic structures and functional properties of layered metal chalcogenides can be manipulated through various strategies.[13,72] Other unknown and novel properties are also expected to be discovered by following these strategies.[68,73] Generally, the 'standard' approaches that have been developed over time for modifying bulk materials fall into the following categories: (i) defect engineering, the manipulation of zero-dimensional (point), one-dimensional (dislocation) and two-dimensional (stacking) defects;[74–76] (ii) intercalation, the insertion of guest species such as metal ions and/or molecules between the layers of metal chalcogenide hosts;[77,78] (iii) substitution, the chemical substitution of constituent elements with other (extrinsic) elements, including cation substitution and anion substitution;[79,80] (iv) layer-number tuning, the exfoliation of bulk layered chalcogenides into specific numbers of layers, *i.e.*, single, double, and multiple layer(s);[70,71] and (v) interfacial interaction, the formation of hetero-interfaces and deposition of chalcogenides on a substrate surface.[81,82]

Layered metal chalcogenides present a great variety of compositions and crystal structures, as exemplified by the dichalcogenides of transition and Group 13–15 metals. Generally, the structural configuration of layered chalcogenides is that the bonding of metal cations and chalcogens leads to the formation of independent monolayers, and the layers are stacked to form stratiform structures *via* van der Waals interactions. For different layered chalcogenides, the structural configuration of individual monolayers and the stacking pattern between layers can be different.

Here, we will take MX_2 (M = transition metal elements, X = S, Se, Te) as examples to illustrate the variance in crystal structure in detail. Each independent monolayer of MX_2 has a 'sandwich' structure consisting of upper and lower layers of X atoms and a middle layer of M atoms in the form of X–M–X stacking.[65,83] The basic MX_2 'sandwich' has two different types: (i) MX_6 trigonal prisms, in which the upper and lower chalcogen layers both form equilateral triangles in the *ab* plane, which are aligned directly above one-another along the *c* axis with the transition metal between them (*i.e.*, located in the centres of each trigonal prism); (ii) MX_6 octahedra, where the upper chalcogen triangle is rotated by 180° in the *ab* plane with respect to the lower (and again M is located between the respective triangles).[83,84] We use three letters to represent the positions of the three layers in the 'sandwich' structure from top to bottom, with capital letters representing chalcogens and lowercase letters representing transition metals. Hence, the trigonal prismatic and the octahedral coordination environments (layers) can be expressed as AbA (upper and lower X layers are similarly aligned) and AbC (upper and lower X layers are differently aligned), respectively.

The adjacent MX_2 layers are connected in the third dimension *via* weak van der Waals interactions. Due to the different 'sandwich' units and the ways in which the layers are stacked, MX_2 exist in a variety of related structural types. These structures are classified as follows: using Arabic numbers to represent the number of independent monolayers contained within a unit cell, and the letters T, H and R to denote either the trigonal, hexagonal, or rhombohedral crystal systems, respectively, of the unit cell.[65,83] By considering this formalism, we can introduce the three major crystal structure types, that is, 1T, 2H and 3R (Figure 1.1). The 1T type structure contains 'sandwich' layers with octahedral configuration that stack with the sequence ...AbC AbC..., possessing a trigonal rotational symmetry.[65,83] As for the 2H type, the unit cell contains two 'sandwich' layers with trigonal prismatic configuration; thus in each 'sandwich' the X atoms in the lower layer are vertically aligned with the X atoms in the upper layer, leading to a stacking sequence of ...AbA BaB AbA BaB... with a hexagonal rotational symmetry.[65,83] The configuration of each 'sandwich' layer in the 3R type is the same as that in 2H type, (*i.e.*, trigonal prismatic) with the main distinction between structures being the stacking pattern of the 'sandwich' layers, which is ...AbA CaC BcB AbA CaC BcB... in the 3R type structure, which possess a rhombohedral symmetry.[65,83]

1.1.3 Nanostructured Metal Chalcogenides

Since the discovery of carbon nanotubes (CNTs) in 1991,[85] nanomaterials have sparked tremendous interest, which can be ascribed to their peculiar physicochemical properties and great potential in advanced miniaturised functional

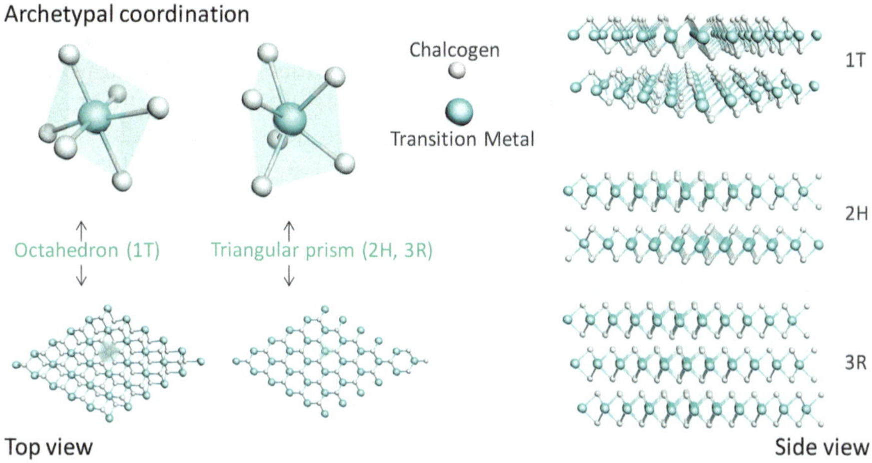

Figure 1.1 Schematic illustration of three structural polytypes of layered metal chalcogenides: 1T, 2H and 3R. Reproduced from ref. 83 with permission from the Royal Society of Chemistry.

devices. The term 'nanostructure' requires that at least one dimension of a material falls in the nanoscale, thus comprising zero-dimensional nanoparticles (*e.g.*, quantum dots), one-dimensional nanostructures (*e.g.*, nanowires, nanotubes, nanorods, nanobelts, *etc.*), two-dimensional nanostructures (*e.g.*, nanosheets) and three-dimensional bulk structures consisting of nanograins and/or incorporating nano-inclusions. The main differences between nanomaterials and bulk materials are the high specific surface area and high surface-to-volume ratios of the former as compared with the latter. With a typical dimension spanning from 1–100 nm, nanomaterials bridge the gap between molecules and bulk materials (such as crystalline solids). Aside from a high specific surface area, materials in this size range exhibit unique and fascinating features compared with their bulk analogues, such as a reduced melting point, improved catalytic activity, enhanced electron transport and a modified band structure.[43,44,86]

The two design strategies for fabricating nanostructured materials are often referred to as 'top-down' and 'bottom-up' methods. Both are of great significance to nanotechnology and modern industry and each is highly applicable to the synthesis of nanostructured chalcogenides. Top-down methods generally obtain desired nanostructures through breaking down bulk materials *via* mechanical, physical or chemical approaches. The major drawback of these methods is the difficulty in maintaining order in extended structures and retaining structural periodicity. Moreover, surface structure imperfections may be introduced into materials in this way.[86–88] The defects that are introduced by top-down approaches can exert a significant influence on the physical properties and surface chemistry of nanomaterials, because the surface-to-volume ratio of nanomaterials is crucial. As the name suggests, bottom-up methods involve building materials from the bottom, such as atom-by-atom, molecule-by-molecule, and cluster-by-cluster.

This procedure is expected to produce nanomaterials with more homogeneous compositions and controllable morphologies, while permitting particle size to be tuned and the surface to be functionalised. A drawback, however, can be the necessary utilisation of toxic precursors or the employment of reagents with difficult or restricted handling requirements.[86,89–91]

Since the first report of the bottom-up syntheses of MoS_2 and WS_2 nanotubes in the 1990s,[63,92,93] other nanostructured metal chalcogenides have attracted tremendous attention due to their distinctive morphological characteristics and promising physicochemical properties. Among chalcogenide nanostructures, nanocrystals or quantum dots of semiconductors consisting of chalcogens and Group 12 and 14 elements, such as CdX, ZnX and PbX (X = S, Se, Te), offer unusual and frequently useful functional properties.[12,94] A few Group 4, 5 and 6 metal dichalcogenides, such as HfS_2, ZrS_2, NbS_2, $NbSe_2$, MoS_2, $MoSe_2$, WS_2 and WSe_2, possess layered structures that are beneficial to nanotube formation.[95] In addition, two-dimensional layer-structured metal chalcogenides, such as Bi_2Se_3 and Bi_2Te_3, normally grow in a thermodynamically favoured manner, which can grow into a nanoplate/nanosheet form even without the assistance of shape-controlling agents.[43,96]

Nanostructured metal chalcogenides and their heterostructures have aroused extensive interest in multifarious applications, such as thermoelectric devices, solar cells, lithium-ion batteries, supercapacitors, catalysis, fuel cells, sensors, light-emitting diodes, and phase-change memory materials.[1,43,44,97] The properties of nanostructured metal chalcogenides can be intentionally regulated depending on their specific applications. For example, Bi_2Te_3-based thermoelectric materials consisting of nanoscale grains and nano-inclusions display reduced lattice thermal conductivity, due to strengthened phonon scattering.[98–100] Many nanostructured metal chalcogenides (*e.g.*, CuS) have become popular options for solar absorber materials.[101] Nanostructured transition metal chalcogenides are hailed as promising non-noble metal catalysts for oxygen reduction reaction and are expected to be effective replacements for precious Pt-based electrocatalysts.[102,103] The use of germanium chalcogenide nanostructures (*e.g.*, $GeSb_2Te_4$) for data storage devices could hopefully pave the way for future large-scale applications[104,105] and finally CdSe can produce any colour across the visible spectrum by adjusting the size of its constituent particles.[106] Hence it can be seen, just from this selection of examples, that many efforts can be made to optimise the fundamental properties of metal chalcogenides by modifying their size, morphology, crystallinity, and defect structure through nanostructuring.

1.1.4 Metal Chalcogenides as Important Materials

In the 21st century, with the exploration of several novel physical phenomena, various new metal chalcogenide systems have been discovered and corresponding application fields have been further expanded.[107–109] A great diversity of multiple-element combinations and stoichiometries have been demonstrated to exist in solid state metal chalcogenides.[1] In addition, non-stoichiometric phases are abundant for metal chalcogenides, especially for those containing transition metals, where variable valences are favoured and electronegativity differences are minimal.[1] In this section, we concentrate on two archetypal metal chalcogenide materials systems, MoS_2 and Bi_2Te_3, for the reason that the diverse chemical and physical properties of these layered chalcogenides have enabled their successful application across different fields.[38,61,65,110–112] Here, we focus on their intrinsic structure and property characteristics and briefly introduce their corresponding fields of application. It should be noted that there are a wide variety of other chalcogenides with many other attractive properties and applications, but these cannot be covered here due to limited space and scope.

1.1.4.1 *Molybdenum Sulfide*

Molybdenum sulfide (MoS_2), with a layered crystal structure and available in nature as the mineral molybdenite, is perhaps the most studied transition metal dichalcogenide. It has been investigated in the fields of dry lubrication,[113] photovoltaics,[114] catalysis[115] and batteries[116] since the 1960s. Recently, few-layer MoS_2 has attracted considerable attention due to its semiconducting characteristics and the potential for its application in electronics and optoelectronics.[65] Different from graphene, which has a single plane of atoms per sheet, mono-layer MoS_2, comprises a plane of Mo atoms sandwiched between two planes of covalently bonded S atoms, lacks the 'wrinkled' morphology of graphene and characteristically forms smoother, more planar sheets at the nanoscale.[117] Each single layer of MoS_2 is about 6.5 Å thick.[65,118] The crystal structure of MoS_2 is composed of stacked S–Mo–S layers held together by weak van der Waals forces[65,118] and the van der Waals gap has been measured to be about 3.49 Å.[119] Within the S–Mo–S layer, the electronegativity difference between Mo and S results in covalent bonds that are partially polarised. Mo relinquishes (primarily d-band) valence electrons to S, and is left in an oxidation state of +4, while the oxidation state of the S anions becomes −2.[120]

Bulk MoS_2 is composed of vertically stacked, weakly interacting S–Mo–S layers, which can be easily sheared by mechanical cleavage and liquid exfoliation.[121] Hence, single- or few-layer MoS_2 can be produced easily and the sheets studied individually.[110] Previous studies revealed that MoS_2 is a polytypic material based on the coordination of S atoms with Mo atoms within the sheets and the manner in which the sheets are assembled,[122] MoS_2 is categorised into three fundamental structural polytypes, that is, 1T-MoS_2, 2H-MoS_2, and 3R-MoS_2,[65,67] the space groups of which are $P\bar{3}m1$, $P6_3/mmc$, and $R3m$, respectively.[123,124] Although the three polytypes have different lattice parameters and cell symmetries, the Mo–S bond lengths are approximately 2.4 Å for all of them.[125]

Among the three polymorphs of MoS_2, the 2H and 3R polytypes are found to be naturally occurring, while the metastable 1T polytype is a purely synthetic phase.[126] Density functional theory calculations[127] and X-ray diffraction experiments[128] have indicated that in fact the 1T structure of MoS_2 should indeed be thermodynamically unstable. 2H- and 3R-MoS_2 possess Mo in trigonal prismatic coordination, while 1T-MoS_2 contains octahedrally coordinated Mo atoms.[126] The alternative stacking arrangements of the trigonal prismatic Mo–S layers lead to the structural difference between 2H- and 3R-MoS_2. Despite some small differences in band structures, 2H- and 3R-MoS_2 have nearly indistinguishable properties, such as n-type semiconducting properties and intrinsic diamagnetism.[129] On the contrary, 1T-MoS_2 is paramagnetic and metallic, consistent with the change in coordination geometry of the Mo atoms.[130] As expected, the complete filling of orbitals (bands) results in semiconducting behaviour while partial filling leads to metallic behaviour.[131] Therefore, as the orbital occupation varies, the electronic properties of MoS_2 gradually change from metal through semiconductor to topological insulator, which provides a diversity of chemical and physical properties for wide varieties of applications.[67,131]

For 2H-MoS_2, the stacking of the trigonal prismatic Mo–S layers yields a thermodynamically-stable structure,[132] which is the most ubiquitous polytype and its widespread application benefits from its plenitude in the Earth's crust.[133] In the majority of studies, it is the 2H-MoS_2 polytype that has been utilised as a precursor to prepare mono- or few-layer MoS_2 nanosheets.[67] Given the stability of the 2H polytype, it is also worth noting that 3R- and 1T-MoS_2 can easily transform into the 2H form.[129] For example, 3R-MoS_2 can be converted to the 2H phase through heating, while 1T-MoS_2 can be converted to the 2H phase by elemental doping, annealing or by applying strain.[134–138] Interestingly, the semiconducting 2H and metallic 1T phases can coexist in chemically exfoliated MoS_2 mono-layers, which was observed utilizing scanning transmission electron microscopy (STEM).[139]

Basically, due to the changes in degree of quantum confinement, interlayer coupling and symmetry elements, exfoliated thin-layer MoS_2 nanosheets exhibit dramatically distinct physicochemical properties from bulk MoS_2.[84] For example, bulk MoS_2 has an indirect band gap of ~1.3 eV (Figure 1.2), a valence-band maximum (VBM) that is highly dependent on the interlayer interactions and is black in colour due to the absorption of all visible light.[140,141] In contrast, mono-layer MoS_2 is a direct band gap semiconductor with a conduction-band minimum (CBM) and VBM coinciding at the K-point (Figure 1.2),[84] meanwhile its band gap increases from ~1.3 eV to ~1.9 eV due to the quantum confinement effect. Accordingly, a strong photoluminescence, which is absent in bulk MoS_2, emerges in single-layer 2H-MoS_2.[142] In addition, the colour of MoS_2 samples, prepared from MoS_2 powders in *N*-methyl-2-pyrrolidone by sonication and centrifugation, varies from black to light-yellow with decreasing layer number and size, indicating the widening of the band gap.[143] Moreover, the crossover from an indirect to a direct band gap semiconductor accounts for the luminescence enhancement in single-layer MoS_2.[67]

Because of the strong in-plane Mo–S bonding and weak out-of-plane van der Waals interaction, the mechanical and functional properties of MoS_2

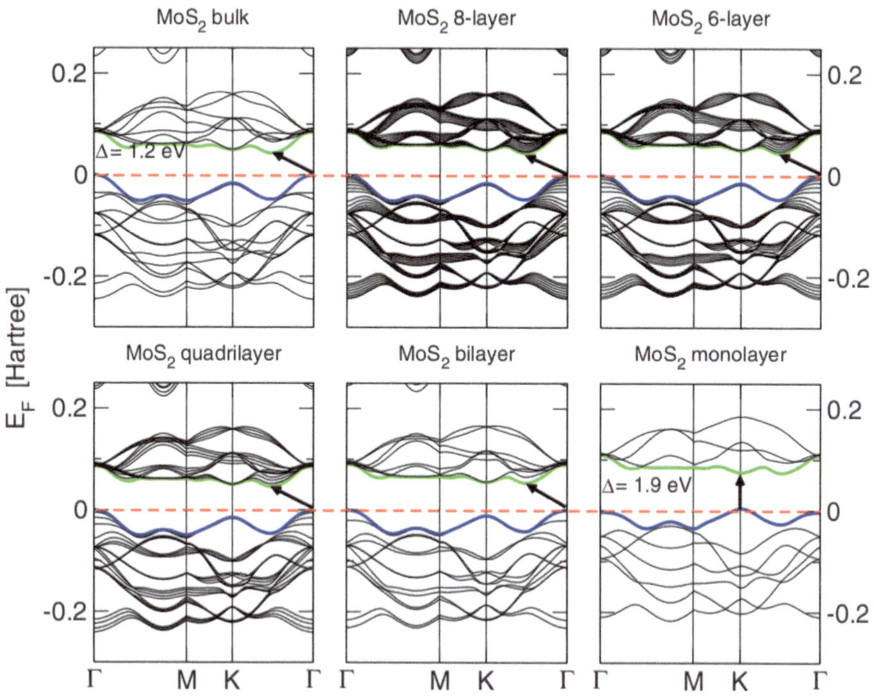

Figure 1.2 Band structures calculated from first-principles density functional theory for bulk and thin-layer MoS_2. The horizontal dashed line indicates the Fermi level. The arrows indicate the functional band gap (indirect or direct). Reproduced from ref. 141 with permission from American Physical Society, Copyright 2011.

are highly anisotropic.[144] For example, thermal expansion in the direction perpendicular to the basal plane (*i.e.*, parallel to the *c*-axis) is more substantial compared with that of the corresponding planar direction(s), while the thermal conductivity is orders of magnitude lower in the *c*-direction, normal to the basal *ab* plane.[120] Moreover, the crystal structure of MoS_2 offers two-dimensional penetrable channels for electron transport, whereas the movement of electrons in the perpendicular direction is hampered by an interlayer potential difference.[65,118] Therefore, the electrical conductivity within layers is higher than that across adjacent layers. Thus, decreasing the layer numbers contributes to less constrained electron transport and improved electrical conductivity. Additionally, MoS_2 possesses structural defects, such as point defects (vacancies), grain boundaries and edges.[145] These defects offer the feasibility of surface modification and functionalisation, which increase the options for the further manipulation of the physicochemical properties of MoS_2.[145] For example, functionalisation of MoS_2 can be achieved *via* the step-by-step covalent assembly of lipoic acid and fluorescein isothiocyanate, contributing to the simultaneous tailoring of fluorescent and electronic properties.[146]

MoS_2 is widely explored in various applications due to its unusual combination of physical properties such as high carrier mobility and tuneable band structure.[147,148] Altering the morphology provides another route by which the physical properties of MoS_2 can be modified; particularly relevant are variations in structural stability, band gap and optical properties.[129] In terms of mechanical properties, MoS_2 has an ultrahigh mechanical strength and a larger Young's modulus than steel.[149,150] Single-layer MoS_2 is marginally stronger than its bulk counterpart.[150] In addition, MoS_2 layers exhibit remarkably high mechanical flexibility. For example, single-layer MoS_2 can be deformed by an effective strain of up to 11% without fracture,[150] such that fabricated MoS_2 thin-film transistors can be bent to a curvature radius of 0.75 mm without obvious damage or denigration of electrical properties.[151] These features emphasise the role of MoS_2 as a promising material for flexible electronics.

In the tribology field, MoS_2 has long been recognised as an exceptional lubricant due to the relative inertness of sulfur basal planes and the weak van der Waals forces between S–Mo–S layers.[152] It is a favoured dry lubricant for aerospace applications; a status which can be ascribed to its outstanding lubricity under high load and *in vacuo*.[153] In terms of energy storage, MoS_2 is a promising electrode material for lithium-ion batteries, where the rather mediocre intermediate insertion voltage (~2 V *vs.* Li/Li^+) can be at least part-compensated by an outstanding reversible capacity (~1290 mA h g^{-1} at 100 mA g^{-1} after 50 cycles) and a good rate capability (*e.g.*, a capacity of ~554 mA h g^{-1} after 20 cycles at 50C).[126] Additionally, with respect to energy conversion, the characteristic tuneable direct band gap endows single-layer MoS_2 with the possibility to serve as a viable semiconductor in photovoltaic applications.[154] In the sensing field, the layered structure and large specific surface area of MoS_2 nanosheets also offer promise for high sensitivity, high-efficiency field effect transistor (FET), optical and electrochemical devices.[117] MoS_2 nanosheets/nanoflakes also exhibit piezoelectric properties, which facilitate their applications in mechanical transducers and piezo-electrocatalysis.[155,156]

MoS_2 is widely feted in various branches of catalysis, for example due to a band gap that couples well with the solar spectrum and on account of the high activity of its edge sites (*i.e.*, the coordinatively-unsaturated sites exposed at the sheet edges).[157] It has also served as a synergistic catalytic support for gold nanoparticles[158] and for nickel–iron alloys utilised in the electrooxidation of hydrazine.[159] Meanwhile, when MoS_2 is grown on the surface of a photocatalyst (*e.g.*, oxygenated monolayer graphitic carbon nitride, O–g-C_3N_4), the chalcogenide's layered S–Mo–S structure is able to suppress agglomeration of the photocatalyst host, which is essential for maintaining the high activity and stability of the photocatalyst.[160] A further useful feature of exfoliated single-layer MoS_2 is the availability of a myriad of S atoms exposed on the surface of the nanostructure. Sulfur is a soft Lewis base exhibiting a high affinity for heavy metal ions that act as corresponding soft Lewis acids.[139,147] MoS_2 nanosheets therefore show both a high selective adsorption capacity for certain metal ions, which can be ascribed to abundant sulfur adsorption sites and fast kinetics due to the easy access to these sites.[135] Hence, single-layer MoS_2 is among the most effective adsorbent materials for removing heavy metal ions such as mercury from industrial waste water.[161]

1.1.4.2 *Bismuth Telluride*

Layered bismuth telluride (Bi_2Te_3) crystallises with a rhombohedral structure (space group $R\bar{3}m$), comprising quintuple layers that are stacked periodically along the *c*-axis (Figure 1.3(a)).[42] As the term suggests, each quintuple layer is composed of five atomic layers that are arranged in the order of –Te(1)–Bi–Te(2)–Bi–Te(1)–, where the numbers in parentheses designate two inequivalent Te sites. Bi atoms are octahedrally coordinated by Te atoms; Te(1) site atoms are bonded with three Bi atoms and three other Te(1) atoms, and Te(2) site atoms are octahedrally coordinated by Bi atoms.[42] A complete unit cell is formed by stacking three such quintuple layers in the [0001] direction.[162] Because the strengths of the ionic and covalent bonds within the quintuple layers are much higher than the van der Waals interactions between them, Bi_2Te_3 is liable to slip along the Te(1) plane during mechanical deformation (*i.e.*, the (0001) acts as a cleavage plane), which is the primary reason for the poor mechanical properties of Bi_2Te_3 single crystals. Bi_2Se_3 and Sb_2Te_3 are isostructural with Bi_2Te_3, and thus the corresponding quintuple

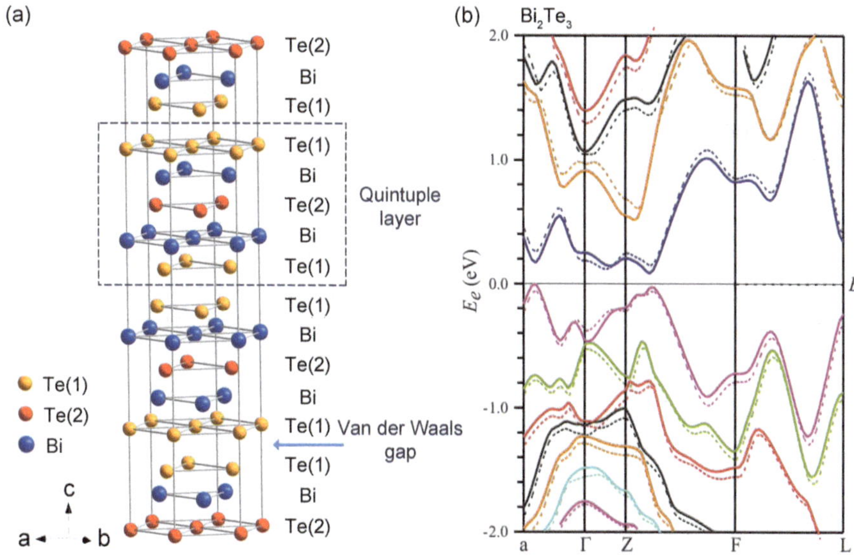

Figure 1.3 (a) Crystal structural model of Bi_2Te_3. (b) The calculated electronic band structure of Bi_2Te_3 along the high-symmetry lines with spin–orbit coupling. The solid and dashed lines are the results with and without the $p_{1/2}$ corrections included. Reproduced from ref. 169 with permission from American Physical Society, Copyright 2008.

layers are –Se(1)–Bi–Se(2)–Bi–Se(1)– and –Te(1)–Sb–Te(2)–Sb–Te(1)–, respectively. These three compounds can form ternary or quaternary continuous solid solutions with tuneable crystal defects, adjustable band gaps and thermoelectric properties that improve upon those of the respective binary compounds. In contrast, Bi_2S_3, Sb_2Se_3 and Sb_2S_3, however, crystallise in an entirely different orthorhombic structure.[163,164] Nevertheless, rhombohedral Bi_2Te_3, Bi_2Se_3, and Sb_2Te_3, can also form solid solutions with the aforementioned orthorhombic binary chalcogenides over limited compositional ranges without altering their rhombohedral ($R\bar{3}m$) structure.[164,165] Beyond these limits rhombohedral and orthorhombic phase separation occurs until a single-phase region of the orthorhombic structure is reached.[164,165] It is worth mentioning that when more electronegative S or Se substitutes for Te, it is the Te(2) sites that are preferentially occupied compared with Te(1).[166,167] Such site-selective ordered occupation leads to the minimum energy structure in the system.[167,168]

The band structure of a semiconductor greatly influences its carrier transport properties.[169] The critical band features, including band shape, band gap, and the number of energy valleys/valley degeneracy (where an energy valley represents the local extreme of the electron energy in the momentum space of the band structure),[170,171] heavily determine the electrical transport properties of materials. From the band structure of Bi_2Te_3 calculated by density functional theory (DFT) (Figure 1.3(b)), it can be clearly observed that Bi_2Te_3 is a typical indirect band gap semiconductor with multiple valleys, and a narrow band gap of ~0.13 eV.[172] The conduction and valence band structure can be represented as a multi-ellipsoidal model with six constant ellipsoids on a reflection plane.[172] Crystal symmetry causes six degenerate carrier pockets in both the conduction band and valence band of Bi_2Te_3, which benefit the higher density of states and effective mass and in turn contribute to the excellent electrical transport properties.[169] Similarly, isostructural Sb_2Te_3 and Bi_2Se_3 also have complex, multi-valley band structures.[42,173]

Manipulating point defects and their interplay makes it possible to regulate the carrier concentration and band structure of Bi_2Te_3 and the point defects in Bi_2Te_3 can be categorised into intrinsic and extrinsic types. The former refers to *anti*site defects and vacancies, while the latter refers to guest atoms introduced generally by doping or alloying. Bi_2Te_3 produced *via* zone-melting as an ingot is a p-type semiconductor due to *anti*site defects denoted as Bi'_{Te} (*i.e.*, Bi atoms occupy the Te positions, providing one hole per *anti*site defect as Bi has one less electron in the outermost shell than Te).[174] Halogens are most often used as donor level dopants to tune the electron concentration of n-type Bi_2Te_3,[175,176] while acceptor levels in p-type Bi_2Te_3 are commonly provided by Ag,[177] Cu,[178] Pb,[179] and Mn dopants.[180]

Due to its distinctive layered structure and heterogenous bonding features, Bi_2Te_3 inevitably exhibits anisotropic electron/phonon transport. For example, Bi_2Te_3 has both higher electrical and thermal conductivity along the direction(s) perpendicular to the *c*-axis (in-plane). Previous studies showed that the in-plane electron mobility and hole mobility of Bi_2Te_3 single crystals reached 1200 $cm^2 V^{-1} s^{-1}$ and 510 $cm^2 V^{-1} s^{-1}$, respectively, at 293 K.[181] These figures are accordingly approximately four times and three times those measured for the out-of-plane direction (along the *c*-axis).[181] Correspondingly, the in-plane electrical conductivity and thermal conductivity are 3–7 times and 2–2.5 times higher than those in the out-of-plane direction.[182,183] Interestingly, it has been observed that polycrystalline n-type Bi_2Te_3 is easier to texture—*i.e.*, align in the (00*l*) direction—than its p-type counterpart under the same preparation conditions.[184] One can hence optimise the anisotropic properties of polycrystalline doped Bi_2Te_3 along a certain direction by regulating the degree of orientation.

It is the small difference in electronegativity between the Bi and Te atoms in Bi_2Te_3 that accounts for the predominantly covalent nature of the Bi–Te bonding. This feature contributes to sharpened electronic bands and in turn the high carrier mobility, ensuring good electrical transport as seen in materials like PbSe.[185] Meanwhile, it is the high atomic masses of both Bi and Te in Bi_2Te_3, combined with the relatively low melting point of the telluride (~858 K), that endows phonon vibration with high anharmonicity, thus leading to low lattice thermal conductivity.[183,186] These qualities combine to make Bi_2Te_3 the most outstanding thermoelectric material currently available for operation near room temperature, and account for its commercialisation in solid-state refrigeration.[112,187,188] The formation of solid solutions with Sb and Se as substituents engenders a high level of control over the band structure and carrier concentration of Bi_2Te_3, leading to a range of excellent p- and n-type Bi_2Te_3-based thermoelectric materials capable of working across different temperature ranges.[60,61,188–190] These solid solutions generally possess better thermoelectric properties and consequently find wider application than the corresponding binary compounds. In addition, Bi_2Te_3 also serves as an example of an important topological insulator with a number of thickness-dependent physicochemical properties.[191] By exploiting its fascinating functional properties, Bi_2Te_3 has already been utilised widely to fabricate various devices for power generation, precision temperature control of microchips, and as a component of self-powered wearable electronics.[192,193]

1.2 Synthesis Methods for Metal Chalcogenides

The development of materials heavily depends on the ability to synthesise them, to explore more cost-effective, energy-efficient synthetic methodologies, and to design novel materials with appropriate compositions and desired functionalities. Synthesis methods make various configurations and microstructures possible for solid materials, which therefore play essential roles in materials innovation. During the past three decades, a variety of synthetic methods have been introduced and implemented in the synthesis of solid-state metal chalcogenides, which include but are not limited to conventional high-temperature melting/heating, hydrothermal (or more broadly, solvothermal) synthesis, vacuum/high-temperature deposition from the gas phase, and emerging techniques such as microwave-assisted synthesis.

Utilising different preparation techniques can generate metal chalcogenides with distinct properties, which depend primarily on chemical composition, crystal structures, microstructure, and crystal defects. The preparation of specific materials requires appropriate regulation and optimisation of synthesis conditions. The following section will discuss the synthetic methodology of metal chalcogenides, which can be divided into three thematic areas: synthesis methods for bulk materials, synthesis of thin films, and nanomaterials synthesis.

1.2.1 Synthesis Methods for Bulk Materials

1.2.1.1 Single Crystals

The Bridgman–Stockbarger technique is widely utilised for single crystal growth and purification.[194,195] Raw materials are loaded into a crucible or boat, melted and then resolidified by applying a positive temperature gradient across the melt. This growth process from the melt can be performed in either a vertical or horizontal configuration,[196] from which the grown crystals generally exhibit either circular or D-shaped cross-sections, respectively.[197] Crystals with circular cross-section from vertical Bridgman growth are preferred for large-scale epitaxial growth and device fabrication, while horizontal growth encourages high crystal quality due to the reduced thermal stress that results from the configuration.[197] This method has been applied for the crystal preparation of various metal chalcogenides, such as MSe (M = Ga, Ge, Cd, Sn),[198–201] MTe (M = Zn, Ga, Cd, Pb),[198,199,202,203] $SnSe_2$,[204] Bi_2X_3 (X = Se, Te),[205] $AgSbTe_2$,[206] Cu_2SnSe_3,[207] $CuInSe_2$,[208] $Cs_2Hg_6S_7$,[209] $Cs_2Cd_3Te_4$,[209] $TlMSe_2$ (M = Bi, Sb),[210] $Tl_2GaInSe_4$[211] and $AgGaGe_3Se_8$.[212]

The Czochralski technique has been also used for the preparation of metal chalcogenide single crystals.[7,197] The apparatus consists of three main components: a crucible, a heater, and a pulling rod positioned axially above the crucible. The growth process starts with melting the required material in a crucible. A seed crystal attached to the end of the pulling rod is then dipped into the melt and slowly lifted while being rotated simultaneously.[197] The melt temperature and pulling speed are regulated to yield a small meniscus at the end of the seed crystal. This is performed in order to eliminate dislocations and reduce the effects of thermal asymmetry on crystal growth.[197] The main advantage of this method is the absence of a container for the crystal and so there is no contact between the rod on which the crystal is grown and the melt-containing crucible during cooling. This arrangement eliminates the thermal stress that is normally generated by the mismatch in the thermal expansion coefficients of the container and the crystal, which is important in the production of dislocation-free crystals.[197] Numerous metal chalcogenide single crystals such as MTe (M = Ge, Sn, Pb),[213,214] M_2Se_3 (M = Sb, Bi)[215] and M_2Te_3 (M = Sb, Bi)[190,215–217] have been successfully prepared using the Czochralski technique.

1.2.1.2 Polycrystalline Bulk Materials

Perhaps the most common way to synthesise polycrystalline metal chalcogenides is by traditional high-temperature solid-state methods, which involves the direct reaction of stoichiometric quantities of high-purity metal and chalcogen element precursors in a sealed ampoule (typically quartz) at relatively high temperatures (*ca.* 400–1000 °C) in a furnace.[1] Briefly, certain proportions of reactants are weighed and ground together, pressed into a pellet and then heated to high temperatures to facilitate interdiffusion and compound formation. Considering the propensity of chalcogenides to convert to their corresponding oxides at high temperatures, even when only trace amounts of oxygen are present, reactions are preferentially conducted in either a vacuum-sealed container or an inert-gas stream. After reaction for hours or days, the ampoule is either quenched or slow-cooled to room temperature. Notably, covering reaction crucibles are beneficial to prevent the evaporation of highly volatile components (such as the chalcogen), thus avoiding significant deviation in the composition of the anticipated product materials.

The method typically yields powders that can initially be multiphasic. Post-annealing of the products for several days under vacuum or inert atmosphere without melting is usually required to obtain higher-quality crystals and/or single-phase thermodynamically stable end-products. Many metal chalcogenides can be prepared using this route, including the binary sulfides, selenides and tellurides of transition and main group metals such as Co, Cd, Hg, Pb, Sn, Sb and Ge,[1,61,218] as well as a variety of complex ternary, quaternary and higher compounds containing elements from all parts of the periodic table (*e.g.*, AgCuTe,[219] $Pt_3Pb_2Se_2$,[220] $K_2MnGe_3S_8$,[221] $EuHgSnS_4$,[222] $EuHgGeSe_4$,[222] $Rb_3Ga_3Ge_7S_{20}$[223]).

1.2.2 Synthesis of Thin Films

Chemical vapour deposition (CVD) is a process in which desired thin films or nanostructures are deposited on solid substrates by a heterogeneous chemical reaction of vapour-phase precursors.[224] In CVD, multispecies of precursors are transported in the gas phase either by evaporation or sublimation through heating, plasma excitation or optical radiation and introduced into a reaction chamber. For example, the synthesis of phase-pure iron pyrite (FeS_2) thin films use di-*tert*-butyl disulfide and $FeCl_3$ as the sulfur and iron precursors, respectively,[225] and the syntheses of MoX_2 (X = S, Se) use $Mo(CO)_6$ and X (S or Se) as the sources of the respective elements.[226] When the gas flows over a heated substrate, surface chemical reactions occur to form a desired thin film or nanostructures on a particular substrate, during which further decomposition of the precursors takes place. Defects present in the substrate can act as nucleation sites from which further growth can occur to form desired materials over time.[197] Surface diffusion and nucleation affect the morphology of the prepared films that can be epitaxial, amorphous or polycrystalline.[197] Additionally, several factors, such as the deposition temperature, substrate, gas pressure and gas flow rate can strongly affect the crystallinity and morphology of materials prepared by CVD.[197,227,228] CVD offers a variety of advantages in the deposition of ordered nanostructured thin films with varied compositions, and the technique is suitable for both batch and semicontinuous operations. A series of metal chalcogenide films have been successfully fabricated, ranging from films of binary and ternary compounds such as PbX (X = S, Se),[229] CuX (X = S, Se),[230] SnX (X = Se, Te),[231] MTe (M = Ge, Cd),[232,233] MoX_2 (X = S, Se),[234] WX_2 (X = S, Se),[234] $ReSe_2$,[235] M_2X (M = Cu, Ag; X = Se, Te),[231] M_2S_3 (M = Cr, Bi),[236,237] Bi_2Se_3,[237] M_2Te_3 (M = In, Sb),[233,238] $CdCr_2S_4$[239] and In_3SbTe_2[240] films, to uniquely structured films such as Fe-intercalated $ZrSe_2$,[241] NbS_2/MoS_2 heterostructures[242] and ReS_2/WS_2[243] heterostructured films.

Physical vapour deposition (PVD) is a commonly utilised technique for the preparation of metal chalcogenide thin films.[197] In contrast to chemical vapour deposition, PVD realises the synthesis of thin films without chemical transitions from precursors to products. In a typical PVD procedure, target materials with desired compositions are evaporated by various physical means such as by the application of a laser, magnetron or electron-beam under vacuum.[86] The generated vapour is deposited onto a heated substrate to form the desired film.[86] The substrate is heated to an appropriate temperature and continuously rotated as needed to achieve improved growth homogeneity.[86] The basic PVD process is generally divided into three categories: evaporation, sputtering and ion plating in terms of the physical mechanisms of deposition.[86,197,244] PVD has been applied in the film preparation of a host of chalcogenides such as SnS,[245] Ag_2S,[246] Sb_2X_3 (X = S, Te),[247,248] M_2Se_3 (M = Bi, As),[247,249] Ba-doped Ag_2S (Ba:Ag_2S),[246] BaS-doped Sb_2S_3 (BaS:Sb_2S_3)[250] and SnS-doped PbS (SnS:PbS)[251] among others.

1.2.3 Nanomaterials Synthesis

Chemistry plays a major role in developing new nanomaterials with novel properties of technological importance.[252] Chemical fabrication routes are generally bottom-up strategies that allow a better control of the morphology and dimensions of synthesised nano/microstructures, since various parameters can be regulated individually or simultaneously during preparation. The primary advantage that chemical methods provide is good homogeneity of samples, because chemical synthesis offers mixing and reaction at the molecular level. Methods can be designed to prepare new materials/crystal structures by understanding how matter is assembled and constructed on an atomic and molecular level and the consequent effects on the material's properties.[106] In this section, several typical chemical synthesis methods of metal chalcogenides are discussed, while other methods for chalcogenide nanomaterials synthesis, such as mechanical alloying and melt spinning, are also introduced.

1.2.3.1 *Hydrothermal and Solvothermal Synthesis*

Hydrothermal and solvothermal methods have been applied for the synthesis of various metal chalcogenide nano/microstructures with specific sizes and shapes.[102,253–259] Hydrothermal synthesis involves chemical reactions in an aqueous medium within a closed vessel at temperatures above 373 K and under elevated autogenous pressure (>0.1 MPa). More broadly, solvothermal synthesis utilises alternative non-aqueous solvents in a similar process. Since such non-aqueous solvents can possess boiling temperatures rather different from water, synthesis can demand modifications to the heating temperature, in turn altering the internal pressure, which provides additional variables towards the successful synthesis of various metal chalcogenides.[260] In addition, solvent properties, such as polarity and viscosity directly impact the solubility, reactivity, and diffusion behaviour of precursors in liquid-phase synthesis, which will further tune the phase, shape, and size of products.[261] Various organic solvents, with very different physiochemical properties, are often utilised as reaction media to replace water. Following the successes of hydrothermal and solvothermal syntheses, a promising derivative approach, known as the mixed solvent method, has evolved, which utilises at least two solvents together. The properties of mixed solvents can be regulated by adjusting the components and volume ratios of solvents to achieve the most appropriate synthetic conditions to obtain certain desired products.[102,258]

Solvothermal synthesis methods typically involve the utilisation of sealable reactors composed of a Teflon™-lined vessel and a stainless-steel autoclave. Inert Teflon™ liners are used to avoid any reaction between the reactants and the reaction vessels and to protect the stainless-steel outer shell of the autoclave from corrosive reagents. Figure 1.4(a) provides a schematic diagram of an autoclave and the solvothermal preparation processes. Specifically, chemical precursors in the appropriate ratios are added to pre-added solvent—serving as the reaction medium—in a Teflon™ liner. The Teflon™ liner is sealed in a stainless-steel autoclave, which is heated to a prescribed temperature above the boiling point of the solvent for a certain duration. In the typical solvothermal syntheses of metal chalcogenides, reaction temperatures are usually in the range of 373–523 K and chalcogen sources can include reagents such as S, Se, Te, Na_2S, Na_2SeO_3, Na_2TeO_3, thiourea, SeO_2, TeO_2 and others.[108] During solvothermal synthesis, pressure is generated in the closed vessel due to solvent vaporisation at elevated temperatures and this increases rapidly with rising temperature.[262] The generated pressure within the reactors also strongly relies on the percentage fill, which is defined in terms of the solvent volume and container volume. The autogenous pressure will rise more rapidly to a higher value when the percentage fill is large.[263] One should be cautious that if the autogenous pressure is higher than the maximum pressure tolerated by the autoclave, it is possible that the autoclave could explode, causing safety issues. As such, the percentage fill in solvothermal synthesis reported previously is generally not higher than 80%.[263] It should be also mentioned that most autoclaves have

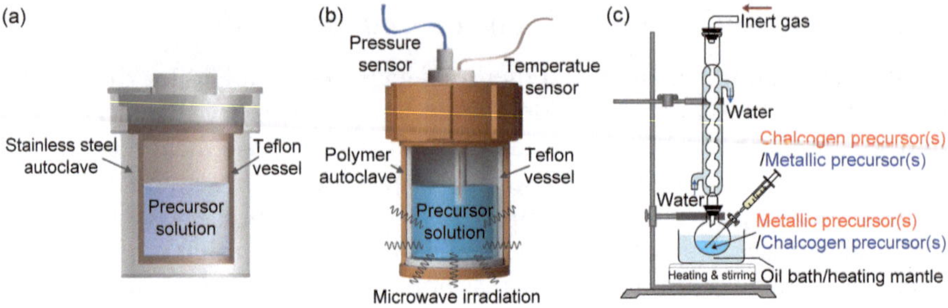

Figure 1.4 Schematic illustration of solvothermal and hot injection methods for metal chalcogenide preparation: (a) solvothermal synthesis, (b) microwave-assisted solvothermal synthesis, and (c) hot injection synthesis.

a bursting disc/pressure release valve to prevent such explosions. The starting materials in solvothermal syntheses should be selected with sufficient solubility in the utilised solvent at elevated temperature, so that reactions can continue until completion. The crystallisation of the desired (insoluble) products then occurs at elevated pressure and temperature. It should be noted that variations of solvothermal syntheses can be performed in a microwave field (Figure 1.4(b)), which will be discussed in detail in the section on microwave-assisted solution synthesis.

When compared to solid state ceramic techniques, for example, solvothermal, and mixed solvent methods have many advantages, such as fast reaction kinetics, short processing times, low energy consumption and the high crystallinity of synthesised samples.[254] These methods can realise effective control on the composition, crystallinity, particle size, and microstructure of products by simply regulating reaction variables such as reagent type, temperature, pressure, the presence of additives (*e.g.*, mineralisers, surfactants), reaction duration, solvent ratio, and pH.[260] Considering pH as an example, in alkaline solution OH^- can realise the disproportionation of Te into $[TeO_3]^{2-}$ and Te^{2-} through $3Te + 6OH^- \rightarrow 2Te^{2-} + [TeO_3]^{2-} + 3H_2O$[264,265] and in turn promote the formation of metal chalcogenide *via* the reaction between metal ions and Te^{2-}. In contrast, acidic conditions would facilitate the formation of metal–halide bonds by ligand replacement, which is vital to achieving the controllable doping of halides into metal chalcogenides.[266,267] Variations of solvothermal methods have achieved great success in the preparation of metal chalcogenide nano/microstructures, such as FeS_2 nanowires,[268] $FeSe_2$ nanoflowers,[269] MnX (X = S, Se) nanorods,[270,271] ZnSe nanobelts,[272] CoTe nanotubes,[273] Bi_2Te_3 nanotubes[274] (Figure 1.5(a)), PbTe nanowires[275] (Figure 1.5(b)), T-shaped Bi_2Te_3-Te heteronanojunctions[276] (Figure 1.5(c)), SnS_2 hexagonal nanoplates[277] (Figure 1.5(d)), nano-string-clusters of Bi_2Te_3[278] (Figure 1.5(e)) and hierarchical flowerlike In_3Se_4[279] (Figure 1.5(f)). In addition, crystalline products are obtained at relatively lower temperature in solvothermal synthesis methods as compared to melting techniques, for example. However, it should be noted that solvothermal

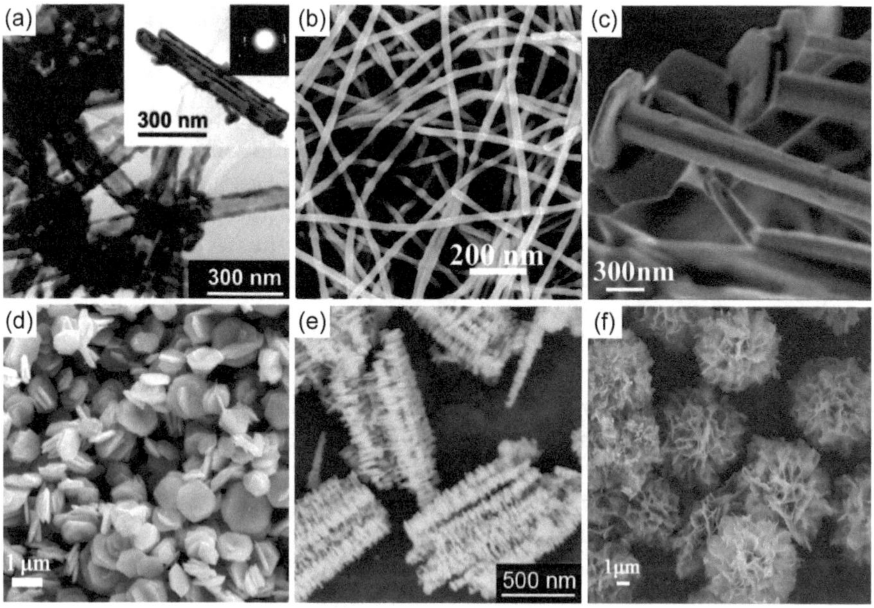

Figure 1.5 Electron microscopy images of some typical metal chalcogenides prepared *via* hydrothermal and solvothermal methods. (a) Bi_2Te_3 nanotubes. Reproduced from ref. 274 with permission from AIP Publishing, Copyright 2005. (b) PbTe nanowires. Reproduced from ref. 275 with permission from American Chemical Society, Copyright 2008. (c) Bi_2Te_3-Te heterojunctions. Reproduced from ref. 276 with permission from American Chemical Society, Copyright 2013. (d) SnS_2 hexagonal nanoplates. Reproduced from ref. 277 with permission from Elsevier, Copyright 2022. (e) Bi_2Te_3 nanostring-clusters. Reproduced from ref. 278 with permission from American Chemical Society, Copyright 2010. (f) In_3Se_4 flowers. Reproduced from ref. 279 with permission from American Chemical Society, Copyright 2013.

synthesis also presents some clear disadvantages. For example, it is difficult to study the nucleation and growth processes of products *in situ* as reactions are performed in closed, non-transparent reactors.

1.2.3.2 Hot Injection

Hot injection is an effective synthetic method for the preparation of metal chalcogenides with good crystallinity and narrow particle size distributions. In a hot injection route, a stock solution containing reactive chalcogen precursors is injected quickly into a pre-heated solution containing metal salts. This induces a high level of supersaturation in the reaction mixtures, which in turn leads to a short burst of nucleation that is followed by subsequent growth by aging at a prescribed temperature.[280,281] Figure 1.4(c) illustrates the hot injection method schematically. The corresponding nucleation takes place instantaneously at a particular temperature.[282] The kinetics of hot injection is such that there is close to negligible time between injection and precipitation. Mechanistic insights have indicated that the separation of the nucleation and growth steps is critical for the formation of monodisperse nanocrystals.[283] The diameter and morphology of synthesised metal chalcogenide nanocrystals can be effectively tuned by judiciously adjusting the experimental parameters such as type and ratios of starting reagents, injection temperature, reaction duration and use of a surfactant.[108,284] For example, inclusion of an appropriate surfactant can reduce the surface tension between two liquids or between a solid and a liquid during hot injection synthesis and can also prevent nanoparticle agglomeration by acting as a capping reagent.[260] Hence, the surfactant enhances colloidal stability so that monodispersed nanoparticles can remain stable in a liquid for a long time. Generally, the polar groups of surfactants are used in a reaction medium to bind to the surface of nanocrystals.[260] However, if the surfactant is not completely removed from nanocrystals during the post-synthesis treatment, the residual surfactant may have an adverse effect on the physiochemical properties (*e.g.*, electrical conductivity) of the products.[285,286]

The hot injection method was first used by Murray *et al.* to prepare monodisperse CdSe nanocrystals in 1993.[94] Since then, many other metal chalcogenide nanocrystals have been successfully synthesised *via* this method, such as CdSe[94] (Figure 1.6(a)), Cu_3SbSe_4[287] (Figure 1.6(b)), ZnS[288] (Figure 1.6(c)), SnSe[289,290] (Figure 1.6(d) and (e)), Cu_5FeS_4[56] (Figure 1.6(f)), CuS,[291] Cu_2S,[292] In_2S_3,[293] SnTe,[294] GeTe,[295] Ag_2X,[296] PbX,[297,298] Bi_2X_3,[299,300] $CuInX_2$,[301–303] $CuGaX_2$ (X = S, Se, Te),[304] SnSe/rGO nanocomposite (rGO = reduced graphene oxide),[305] Cu_2SnS_3,[306] $Cu_2ZnSnSe_4$,[307] $Cu_2CdSnSe_4$[308] and $Cu_2ZnGeSe_4$[309] nanocrystals. The particle size and morphology of the metal chalcogenide nanostructures prepared this way are diverse, ranging from nanoparticles and nanorods to nanoplates and nanoflowers. The dominant form depends on the crystal structures of the chalcogenides as well as the synthetic conditions, such as reaction temperature, reaction duration and types of precursors/solvents/surfactants.[80,289,310–313]

Physical properties can also be modified by variations in the hot injection protocol. For example, Han *et al.* could regulate the type of charge carrier in SnSe by changing the Sn precursor and surfactant using an aqueous hot injection method.[289,290] Following this demonstration, p-type SnSe/rGO nanocomposites were also successfully fabricated through a facile *in situ* hot injection synthesis method.[305] From these principles, a general surfactant-free, hot injection method was subsequently developed for synthesising binary silver chalcogenides Ag_2X (X = S, Se, Te).[296] Among other examples, Ithurria and Dubertret pioneered the synthesis of ultrathin CdSe nanoplates *via* hot injection and achieved a precise control of the geometrical parameters of the product crystals (*i.e.*, thickness, crystal shape, lateral size),[314] while Ibáñez *et al.* exerted morphological control in their synthesis of $Cu_2CdSnSe_4$ to obtain either tetrahedral or penta-tetrahedral shapes.[308] Allen and Bawendi have also reported the controllable hot injection synthesis of Cu–In–Se quantum dots of variable composition and size.[303] The nanocrystals exhibit red

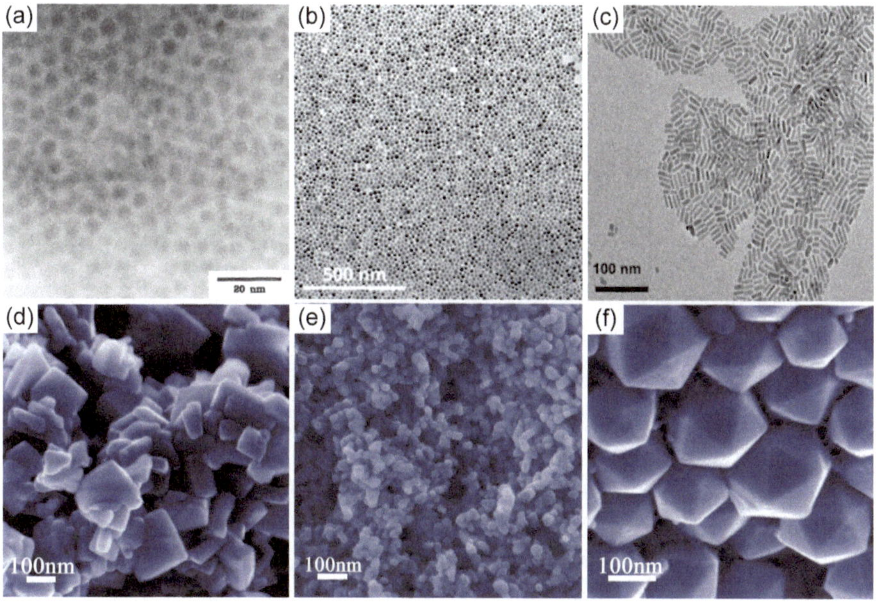

Figure 1.6 Electron microscopy images of some typical metal chalcogenides prepared through hot injection method. (a) CdSe nanocrystals. Reproduced from ref. 94 with permission from American Chemical Society, Copyright 1993. (b) Cu_3SbSe_4 nanocrystals. Reproduced from ref. 287 with permission from the Royal Society of Chemistry. (c) ZnS nanorods. Reproduced from ref. 288 with permission of American Chemical Society, Copyright 2005. (d) SnSe nanoplates. Reproduced from ref. 289 with permission of John Wiley & Sons, Copyright 2016. (e) SnSe nanoparticles. Reproduced from ref. 290 with permission of John Wiley & Sons, Copyright 2017. (f) Cu_5FeS_4 icosahedrons. Reproduced from ref. 56 with permission from John Wiley & Sons, Copyright 2018.

to near-infrared photoluminescence with a direct dependence on both the composition and size of the dots.[303] The above are just a few examples that demonstrate how the hot injection method can be applied across multiple systems at multiple scales while also providing significant synthetic flexibility in the preparation of metal chalcogenide nano/microstructures.

1.2.3.3 Microwave-assisted Solution Synthesis

The utilisation of microwaves in liquid-phase inorganic synthesis was first explored by Komarneni and Roy in 1985.[315] Microwaves are electromagnetic waves of frequencies between 0.3–300 GHz, flanked by the infrared and radiofrequency regions of the electromagnetic spectrum. To avoid interference with cable communication and mobile phone frequencies, microwave reactors for chemical synthesis predominantly operate at a specific regulated frequency of 2.45 GHz.[260,316] This equates to an energy that is insufficient in itself to break typical chemical bonds but is capable of inducing molecular rotations.[260] After decades of development, microwave-assisted synthesis has been widely applied in the fabrication of metal chalcogenides, which is considered as a green synthesis technique. Figure 1.4(b) illustrates the microwave-assisted solvothermal method schematically. Microwaves are non-ionising and only act so as to activate a materials system thermally.[260] Notably, microwave radiation is directly applied to reactants not to reaction vessels. By design, the vessel is effectively transparent to microwave energy, which is absorbed by the reactants and solvents, resulting in efficient heat transfer, principally through dielectric heating. The heating effect of microwaves is triggered by dipolar polarisation, ionic conduction or electrical conduction mechanisms depending on the materials used for reaction.[260,316] Microwave heating *via* electromagnetic waves can result in faster reaction kinetics and shorter reaction times (often by orders of magnitude) compared to conventional heating.

The intensity of microwave heating in a reaction mainly depends on the physical properties of the substance placed in the microwave field.[260] The ability of a specific substance to convert microwave energy into heat is determined by the so-called loss tangent ($\tan\delta$).[260] A high $\tan\delta$ material usually has a high dielectric loss, which represents the microwave energy converted to heat.[260,261] In fact, the microwave-assisted synthesis of bulk chalcogenides can be performed both in the liquid phase and the solid state, as long as the solvent, liquid or solid reactants possess high $\tan\delta$ values. Microwave-assisted solution synthesis, however, is more appropriate for the syntheses of chalcogenide nanomaterials specifically. Solvents that possess high $\tan\delta$ values typically also demonstrate good microwave absorption capacity and thus are favoured in the exploitation of the thermal effects achievable using microwave radiation.[260,261,317] Consequently, polar solvents such as water and alcohol are considered as good media for the microwave-assisted syntheses of inorganic nanostructures in solution. With a high boiling point (471 K), a chemically reducing capability, and a high value of $\tan\delta$, ethylene glycol finds wide application in the microwave-assisted syntheses of metal chalcogenides.[260] Generally, therefore, solvents containing dipolar species couple most effectively with microwaves, leading to a rapid rise in temperature and reaction rate.

Microwave heating offers the possibility of rapid chemical synthesis in solution in a very short duration compared to conventional heating, resulting in the potential advantages of lower cost, higher yields, side reaction suppression, and energy saving for chalcogenide nanomaterials production.[108,260,261] Microwave-assisted solvothermal synthesis employs autoclaves made of high-strength polymeric materials that are transparent to microwaves.[318] Apart from the incident microwave power, other reaction parameters, such as the reaction temperature, reaction duration, solvent type and pH also affect microwave-assisted solution synthesis. Through rational design of these parameters, metal chalcogenides with desired chemical composition, diameter, and morphology can be effectively synthesised.

For example, Mehta *et al.* developed a rapid and scalable microwave-assisted solvothermal synthesis of sulfurised Sb_2Se_3 nanowires and nanotubes.[319] By adjusting the microwave dose, which represents the product of microwave power and time, the diameter and length of these 1D nanostructures can be controlled. Panda *et al.* also reported a similar morphology control for metal chalcogenide nanostructures in the microwave synthesis:[320] by simply tuning the microwave radiation times, the stepwise formation of the ZnS and ZnSe nanowires from small spherical or elongated particles through short rods to long nanowires was observed. Up to now, a variety of metal chalcogenides have been successfully prepared by microwave-assisted solution methods, including binary materials such as HgS,[321] $MoSe_2$,[322] SnS_2,[323] CdX,[324–327] ZnX,[327–330] SnX,[331–333] PbX,[327,334,335] CuX,[336–339] Cu_2X,[340–342] Ag_2X,[343–345] Sb_2X_3,[346–348] and Bi_2X_3[349–351] (X = S, Se, Te), as well as ternary and quaternary metal chalcogenides such as $AgInS_2$,[352] $AgIn_5S_8$,[353] $CuInX_2$ (X = S, Se),[354] $CuIn_2S_4$,[355] Cu_3SbSe_4[356] and Cu_2ZnSnS_4.[357]

1.2.3.4 Liquid Exfoliation

Liquid exfoliation is regarded as a highly scalable and achievable solution-based method for producing large quantities of thin-layer metal chalcogenides with good crystallinity using mild conditions.[358–360] This method utilises high shearing or ultrasonication of crystal layers dispersed in a stabilising solvent (such as polymer solutions, ionic liquids, *etc.*) to produce nanosheets. The thin flakes are then separated from the solvent by centrifugation. Further, two-dimensional metal chalcogenides can be intercalated between layers by various guest species[361] and subsequently exfoliated into mono- and few-layer nanosheets by simple ultrasonication.[135,362–365] A typical chemical liquid exfoliation procedure involves submerging bulk powders in a solution of an oxidisable Li-containing compound (*e.g.*, *n*-butyl lithium) for a specific duration to allow the Li ions to intercalate. If this procedure is followed by exposing the intercalated materials to water, then the water will react with the Li between layers, resulting in H_2 generation and in turn the separation of the layers.[135,363]

The type of solvent used in liquid exfoliation directly affects the layer number, thickness, yield, stability and defect chemistry of the exfoliated material.[244] To inhibit the reaggregation of exfoliated materials, the selected solvent must have a surface tension comparable with the surface energy of the layered materials.[244] Coleman *et al.* reported a universally effective liquid exfoliation method to produce various nanosheets by the dispersion and ultrasonication of a series of layered metal chalcogenides in dozens of solvents with varying surface tensions.[360] Studies of the corresponding correlations were performed using optical absorption spectroscopy to measure the quality of the dispersed materials. Based on the application of theory to the analysis, they proposed that ideal solvents are those that minimise the energy of exfoliation.[360] For example, *N*-methyl-pyrrolidone and isopropanol emerge as promising solvents for exfoliating numerous layered compounds.

Through this method, layered compounds, such as $NbSe_2$,[360,366,367] $NiTe_2$,[360] TiS_2,[368] MoX_2 (X = S, Se, Te),[360,366,367,369] WX_2 (X = S, Se),[360,366,367,369] TaX_2 (X = S, Se),[360,367] M_2Se_3 (M = Bi, In)[366] and M_2Te_3 (M = Sb, Bi),[360,366] can be successfully dispersed, exfoliated and deposited as individual flakes or films.

1.2.3.5 *Mechanical Alloying*

Mechanical alloying, which is a process of alloy formation in the solid state through the ball milling of high-purity elemental components, has proven to be an efficient method for the preparation of metal chalcogenide nanoparticles.[370–373] In a typical process, a certain proportion of the respective reactant elemental pieces or powders are loaded into a jar/vessel made of cemented carbides (such as WC), steel or agate, together with a specific number of grinding balls (also manufactured from the equivalent jar material). The tightly sealed jars containing the elemental mixtures are rotated at a predefined frequency, so that the grinding balls experience repeatable mechanical forces resulting from collision, friction, compression, shearing, and grinding.[370,374] These processes impart mechanical energy to the powders, which makes the powders undergo continuous fracture and welding. The high-intensity, long-duration effects of ball milling ensure that the products are homogeneous, usually resulting in evenly dispersed, ultrafine alloyed particles.[370]

Factors directly affecting the degree of alloying include the type of ball mill, the composition of the grinding balls and the milling conditions employed (*e.g.*, time, speed, ball–powder ratio).[375] The types of instruments for mechanical alloying can vary in the way that energy is imparted and include vibrating, planetary, rotational and high-energy (vibrating–rotational) machines.[263] It is also possible to mill cryogenically and under different gaseous atmospheres. Tungsten carbide, agate, zirconia, and hardened stainless steel are usually utilised for the construction of jars and the grinding media. The combination of high-energy ball milling and hardened-steel grinding balls generally provides optimal mechanical alloying strength. A reasonable selection of technological parameters, such as the materials of manufacture, the size/dimension of the vessels and the ratio of grinding balls, can improve the ball milling efficiency.[370] As the materials used to manufacture the milling jars and media directly determine key characteristics such as density and hardness, it is necessary to choose the appropriate jars and balls in accordance with the intrinsic properties of the samples to be milled and the desired features of products. The matching of grinding ball material to the jars/vessel material helps avoid erosion and the introduction of unintended impurities into milled mixtures. Additionally, the ball to powder weight ratio is also crucial.[370] Usually, the ball-milling energy increases with the increase of ball to powder weight ratio. However, when the weight ratio exceeds a certain threshold, the limited acceleration space will reduce the collision energy.[370] Ball milling duration also has a significant impact on the phase and microstructure of products; therefore, selecting an appropriate ball milling time is critical for highly efficient preparation and controlled energy consumption. By combining mechanical alloying with various phase and microstructure characterisation techniques, the optimal milling parameters can be tuned to specific metal chalcogenide systems.

Mechanical alloying allows for solid-state reactions without high treatment temperatures. The structural defects of milled products, which are generated by the strong mechanical collisions between balls and between balls and the walls of the container, present plentiful possibilities to obtain various metastable and defect-rich materials with functionalities modulated by these features.[100,370] For example, p-type nanocrystalline BiSbTe bulk fabricated as a thermoelectric through hot pressing the ball-milled nanopowders, achieved a peak dimensionless figure of merit of 1.4 at 373 K owing to the low thermal conductivity caused by the increased phonon scattering by grain boundaries and defects.[100] Given the propensity of the method to shift away from equilibrium conditions, milling can enable the generation of alloys that are difficult to obtain *via* solidification processes due to unfavourable thermodynamics (as depicted by phase diagrams).[373] A tremendous variety of metal chalcogenide nanomaterials, including stoichiometric compounds such as PbTe,[376] SnS_2,[377] SnX,[378–380] Ag_2X,[381] AgCuTe,[382] $CuInX_2$,[224] $CuCrS_2$ [383] and $CuCr_2X_4$ (X = S, Se, Te),[384] as well as solid solutions and composite architectures such as $Sn(S,Se)_2$,[377] $(Bi,Sb)_2Te_3$ [100] and CNTs/Cu_2Se[385] have been prepared this way.

1.2.3.6 Melt Spinning

Solidification is a conventional method to fabricate metals and alloys, which involves heating a material above its melting point and then allowing the melt to cool to room temperature at various speeds.[386] The heating sources for melting are versatile, such as electrical resistance heating, induction heating, arc melting, and levitation melting.[387,388] In particular, the melt spinning technique, involving the rapid cooling of melts, has been widely applied to prepare amorphous alloys.[389,390] Figure 1.7(a) presents a schematic illustration of melt spinning, revealing that in a typical process, a thin stream of melt is ejected onto a high-speed rotating Cu wheel (that is cooled internally, usually by water) under a protective Ar atmosphere. Due to the ultra-high cooling rate of 10^4–10^7 K s^{-1},[391] the melt contacting the Cu wheel will cool relatively quickly and form a nanostructured ribbon-like material with a relatively uniform grain size distribution or even an amorphous material.[386,392] This method has been applied to prepare a variety of nanostructured metal chalcogenides, such as Bi_2Te_3–,[393–395] $AgSbTe_2$–,[396,397] SnTe–,[398] and PbTe–based materials.[399]

The microstructures generated in melt spun materials are sensitive to their positions relative to the rotating Cu wheel.[398,400,401] In general, melt spinning will create ribbons with two entirely different surfaces: the surface that contacts the Cu wheel is defined as the contact surface, while the opposing one is called the free surface. These two surfaces have significantly distinct cooling rates, temperatures, critical radii of nucleation and growth rates of nucleation sites, leading to the formation of multiscale microstructures.[391,400] Progressing from the contact surface towards the free surface of the ribbons, one can observe an evolution from amorphous microstructures, through

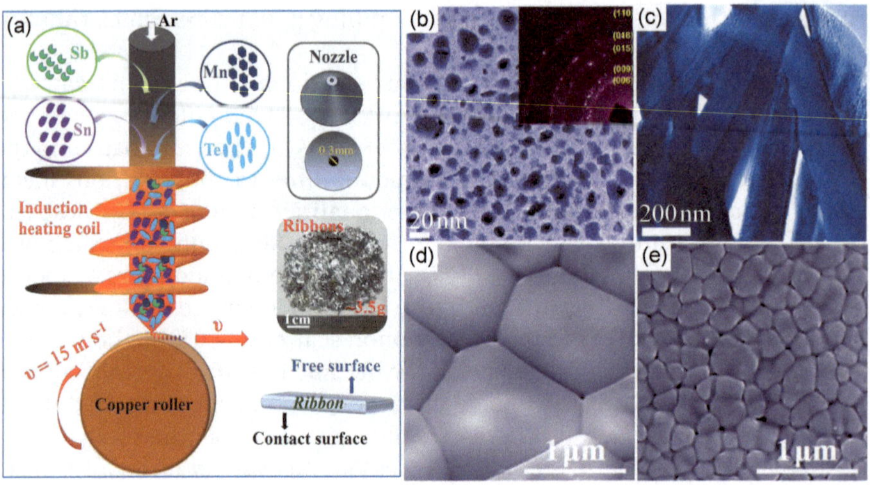

Figure 1.7 (a) Schematic illustration of melt spinning in the synthesis of SnTe-based ribbons. Reproduced from ref. 398 with permission from Elsevier, Copyright 2021. (b) and (c) TEM images of the contact surface and free surface, respectively of BiSbTe melt-spun ribbons. Reproduced from ref. 400 with permission from American Chemical Society, Copyright 2010. (d) and (e) SEM images collected from the free surfaces of SnTe and $Sn_{0.84}Sb_{0.16}Te$ melt-spun ribbons, respectively. Reproduced from ref. 398 with permission from Elsevier, Copyright 2021.

fine nanocrystalline regions, to larger nano/micro-sized crystals.[263,400] This is what occurs in Bi_2Te_3-based materials for example (Figure 1.7(b) and (c)) and the contact surface of a $Bi_{0.48}Sb_{1.52}Te_3$ ribbon features an amorphous structure embedded with 5–10 nm nanocrystalline domains, while the free surface is characterised by micro-sized dendritic crystals with nanoscale features.[400] Significantly, the nanostructures of melt-spun ribbons are determined not only by the melt spinning process but also by the properties of the starting materials, so different materials could possess totally different nano/microstructures.[398,402] For example, Yan *et al.* controllably synthesised SnTe ribbons consisting of columnar grains *via* melt spinning (Figure 1.7(d) and (e)) and found that Sb doping effectively hindered crystal growth during solidification, leading to a decrease in the diameters of the SnTe-based columnar grains.[398] To summarise, adjusting the chemical compositions of the raw materials and the cooling conditions during melt spinning can effectively dictate the growth of different nanostructured chalcogenide materials.

1.3 Energy-related Applications of Metal Chalcogenides

Energy serves as the backbone for modern society. The accomplishments of civilisation have mainly been achieved through the increasingly extensive and efficient harnessing of multi-form energy to extend human ingenuity and capability. The world's routine consumption of energy is confronted by a series of serious challenges in the context of long-term sustainability, which include the depletion of traditional non-renewable fossil fuel reserves, rapidly surging energy demands, global warming, and other energy-relevant environmental problems.[403,404] As a sustainable energy supply is essential for global development, the exploration of novel high-efficiency strategies for clean, renewable energy storage and conversion is becoming increasingly urgent.[404,405] Among different types of materials, metal chalcogenides, exhibiting rich chemical compositions, diverse crystal structures, tuneable nano/microstructures, and fascinating physicochemical properties, have been considered as promising building blocks for renewable energy conversion and storage devices.[6,72,108,406–408] In addition, many metal chalcogenides possess several other advantages with respect to the selection of materials for devices, such as relatively low cost and high Earth abundance, each of which are of great significance for practical (commercial) application.[108] Metal chalcogenides have attracted wide attention in a variety of promising energy-related applications, such as lithium-ion batteries,[409–412] supercapacitors,[413–418] thermoelectric conversion,[109,188,419–425] solar cells[9,58,298,426,427] and catalysis.[428–432] This section will focus on these typical applications of metal chalcogenides.

1.3.1 Energy Storage Applications

1.3.1.1 Lithium-ion Batteries

Efficient energy storage is a long-standing scientific and technological issue that has global effects on humanity. Electrochemical techniques with the advantages of environmental friendliness and high efficiency are regarded as ones of the most promising candidates for clean energy storage.[433–435] Secondary batteries, also called rechargeable batteries, comprise cells constructed for storing and releasing electrical energy in which each cell, after discharge, can be restored to its original charged state. Metal chalcogenides feature in various battery chemistries, including lithium-ion, sodium-ion and lithium–sulfur batteries among others.[436–440]

A lithium-ion battery (LIB) refers to a type of secondary battery, which utilises internal lithium ions and external electrons to move between the positive and negative electrodes to convert chemical energy into electricity and *vice versa*. LIBs entered the energy stage in the early 1970s and have been quickly developed since their first commercialisation by the Sony Corporation in 1991.[441,442] The main components of a LIB include a positive electrode (*i.e.*, cathode), a negative electrode (*i.e.*, anode), an electrolyte and a separator. The operation principles of LIBs can be summarised as follows (Figure 1.8): (1) when LIBs are charging, Li^+ ions de-intercalate under an applied potential from the cathode and then intercalate into (react with) the anode; (2) when discharging, Li^+ ions transport through the electrolyte to intercalate back into (react with) the cathode, generating a current.[443,444] During such charge/discharge cycles, Li^+ ions shuttle back and forth between the cathode and

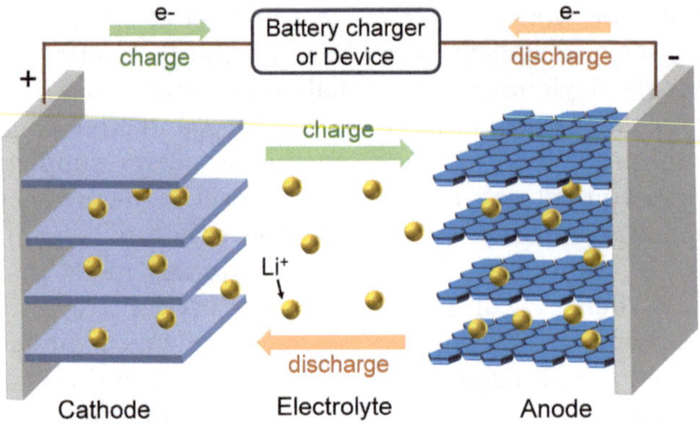

Figure 1.8 Schematic illustration of the working principle for LIBs.

anode, realising the interconversion between chemical energy and electrical energy and enabling electrochemical energy storage within a battery. Accordingly, LIBs are also referred as 'rocking chair batteries', as originally proposed by Armand *et al.* in 1980.[445]

LIBs have already broadly served as the predominant power supply for consumer electronics and electric vehicles for several decades, owing to their relatively high energy density, low self-discharge rate and good cycle life.[446–451] Commercial LIBs generally utilise graphite as anodes and oxide Li-ion intercalation host materials (*e.g.*, $LiCoO_2$) as cathodes. However, these electrode materials experience several limitations. For example, the relatively low theoretical capacity (372 mA h g^{-1}) and slow Li^+ diffusion rate (10^{-8} cm^2 s^{-1}) of the graphite anode impose restrictions on the energy density and power density of LIBs, respectively.[108] Moreover, the growing demands for lighter and thinner LIBs with higher capacity has boosted ongoing research for new electrode materials with superior properties. Therefore, tremendous efforts have been devoted to the exploration and design of alternative electrode materials with high energy storage performance.[452–455] Among them, metal chalcogenides especially layered ones have been regarded as electrode materials of immense potential for LIBs due to the possibilities of high capacity and long cycling life.[409,456]

Typically, MoS_2 has been widely investigated as an anode material for rechargeable batteries since its layered structure and large specific surface area can provide plentiful active sites for the reversible electrochemical insertion of guest ions.[457–459] Additionally, MoS_2 exhibits good capacity retention given the high reversibility of redox reactions with Li ions.[436] A high specific capacity of 750 mA h g^{-1} has been reported when using restacked MoS_2 single layers as electrodes.[460] However, MoS_2 has some intrinsic shortcomings: for example, the electrochemical performance is downgraded by the unfavourable reaction kinetics caused by its low electrical conductivity and the relatively large volume changes that occur during the charge and discharge processes.[461] Fortunately, the above problems can be mitigated *via* the modification of MoS_2 using a number of different strategies. Introducing defects—defect engineering—has been shown to endow MoS_2 with beneficial characteristics, such as providing additional electrochemically active sites for Li^+, resulting in optimised electrochemical performance.[410,412,462] For example, vacancies introduced into MoS_2 break the periodic crystal structure, leading to a much decreased interlayer diffusion energy barrier;[462] Qin *et al.* reported N-doped mesoporous MoS_2 nanosheets with enhanced Li^+ ion storage performance (a consistent capacity of 998 mA h g^{-1} after 100 cycles), which is a consequence of their large surface area, porous microstructure and enlarged interlayer distance.[410]

Additionally, the interplanar distance can play a decisive role in facilitating Li ion intercalation and diffusion in an MoS_2 anode. Liu *et al.* prepared a highly ordered mesoporous MoS_2 material with an expanded (002) interplanar spacing of 0.66 nm, providing sufficient space for ultra-fast Li^+ ion intercalation.[463] In fact, the three-dimensional diffusion of Li^+ is of vital importance to achieve fast charge/discharge kinetics. Tao *et al.* synthesised defect-enriched few-layer MoS_2 nanosheets for Li^+ ion storage and determined that the energy barrier for Li^+ ions to pass through the double-vacancy MoS_2 layer is only 0.238 eV, which is dramatically lower than the zero-defect (3.14 eV) and single-defect (7.97 eV) MoS_2.[411] In addition, MoS_2 can be incorporated with other capacitive and conductive materials to yield MoS_2-based hybrid LIB electrodes with improved electrochemical performance and stability.[412,464–466] For example, MoS_2/graphene composites exhibited a high reversible capacity of 1100 mA h g^{-1} at a current density of 100 mA g^{-1} and good cycling stability when applied as an anode in an LIB.[465] Similarly, MoS_2/CNT hybrids achieved enhanced ionic transfer, thus yielding an extraordinary capacity of 1320 mA h g^{-1} at a current density of 100 mA g^{-1} and an impressive cycling stability of up to 1000 cycles without appreciable capacity fade.[466]

Apart from Mo_2S, a variety of metal sulfides, selenides, and tellurides have also been explored and developed as high-performance LIB electrode materials. These include: MS (M = Zn, Mn, Fe, Sn, Cu, Ni, Co),[467–473] MSe (M = Sn, Zn, Fe),[474–476] SnTe,[477] MS_2 (M = Mo, W, Sn, Zr, Ti, Fe, Ni, V, Co),[478–485] MSe_2 (M = Mo, W, Co, Ni, Fe, Sn),[474,475,478,486,487] Cu_2S,[488] M_2S_3 (M = In, Sb),[489,490] M_2X_3 (M = Sb, Bi; X = Se, Te),[491–495] VS_4,[496] In_3Se_4,[497] Ni_3S_4[498] and Co_9S_8.[499]

1.3.1.2 Supercapacitors

The supercapacitor (SC) is considered as one of the most promising energy storage systems due to characteristic features such as high power density, fast charge/discharge, a long cycling lifespan, simple fabrication procedures and relatively low cost. This combination of features yields a device that bridges the gap between conventional capacitors and batteries.[500–505] Current commercial SCs are mainly based on carbon materials, which normally have limited energy densities (less than 10 W h kg^{-1}).[506] A higher energy density is certainly desirable for almost all practical applications of SCs, so intensive efforts have been devoted to developing alternative electrode materials with higher specific capacitance.[414,507,508] Owing to the amalgamation of several intrinsic physicochemical properties coupled with good electrochemical performance, several metal chalcogenides have been proposed as electrode materials for SCs.[415] For example, two-dimensional layered architectures comprising metal chalcogenides and other conducting materials (*e.g.*, chalcogenide/graphene) have been regarded as among the most promising electrode materials for high performance SCs.[509]

A SC consists of two working electrodes (composed of active materials deposited on current collectors), an electrolyte, and a separator that electrically isolates the two working electrodes. According to the different possible charge storage mechanisms, SCs can be classified into three types: electric double-layer capacitors (EDLCs), pseudo-capacitors and hybrid supercapacitors.[414,508,510,511] The function of EDLCs is controlled by the reversible adsorption/desorption of electrolyte ions at the electrode/electrolyte interface, which forms an electrical double layer region.[512,513] The energy storage achievable through this physical process is relatively low and highly dependent on the accessible surface area of the electrode materials for the adsorption/desorption of ions; high surface area carbon-based materials are typically excellent candidates to serve as EDLC electrodes.[514] In contrast, pseudo-capacitance is based on a quick and reversible faradaic charge transfer through redox and/or intercalation reactions in active materials.[515] These processes endow SCs with much higher charge storage capacities than EDLCs. The charge storage made possible *via* faradaic reactions in metal chalcogenides generates appreciably higher energy density in pseudo-capacitors when compared with carbon-based materials.[516] Hybrid SCs combine the concepts exploited in both EDLCs and pseudo-capacitors and contain two electrodes with very different characteristics; the electrode on one side demonstrates electrostatic capacitance, while the electrode on the other side functions *via* pseudo-capacitance.[413,511,517] Similar to any other energy storage system, the performance of an SC is mainly evaluated in terms of the specific capacity, the energy density and the power density.

Two-dimensional layered MoS_2, which comes with the advantages of an ultrahigh surface area, excellent mechanical properties, high ionic conductivity and high theoretical capacity, is regarded as one of the most outstanding candidates for SCs.[418,518–520] For example, Fan *et al.* successfully prepared flower-like MoS_2/C nanospheres with a high specific capacitance of 201.4 F g^{-1} at a current density of 0.2 A g^{-1}.[520] Acerce *et al.* reported that chemically exfoliated MoS_2 nanosheets containing a high proportion of metallic 1T phase can electrochemically intercalate ions such as H^+, Na^+ and K^+, which therefore obtain high capacitances ranging from 400 to 700 F cm^{-3} in different aqueous electrolytes.[418] The resulting 1T nanostructured MoS_2 film electrode was operated at a high voltage of 3.5 V in organic electrolytes, exhibiting favourable volumetric energy and power densities, as well as a high Coulombic efficiency of 95% after 5000 cycles. The excellent electrochemical performance is ascribed to the hydrophilicity and high electrical conductivity of 1T MoS_2, as well as the ability of the exfoliated layers to dynamically expand and intercalate various ions.[418] Further, the synergetic effects that exist between MoS_2 and carbon materials can facilitate charge transport and electrolyte diffusion in C/MoS_2 composites and relieve the volume expansion/contraction of the electrode during cycling.[509,521,522] For example, da Silveira Firmiano *et al.* reported that an rGO/MoS_2 hybrid electrode achieved an energy density of 63 W h kg^{-1} and an exceptional cycling stability with 92% specific capacitance remained after 1000 cycles.[509] Additionally, the coating of transition metals/metal oxides/metal sulfides on MoS_2 works so as to both improve the structural stability and electrochemical performance of MoS_2.[417] For example, Ti/TiO_2/MoS_2 coaxial hybrids demonstrated a capacitance of 230.2 F g^{-1} and an energy density of 2.70 W h kg^{-1} when adopted in a symmetrical SC, in addition to possessing both mechanical flexibility and stability.[417]

Alternatively, $NiCo_2S_4$ with a high redox activity has become very attractive as an SC electrode due to its high electrical conductivity, electrochemical activity, and capacity compared to mono-metallic sulfides.[416,523] Chen *et al.* reported that sea-urchin-like $NiCo_2S_4$ exhibited dramatically improved electrochemical properties when compared to the sulfides of either of the individual metals, Ni or Co.[524] The $NiCo_2S_4$ electrodes also exhibited high storage stability and long lifetimes with a 91.4% specific capacity retention after 5000 cycles at a current density of 20 A g^{-1}.[524] Later, Pu *et al.* successfully prepared hollow hexagonal nanoplates of the same chalcogenide, $NiCo_2S_4$, which displayed a high specific capacitance of 437 F g^{-1} at a current density of 1 A g^{-1} using 3 mol L^{-1} KOH electrolyte solution.[525] Crucially, it was Chen *et al.* that showed that the electrochemical performance of $NiCo_2S_4$ was strongly associated with the Ni/Co atomic ratio and that the best performance was found in a composition with an Ni/Co atomic ratio of 1.[526] In addition, Ouyang *et al.* reported that an S-Co_3O_4@$NiCo_2S_4$ hierarchical structure exhibited battery-like behaviour with a remarkably large areal capacity of 10.9 mA h cm^{-2} at a current density of 8 mA cm^{-2}, together with an outstanding cycling stability, demonstrating a capacity retention of 97.3% after 5000 cycles.[523] The excellent electrochemical performance of this hierarchical composite was attributed to plentiful electrochemically active sites, fast electron transport paths and the expanded surface area provided by the binder-free core–shell structure.[523]

1.3.2 Energy Conversion Applications

1.3.2.1 Thermoelectric Conversion

A large proportion of energy generated worldwide is lost as waste heat. Effective recovery of waste heat is crucial for solving energy and environmental problems by providing supplemental electrical power without burning additional fossil fuels. Fortunately, thermoelectric technology offers a reliable and environmentally friendly solution to harvest electrical energy from the recycling of waste heat.[109,405,527,528] Additionally, thermoelectric technology can be utilised to convert heat produced by concentrated or unconcentrated sunlight into electricity.[529,530] This is important because infrared radiation with photon energy below the band gap of photosensitisers is not absorbed in conventional photovoltaic cells and generates only waste heat.[529] The three thermoelectric phenomena have been known since the respective discoveries made by Thomas Johann Seebeck, Jean Charles Athanase Peltier, and William Thomson in the 19th century and enable the direct and reversible conversion between heat and electricity in an all-solid-state system without noise, vibration, or emissions of greenhouse gases.[531–533] Solid-state power generation and refrigeration systems, based on the Seebeck and Peltier effects, respectively, have received intensive and sustained attention.[527,534–536] Figure 1.9(a) and (b) provide schematic diagrams of the Seebeck and Peltier effects, respectively. Thermoelectric devices possess many advantages, such as no moving parts and long lifespans that require no maintenance for long periods. These traits favour their use in fields such as remote power supplies (*e.g.*, radioisotope thermoelectric generators),[537–539] the automotive industry,[540,541] temperature sensing,[542] temperature control devices,[543] and implantable/wearable devices.[544–546]

In a typical thermoelectric device, both n- and p-type thermoelectric materials are needed and assembled with electrodes and bonding materials. A junction is formed by a pair of p- and n-type semiconductors and a number of these junctions are connected electrically in series and thermally in parallel to form a thermoelectric device, as illustrated in Figure 1.10. The basic working principles are then as follows:[547] when applying a temperature gradient across the two ends of a thermoelectric material or device, the charge carriers will migrate from the hot side of the material to the cold side, creating a potential difference to power an external load (*via* the Seebeck effect, resulting in a voltage induced by a temperature gradient); conversely, if a thermoelectric material or device is connected to an external electric power source, it can function in an inverse fashion as a thermoelectric refrigerator

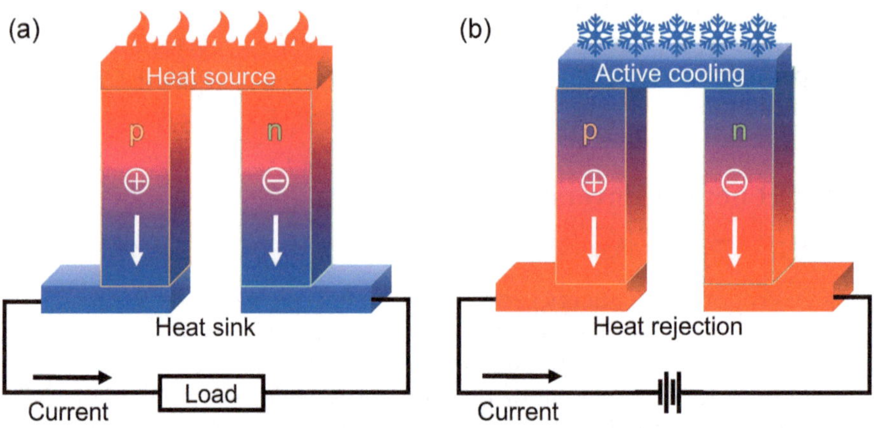

Figure 1.9 Schematic illustration of thermoelectric effects: (a) Seebeck effect and (b) Peltier effect.

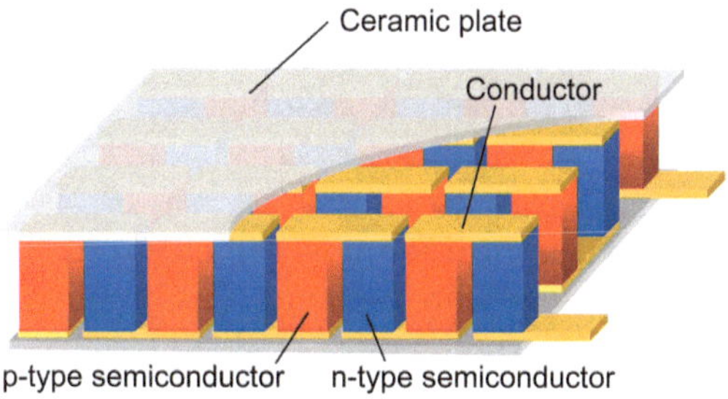

Figure 1.10 Schematic illustration of a thermoelectric device.

(*via* the Peltier effect, generating a temperature gradient induced by an external voltage). The maximum energy conversion efficiency (η_{max}) of a thermoelectric power generation device is given in eqn (1.1):[547]

$$\eta_{max} = \eta_c \times \frac{\sqrt{1+ZT_{avg}}-1}{\sqrt{1+ZT_{avg}}+T_c/T_h} \tag{1.1}$$

where ZT_{avg} is the average figure of merit of thermoelectric material between T_h and T_c, where T_h and T_c are the temperatures of the hot and cold sides, The variable η_c represents the Carnot efficiency, which is expressed in eqn (1.2):[547]

$$\eta_c = 1 - \frac{T_c}{T_h} \tag{1.2}$$

The η_{max} value of a thermoelectric device is strongly dependent on the corresponding thermoelectric performance of the materials employed, which is evaluated (eqn (1.3)) by the dimensionless figure of merit (ZT):[547]

$$ZT = \frac{S^2\sigma T}{\kappa_e + \kappa_l} \tag{1.3}$$

where S, σ, T, κ_e and κ_l are the Seebeck coefficient, electrical conductivity, absolute temperature, electronic thermal conductivity and lattice thermal conductivity, respectively.

High thermoelectric efficiency requires materials with high dimensionless figures of merit—high ZT materials—which are essential to make thermoelectric generation technologies competitive with other energy conversion methods that utilise heat as the energy source.[109] From eqn (1.3), we can intuitively see that an excellent thermoelectric material simultaneously requires a high power factor $S^2\sigma$ (representing good electrical transport) and a suppressed thermal conductivity κ (representing hindered thermal transport). The expressions defining σ, S and κ_e are given in eqn (1.4)–(1.6), respectively:[548–550]

$$\sigma = ne\mu \tag{1.4}$$

$$S = \frac{8\pi^2 k_B{}^2 T}{3eh^2} m^* \left(\frac{\pi}{3n}\right)^{\frac{2}{3}} \tag{1.5}$$

$$\kappa_e = L\sigma T \tag{1.6}$$

where n is the carrier concentration, e is the electron charge, μ is the carrier mobility, k_B is the Boltzmann constant, T is the absolute temperature, h is the Planck constant, m^* is the density of states effective mass, and L is the Lorenz number. Note that the expression for S (eqn (1.5)) is valid for a degenerate semiconductor with a parabolic band dispersion. It can be seen that these three parameters (σ, S and κ_e) are coupled strongly through the carrier concentration n, which makes it difficult to optimise one parameter without (adversely) affecting the other parameters. Therefore, it is necessary to explore strategies that can realise either independent or synergistic regulation of electron/phonon transport in order to optimise thermoelectric performance. After long-term development, researchers have learnt to exploit numerous strategies to improve thermoelectric performance. This might be achieved by enhancing $S^2\sigma$ *via* energy band engineering,[170] optimising carrier concentration,[551] introducing resonant states[552,553] and/or utilising energy filtering effects.[554] Alternatively, performance could be improved *via* suppressing κ_l through nanostructure engineering,[555] defect engineering[556,557] and/or by forming nanocomposites[305,385] and constructing multi-scale hierarchical structures,[558] *etc*. The current priority for developing high performance materials in the thermoelectric field remains to optimise the property parameters synergistically and to integrate all the above strategies effectively.

Generally, a good thermoelectric material behaves as a 'phonon-glass electron-crystal' (PGEC), with the excellent electrical properties typical of a crystalline solid and the ultralow thermal conductivity typical of a glass.[559,560] To date a number of thermoelectric materials, including metal chalcogeni des,[100,108,424,425,561–566] skutterudites,[567–569] clathrates,[570–572] Zintl phase compounds,[573–576] complex alloys,[422,577,578] conductive polymers,[579–581] and some oxides,[582–584] have been identified as promising systems. Among them, metal chalcogenides have already demonstrated outstanding thermoelectric performance and with the huge structural and compositional flexibility that is manifest with chalcogenides, they offer a good platform for designing new and improved materials.[181,263,547,560] For example, Bi_2Te_3 and its solid solutions with Se or Sb are archetypal thermoelectric materials for operation in devices near room temperature,[112,187,193] and PbTe-related materials are prime candidates for medium-temperature applications;[419,585–587] both these systems are representative of traditional and well-established thermoelectric materials. In recent years, many new high performance thermoelectric chalcogenide materials have been identified and widely explored. SnSe- and Cu_2Se-based materials are typical of the new focus for thermoelectrics providing not only some of the highest ZT values in existence, but also meeting ecological requirements of Earth-abundant component elements and low environmental impact.[385,421,425,588–590] We now focus principally on these four metal chalcogenide systems (Bi_2X_3, PbX, SnX and Cu_2X), their fundamental chemistry and their progress to date as bulk thermoelectrics.

1.3.2.1.1 Bi_2Te_3-based Thermoelectric Materials. As state-of-the-art near-room-temperature thermoelectric materials, $(Bi,Sb)_2(Te,Se)_3$ were first commercialised and utilised in thermoelectric refrigerators several decades ago.[547] Across all the possible solid solutions, the compounds $Bi_2Te_{2.7}Se_{0.3}$ and $Bi_{0.5}Sb_{1.5}Te_3$ correspond to the optimal compositions of n- and p-type high-performing thermoelectric materials, respectively.[530] Layer-structured Bi_2Te_3 demonstrates anisotropic thermoelectric properties with higher *ZT* values in the *ab* plane than along the *c*-axis.[182,183] Therefore, one way to enhance the *ZT* of Bi_2Te_3-based compounds is to influence the orientation of crystals. For example, hot deformation has been proven to be an effective method to increase the degree of texturing (preferred orientation) and to optimise the concentration of defects in polycrystalline Bi_2Te_3.[591–594] Consequently, the *ZT* values of the corresponding n-type and p-type Bi_2Te_3-based materials can reach levels of 1.2 (445 K)[591] and 1.3 (380 K),[595] respectively. Nanostructuring can also enhance thermoelectric performance by way of reducing lattice thermal conductivity through introducing high-density grain boundaries.[100,112,596–598] For example, Poudel *et al.* achieved significantly improved *ZT* values (a peak *ZT* of 1.4 at 373 K) in a p-type Bi–Sb–Te nanostructured bulk material prepared by ball milling followed by hot pressing.[100] The magnitude of the figure of merit was mainly ascribed to the low lattice thermal conductivity resulting from the intensified phonon scattering by grain boundaries and defects;[100] similarly, p-type nanostructured $Bi_{0.5}Sb_{1.5}Te_3$ prepared by combining melt spinning and spark plasma sintering reached a maximum *ZT* of 1.56 at 300 K.[598] The thermoelectric performance of Bi_2Te_3-based alloys can also be drastically improved through the introduction of nanoscale secondary phases into matrices.[599] For example, $(Bi,Sb)_2Te_3$ composites embedded with nanostructured SiC achieved a *ZT* of 1.33 at 373 K.[599] Additionally, anti-site defects and vacancies in Bi_2Te_3 can play decisive roles in the conduction type and carrier concentration.[561,597,600,601] Hence, defect engineering can be used very successfully to tune the thermoelectric performance of Bi_2Te_3-based compounds. A further example of this approach was provided by Kim *et al.* who adopted a liquid-phase compaction method to create dislocation arrays at grain boundaries in $Bi_{0.5}Sb_{1.5}Te_3$.[600] In this way they were able to obtain very obvious reductions in lattice thermal conductivity, leading to a maximum *ZT* of 1.86 at 320 K.[600]

1.3.2.1.2 PbTe-based Thermoelectric Materials. Lead chalcogenides, especially PbTe-based materials, are traditional high-performance thermoelectric materials for mid-range temperature (500–900 K) applications.[585,586,602,603] The compounds crystallise with the face-centred cubic NaCl crystal structure (space group $Fm\bar{3}m$). Various dopants have been applied to tune the conduction type and carrier concentration. For example, acceptor doping (p-type) can be achieved by placing Li, Na and K onto the Pb sites, while donor doping (n-type) is possible by either introducing Ga, In, La, Sb, Al, Bi onto the Pb sites or by doping Cl, Br, I on the Te sites. Among these options, Na has proved to be a very effective dopant for p-type PbX (X = S, Se, Te) systems, possessing the highest solubility in PbS (2 mol%) and somewhat lower solubilities in PbSe (0.9 mol%) and PbTe (0.5 mol%).[604] Na doping also introduces point defects and precipitates, leading to a reduced lattice thermal conductivity.[604] Regarding n-type PbX, the best performing materials have been proven to be Al-doped PbSe[605] (with a *ZT* of 1.3 at 850 K) and I-doped PbTe[606] (*ZT* = 1.4 at 700–850 K). Creating nanoprecipitates and/or building hierarchical structures can additionally significantly reduce the lattice thermal conductivity of PbX.[558,602,607–610] Wu *et al.* reported that 3% Na-doped $(PbTe)_{0.8}(PbS)_{0.2}$ achieved a maximum *ZT* of 2.3 at 923 K due mainly to the extremely low thermal conductivity arising from the phase boundaries,[610] whereas Biswas *et al.* achieved significantly reduced lattice thermal conductivity and dramatically enhanced thermoelectric performance (a *ZT* of 2.2 at 923 K) in PbTe-based materials by constructing multi-length scale hierarchical architectures that include point defects, nanoscale endotaxial precipitates and mesoscale grain boundaries in the same material.[558] Such architectures are capable of scattering phonons across different wavelengths.[558] Jiang *et al.* realised a superior *ZT* of 1.8 at 900 K in an n-type PbSe-based high-entropy material that consisted of each of Pb, Sb, Sn, Se, Te and S in a chalcogenide of chemical formula of $Pb_{0.89}Sb_{0.012}Sn_{0.1}Se_{0.5}Te_{0.25}S_{0.25}$.[611] Such a high *ZT* was attributed to first, the consistent electronic transport properties that result from a stabilised single-phase structure and second, the ultra-low lattice thermal conductivity that arises from the strong phonon scattering across a highly strained lattice.[611]

1.3.2.1.3 SnSe-based Thermoelectric Materials. Tin selenide (SnSe) has a layered orthorhombic structure derived from a three-dimensional distortion of the rock-salt structure and possesses a strong anisotropy.[425,588] SnSe undergoes a second-order displacive phase transition from the *Pnma* space group to a higher-symmetry *Cmcm* space group when the temperature increases above ~750 K.[588] SnSe is one of the most promising Te-free thermoelectric candidates emerging in the recent decade and meets requirements of environmental friendliness and low cost in addition to outstanding thermoelectric performance.[425,589,612] Specifically, the low lattice thermal conductivity is due to a strong anharmonicity, and the excellent electrical transport properties result from multiple electronic valence bands.[425,551,589] SnSe-based materials have garnered great attention and experienced vigorous development over a short period of time.[420,613] For example, high *ZT* values of ~2.6 at 923 K[425] and ~2.8 at 773 K[612] were reported in p-type Na-doped SnSe and n-type Br-doped SnSe single crystals, respectively, which originate from both a very low lattice thermal conductivity and the high magnitude of the power factor. Meanwhile, the relatively less impressive thermoelectric performance of polycrystalline SnSe has also been improved by various strategies, such as doping,[378,614–616] forming solid solutions with other chalcogens,[80,617–619] nanostructuring,[614,620,621] obtaining higher purity samples with minimal oxidation[331,622] and composite engineering.[305,340,623–626]

1.3.2.1.4 Cu_2Se-based Thermoelectric Materials. p-Type Cu_2Se-based materials constitute the other main group of candidate thermoelectrics that have been explored extensively in the past decade and their popularity originates from their phonon-liquid electron-crystal (PLEC) behaviour.[385,423,424,590,627] Cu_2Se has a unusual cubic structure at elevated temperature, where ordered Se anions form a rigid sublattice and provide pathways for charge carriers, while Cu cations are randomly distributed around the static Se.[424] The mobile Cu ions in Cu_2Se can scatter phonons significantly and eliminate partial transverse vibrational modes, which leads to extremely low lattice thermal conductivity while preserving good electrical conductivity.[424] Liu *et al.* first reported this liquid-like Cu^+ transport behaviour in non-stoichiometric $Cu_{2-x}Se$ and realised a high *ZT* of 1.5 at 1000 K inspiring systematic investigations of the selenide's PLEC phenomenon and encouraging multiple efforts to enhance thermoelectric performance further.[424] Ball milling and hot pressing was found to induce an extremely low lattice thermal conductivity (0.4–0.5 W m^{-1} K^{-1}) and a *ZT* of ~1.6 at 973 K;[559] as with other chalcogenide thermoelectrics, substitution and nanostructuring strategies also proved extremely successful.[628] $Cu_{1.94}Se_{0.5}S_{0.5}$ solid solutions achieved a remarkable *ZT* of 2.3 at 1000 K,[628] whereas formation of a $Cu_2Se/CuInSe_2$ nanocomposite doped with 1 mol% In improved chemical stability of the selenide and yielded an extraordinary peak *ZT* value of 2.6 at 850 K.[627] At a likely different length scale, Cu_2Se/CNT hybrids achieved a remarkable high *ZT* of 2.4 at 1000 K due to the homogeneously dispersed CNTs inside the Cu_2Se matrix and correspondingly formed Cu_2Se/CNT hybrid interfaces.[385] Apart from Cu_2Se, other Cu-based PLECs, such as Cu_2S,[629] Cu_5FeS_4,[55,56,630,631] and CuAgSe,[632] have also attained excellent *ZT* due to low lattice thermal conductivity and improved electrical transport properties.

1.3.2.1.5 Other Chalcogenide Thermoelectric Materials. In addition to chalcogenides of Bi, Pb, Sn and Cu, Ag_2X (X = S, Se, Te),[296,633–637] GeX (X = Se, Te)[638,639] and various indium selenides (*e.g.*, In_4Se_3, In_3Se_4)[7,562] have also been explored for thermoelectric applications. Orthorhombic-structured Ag_2Se, for example, demonstrates high *ZT* near room-temperature due to its large carrier mobility and low lattice thermal conductivity.[296,636,637] The former is related to the low band effective mass and the latter is ascribed to the low cut-off frequency and group velocity as well as the hybridisation of the optical and acoustic phonon branches.[636] Recently, Ag_2Se/CNT composites were synthesised by *in situ* solution synthesis and spark plasma sintering; due to the interfacial phonon scattering and energy filtering effect, the optimal Ag_2Se/CNT sample registers a remarkably low κ_l of 0.27 W m^{-1} K^{-1} and a slightly enhanced power factor of 2.71 mW m^{-1} K^{-2} at 375 K, leading to an exceptional peak *ZT* of 0.97 at 375 K.[636] Similarly, the $AgSbSe_2$ component in $Ag_2Se/AgSbSe_2$ composites enables intensified phonon scattering and impeded crack propagation, leading to an excellent average *ZT* of ~0.89 between 300 and 375 K and a compressive strength of 197.3 MPa.[637] Undoubtedly, metal chalcogenides have established themselves as most promising thermoelectrics, particularly in the low and medium temperature ranges.

1.3.2.2 Solar Cell Applications

Sunlight is the most widely and abundantly distributed energy source and the solar energy that strikes the Earth in a day is much more than the total energy consumption of the world in a year.[640,641] Energy harvesting directly from sunlight provides a desirable solution for fulfilling the needs of clean energy without negative environmental impact. Solar cells or photovoltaics can capture sunlight and convert solar energy into electricity. These are therefore excellent devices for solar power generation and have already been commercialised, with most of the solar cells that have been utilised based on crystalline silicon.[427,642] High-purity silicon is expensive and so during the past few decades substantial efforts have been devoted to exploring and tailoring various semiconductors for solar cells with lower manufacturing costs compared to silicon without incurring a significant loss of conversion efficiency.[643–647] Metal chalcogenide semiconductors such as cadmium telluride (CdTe) and copper indium gallium selenide, $Cu(In,Ga)(Se,S)_2$, possess outstanding photovoltaic performance but can be produced relatively cheaply compared to silicon.[426,648] The chalcogenide materials have therefore been considered as key materials in the development of high-efficiency solar cells.[649,650]

A solar cell is a semiconductor-based device that serves to convert light energy (certain wavelengths) into electricity *via* the photovoltaic effect. The key indicator in assessing the performance of a solar cell is the amount of electrical power that can be extracted from the light available to it. The size (energy) of the band gap of the semiconducting photoelectrode is inherently important since it should be suitable for the utilisation of the maximal portion of the solar radiation spectrum.[651] The thickness of the semiconducting photoelectrode needs to be large enough to absorb the maximum number of photons of incident solar radiation. The working principle of energy conversion in a solar cell involves three basic processes,[643,652,653] as illustrated in Figure 1.11: (1) absorption of light, which generates electron–hole pairs

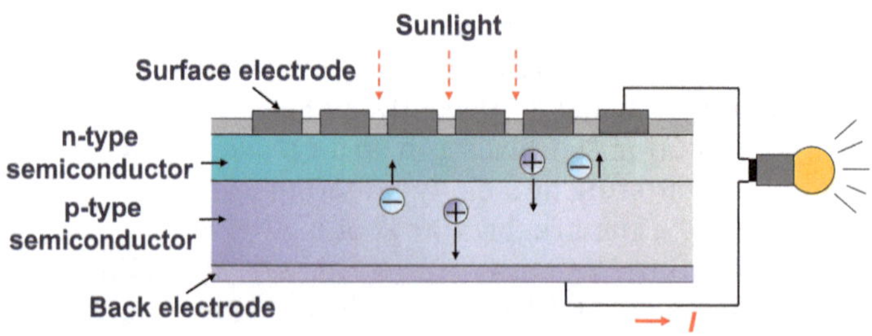

Figure 1.11 Schematic illustration of the working principle of a solar cell.

in the semiconductor; (2) separation of charge carriers of opposite types; (3) separate withdrawal of these carriers to an external circuit. If one considers a p–n junction under illumination for example, the operation of the semiconducting solar cell can be interpreted as follows: the fundamental feature of the junction is the existence of a strong electric field and in equilibrium the electrochemical potentials on the two sides of the junction are equal, so no net electric current is generated; under illumination, however, electron–hole pairs are generated in the semiconductor and subsequently separated by the electric field effect of the junction.[654,655]

CdTe is an important candidate for solar cell applications because of its high optical absorption coefficient and direct band gap (1.45 eV) that closely matches to the solar spectrum for optimum conversion efficiency.[656] For example, a CdTe thin film with a thickness of ~2 μm can absorb nearly 100% of incident solar radiation.[657] Various methods can be utilised for the deposition of CdTe thin films, such as electrodeposition and metal organic chemical vapour deposition.[658–661] In general, the most common CdTe-based solar cell configuration is composed of a p-type CdTe absorber layer and an n-type CdS window layer.[662] A key reason for choosing a CdS/CdTe heterojunction is that the two compounds are miscible and thus an interdiffusion reaction between these two materials will lead to the formation of a $CdS_{1-x}Te_x$ interlayer, which benefits the solar cell performance.[663] CdS/CdTe thin-film solar cells possess an ideal band gap and a high absorption coefficient meaning that a high energy conversion efficiency above 20% can be achieved.[664]

Chalcopyrite $Cu(In,Ga)(Se,S)_2$—known in shorthand as CIGS—has emerged as one of the most promising photovoltaic materials for low-cost, high-efficiency thin film solar cells. Its utility arises from a compositionally-tuneable band gap, high stability under extended periods of excitation and high optical absorption coefficients resulting from the direct band gap character.[665] A high efficiency of 20% for CIGS solar cells has been recorded,[666] which highlights them as probably the most viable substitute for Si in solar cells. The band gap in CIGS can approach ideal values by fine tuning the chemical composition; for example, the band gaps of $CuInSe_2$ and $CuGaSe_2$ are ~1.0 and ~1.7 eV, respectively.[667,668] Generally, the preparation of CIGS-based solar cells starts with the deposition of an absorber material on a Mo-coated glass substrate.[669] The absorber material yielding the highest efficiency is CIGS with a Ga/(Ga+In) ratio of approximately 20% prepared through co-evaporation from elemental sources.[44] Overall, the CIGS solar cell has become a relatively mature, highly efficient energy conversion technology that is commercially produced widely.[670–672]

1.3.2.3 Electrocatalysis for Water Splitting

The hydrogen economy has become a very promising alternative to the current hydrocarbon economy and green hydrogen generation typically involves the process of splitting water into hydrogen and oxygen to harvest renewable energy. The generated hydrogen can then be utilised as a fuel.[428]

Metal chalcogenides have arisen as prime candidates for water splitting *via* catalytic reactions because of highly relevant features such as catalytic edge effects and tuneable band gaps.[673–676] Such catalytic processes are mainly driven by external electrons from either applied current (electrocatalysis) or light irradiation (photocatalysis).

Typically, in the electrocatalytic process of water splitting, the two half reactions—the hydrogen evolution reaction (HER) and the oxygen evolution reaction (OER)—occur at the cathode and anode of an electrolytic cell to produce gaseous molecular hydrogen and oxygen,[677] respectively, as illustrated in Figure 1.12. In electrocatalytic water splitting, the HER generally proceeds *via* two successive steps. The first step of the HER is the adsorption of H *via* the transfer of a proton and its combination with an electron (the Volmer reaction). The equivalent reaction step in alkaline media requires water dissociation to provide the proton (an energetically more demanding process). For the second step, there are two possibilities: one is the Heyrovsky reaction which tends to occur when the concentration of adsorbed H is relatively low. Here the adsorbed H atom combines with a proton from the electrolyte and an electron transferred from the electrode surface to form one equivalent of H_2; the other possibility is the Tafel reaction, in which two adsorbed H atoms combine with each other to form one molecule of H_2. The determination of which reaction step is rate-determining can be simply discerned by measuring the linear Tafel slope from the polarisation curves.[677] The Tafel slope (from a plot of the overpotential *vs.* log current density) represents the response of the reaction rate (current density) to the potential change. It reflects the difficulty of the electrode process; the higher the slope, the more difficult the

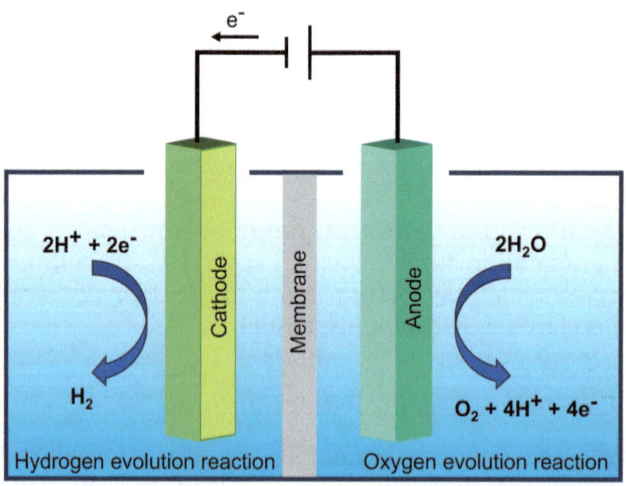

Figure 1.12 Schematic illustration of the process of electrocatalysis for water splitting in acidic media.

electrode process.[428] The overall reaction pathways for the HER are given in eqn (1.7)–(1.10):[428,678]

In acidic media:

$$H^+ + e^- + * \rightarrow H_{ads}\,(\text{Volmer}) \tag{1.7}$$

$$H^+ + e^- + H_{ads} \rightarrow H_2\,(\text{Heyrovsky}) \quad \text{or} \quad 2H_{ads} \rightarrow H_2\,(\text{Tafel}) \tag{1.8}$$

In alkaline media:

$$H_2O + e^- \rightarrow H_{ads} + OH^-\,(\text{Volmer}) \tag{1.9}$$

$$H_2O + e^- + H_{ads} \rightarrow H_2 + OH^-\,(\text{Heyrovsky}) \quad \text{or} \quad 2H_{ads} \rightarrow H_2\,(\text{Tafel}) \tag{1.10}$$

where the symbol '*' represents the active sites on the catalyst surface, and the 'ads' represents adsorbed state of intermediates (H_{ads}).

Compared to the HER, the anodic OER is a complicated multistep electron–proton transfer process accompanied by the absorption of different oxygen intermediates.[679] The generally accepted hypothesis behind the OER mechanism is that the oxygen precursor undergoes the stepwise adsorption–deprotonation–coupling–desorption process (namely, $OH_{ads} \rightarrow O_{ads} \rightarrow OOH_{ads} \rightarrow O_{2ads}$), which is typically named the absorbate evolution mechanism (AEM).[679] Generally, the AEM is proposed to involve several concerted proton–electron transfer reactions.[679] Taking the alkaline OER as an example, OH^- is initially chemisorbed on the coordinatively unsaturated metal site and then effectively deprotonates to generate O_{ads} species (and water + an electron).[679] Afterwards, O–O coupling occurs through nucleophilic attack of OH^- to form an OOH_{ads} intermediate.[679] Finally, the O_2 molecule desorbs from the catalyst surface along with a second deprotonation (forming water), releasing the active metal site and looping the reaction.[679] The overall reaction pathways for the OER in acidic and alkaline media are given in eqn (1.11)–(1.20):[677,679]

In acidic media:

$$* + H_2O \rightarrow OH_{ads} + H^+ + e^- \tag{1.11}$$

$$OH_{ads} \rightarrow O_{ads} + H^+ + e^- \tag{1.12}$$

$$O_{ads} + H_2O \rightarrow OOH_{ads} + H^+ + e^- \tag{1.13}$$

$$OOH_{ads} \rightarrow O_{2ads} + H^+ + e^- \tag{1.14}$$

$$O_{2ads} \rightarrow * + O_2 \tag{1.15}$$

In alkaline media:

$$* + OH^- \rightarrow OH_{ads} + e^- \tag{1.16}$$

$$OH_{ads} + OH^- \rightarrow H_2O + O_{ads} + e^- \tag{1.17}$$

$$O_{ads} + OH^- \rightarrow OOH_{ads} + e^- \tag{1.18}$$

$$OOH_{ads} + OH^- \rightarrow O_{2ads} + H_2O + e^- \quad (1.19)$$

$$O_{2ads} \rightarrow * + O_2 \quad (1.20)$$

where the symbol '*' represents the active sites on the catalyst surface, and the 'ads' represents adsorbed state of intermediates (OH_{ads}, O_{ads}, OOH_{ads}, and O_{2ads}).

The different mechanistic steps of the HER depend on the nature of the electrode materials and the electrolyte media under investigation.[680,681] Notably, since the dissociation of water is required for HER in neutral and alkaline electrolytes, the reaction in such electrolytes is normally two to three orders of magnitude slower than the equivalent process in an acidic medium.[682,683] Overall, H_2 production is strongly dependent on the adsorption/desorption processes, which in turn is governed by the corresponding Gibbs free energy.[678]

Currently, Ru/Ir-based materials work most efficiently for catalysing the OER and Pt-group metals possess the highest electrocatalytic HER activity. However, the high costs and scarcity of both types of catalyst hinder large-scale applications.[684,685] Over the past few decades, researchers have devoted intensive efforts to exploring and designing novel OER/HER catalyst candidates with high Earth abundance, reduced manufacturing cost, and excellent performance.[686–688] Among these candidates, metal chalcogenides are considered as among the most promising electrocatalysts, owing to their low cost and impressive catalytic performance.[689] There are generally two strategies to enhance the activity of an electrocatalyst: by improving the intrinsic catalytic activity and by increasing the number of exposed active sites.[690] As to the former, approaches such as phase engineering,[691,692] defect engineering,[693–695] crystal facet engineering,[696] and multi-metal engineering[697,698] have been developed. Meanwhile, processing a given catalyst by nanostructuring, hybridising with conductive carbon-based materials, and growing on three-dimensional substrates can effectively expose more active sites.[699–704] These strategies can be combined synergistically to achieve even better catalytic performance.

Among the recently developed OER catalysts based on metal chalcogenides, Ni-based and Co-based chalcogenides have been investigated extensively and exhibit excellent activity and stability.[705–708] Zhou *et al.* reported that Ni_3S_2 nanorod arrays had a notable OER activity with a low η_{10} value of 187 mV (η_{10} is used to evaluate the potential difference between the potential reaching a current density of 10 mA cm^{-2} and the thermodynamic potential of HER/OER).[707] Gao *et al.* reported that orthorhombic $CoTe_2$ had better OER activity and stability compared with hexagonal CoTe, which is ascribed to its large active surface, high electrical conductivity and the low adsorption energy of the *OOH intermediate.[708] Furthermore, the incorporation of suitable foreign ions into metal chalcogenides can function well in terms of improving the OER activity.[431,705] For example, Gu *et al.* synthesised Fe-doped $NiSe_2$ ultrathin nanowires with a small η_{10} of 268 mV in 0.1 mol L^{-1} KOH, which was 99 mV and 121 mV lower than that of undoped $NiSe_2$ and commercial RuO_2, respectively.[431] Additionally, the surface properties of metal chalcogenides can be tuned by forming composites with other materials, such that the OER catalytic activity can be enhanced.[709–712] A representative case is the employment of lamellar $CoSe_2$ nanobelts,[709] which have been investigated in some depth as a platform for the grafting of foreign materials successfully giving rise to high performing OER catalysts.[710–712] It was observed that chemically grafting an appropriate material (*e.g.*, graphene, Mn_3O_4, and CeO_2) onto $CoSe_2$ significantly improved its OER activity.[710–712] For example, the obtained CeO_2/$CoSe_2$ composite catalyst had a much improved OER activity, with a small η_{10} of 288 mV and a small Tafel slope of 44 mV dec^{-1} in 0.1 mol L^{-1} KOH, when compared to pure $CoSe_2$ and RuO_2 catalysts. It was hypothesised that the CeO_2 nanoparticles contain highly mobile oxygen vacancies that could bind adsorbates much stronger than normal oxide sites and additionally act so as to transfer electrons to $CoSe_2$. Much of the activity of the catalyst is envisaged to originate from the Co(IV) in $CoSe_2$, which facilitates the formation of OOH and subsequent oxidation to O_2. Therefore, the introduction of CeO_2 appears beneficial in forming OOH_{ads} on the composite surface and given the high mobility of oxygen vacancies at the nanoscale, promotes water oxidation to produce O_2 on the CeO_2/$CoSe_2$ composite.[712]

The research on chalcogenide HER catalysts begins with MoS_2, which has been identified as an efficient HER catalyst both theoretically and experimentally.[713,714] The HER activity of MoS_2 arises from exposed edge surfaces that possess an optimised H* adsorption behaviour similar to Pt.[713–715] This motivated the development of numerous MoS_2 nanostructures to maximise the exposure fraction of edge sites,[157,685,692,715–717] leading to a substantially reduced η_{10} value of ~100 mV in acidic electrolytes. However, the HER performance of MoS_2 in alkaline electrolytes is relatively poor due to unfavourable water adsorption and a slow water dissociation step.[718,719] Doping suitable transition metals (*e.g.*, Ni, Co) into MoS_2 has been shown to accelerate the water dissociation process, thus enabling improved HER activity in alkali electrolytes.[718–720] Nevertheless, substantial efforts have also been devoted to exploring other metal chalcogenide catalysts for the HER in alkaline electrolytes.[429,430,704,721,722] For example, Miao *et al.* reported a mesoporous FeS_2 material with a η_{10} of merely 96 mV and a low Tafel slope of 78 mV dec^{-1} in 0.1 mol L^{-1} KOH.[721] Alternatively, Yin *et al.* studied the influence of various metal dopants (Co, Fe and Cu) on the HER activity of NiS_2, and found that Co-doped NiS_2 had the best HER activity and excellent long-term stability.[722] Intriguingly, it has also been reported that P doping can have beneficial effects on dichalcogenide HER catalysts, for example, inducing a phase change of $CoSe_2$ from cubic to orthorhombic, which resulted in the orthorhombic-$CoSe_2$|P (8 wt% dopant) exhibiting a much enhanced HER activity (η_{10} of 104 mV in 1 mol L^{-1} KOH) as compared to the cubic form of $CoSe_2$ (η_{10} of 330 mV).[429] Similarly, N doping can have a positive effect, as demonstrated in $NiCo_2S_4$.[430] The presence of N was proposed to accelerate the desorption of H* from the S sites and promoted water dissociation, delivering a η_{10} value of 41 mV—a significant decrease from the value of 104 mV for undoped $NiCo_2S_4$.[430]

1.3.2.4 Photocatalysis for Water Splitting

As seen from previous sections, the effective utilisation of solar energy is essential for the achievement of sustainable energy systems. Solar energy collection, conversion and storage are three key processes for practical applications, and these processes can be integrated in a single monolithic photocatalytic water splitting cell, within which solar energy is converted and then stored in simple H–H chemical bonds.[723] Since Fujishima and Honda's ground-breaking work in 1972 on hydrogen generation utilising an n-type TiO_2 electrode,[724] the research in the photocatalysis field has concentrated on the exploration of suitable catalytic materials.[725–743] A number of oxide photocatalysts reported show high activities for water splitting, including TiO_2,[725–730] ZnO,[731–736] Cu_2O,[737,738] and WO_3.[739–743] However, without doping or other optimisation, these oxide photocatalysts are generally active only under ultraviolet light, which accounts for only 5% of the solar spectrum at the Earth surface.[744] As a large fraction of solar light is visible light, it is indispensable to explore and develop novel visible-light-driven photocatalysts for water splitting. In this respect, metal chalcogenides have received considerable attention as candidates for visible-light-driven photocatalysts for water splitting, owing to their suitably narrow band gaps and the position of their valence bands at relatively negative potentials compared to oxides.[744]

While the source of electrons is supplied by an applied external current in electrocatalytic reactions, photocatalysis is driven purely by light and involves five successive processes, including light-material interaction, photogenerated electron–hole excitation and separation, molecular adsorption, surface redox reaction, and product desorption.[745,746] When a semiconductor is irradiated by light (where the incident energy is larger than the band gap), electrons and holes are generated in the conduction band (CB) and valence band (VB), respectively. The photogenerated electrons and holes cause redox reactions similar to electrolysis and water molecules are oxidised by the holes to form oxygen and are reduced by the electrons to form hydrogen for overall water splitting.[744,747] Efficient photocatalytic water splitting relies on the band gap and position of the CB and VB in the electronic structure of the photocatalyst.[748] The band gap determines the energy of the absorbed photons and the produced excitons; *viz.* a narrower band gap implies a wider optical absorption region. In addition, the bottom of the CB must be more negative than the redox potential of H^+/H_2, while the top of VB should be more positive than the redox potential of O_2/H_2O.[749] These criteria indicate that the band gap should be wider than 1.23 eV.[749]

Cadmium chalcogenides (such as CdS) are among the most popular visible-light-driven photocatalysts.[744,750–753] CdS is a well-established photocatalyst with a band gap of 2.4 eV and demonstrates high activity for hydrogen production under visible-light irradiation.[751–753] Nanostructured, high surface area cadmium chalcogenides have proved to be extremely effective photocatalysts for several decades. For example, CdS nanoparticles within a narrow particle size envelope (6–12 nm) and with higher hydrogen evolution rates than bulk materials can be prepared by a precipitation process using different zeolite matrices as templates;[752] CdS nanostructures containing nanopores with a diameter of ~3 nm are able to yield a very large BET surface area of ~112.8 m^2 g^{-1} and delivered a hydrogen yield of about 4.1 mmol h^{-1} under visible-light irradiation.[753] Nanocomposite photocatalysts formed with other metal chalcogenides have also been shown to be successful.[751,754,755] In one case, MoS_2–CdS and WS_2–CdS nanocomposites could be fabricated from few layers of MoS_2 or WS_2 grown upon the surface of CdS, and the resulting nanohybrids rapidly produced astonishingly high yields of hydrogen (1472 mmol h^{-1} g^{-1} for MoS_2–CdS, 1984 mmol h^{-1} g^{-1} for WS_2–CdS) suggesting that the surface of the nanocomposites contained a higher concentration of active sites than CdS itself.[755] In fact, MoS_2 and WS_2 are increasingly finding application more widely as components in composites and heterostructures formed with photocatalytically-active oxides such as TiO_2 and other chalcogenides such as $ZnIn_2S_4$. By forming the composites increased visible light absorption and enhanced charge carrier mobility have been observed.[756,757] All these efforts indicate the immense potential of non-precious metal chalcogenide-based catalysts in terms of the photocatalytic production of hydrogen.

1.4 Summary

This chapter aims to serve as a valuable introduction to the diverse and exciting contemporary solid-state materials chemistry of chalcogenides. In the opening sections it seeks to summarise the basic structures and properties of metal chalcogenides, focuses on the structural characteristics of typical layered metal chalcogenides, briefly captures the development of nanostructured metal chalcogenides and then introduces MoS_2 and Bi_2Te_3, two typical chalcogenide materials with a wide range of applications. The chapter subsequently considers some of the principal synthesis methods that are available to produce single crystals, polycrystalline bulk powders, thin films and nanostructures of metal chalcogenides. Finally, the chapter concentrates on some of the most important fields of development in chalcogenide chemistry today, emphasising the energy storage and energy conversion applications of metal chalcogenides, including the basic principles behind these technologies, some of the foremost metal chalcogenides in use and some of the properties that lie behind their performance.

Chalcogenides with diverse structures and rich compositions exhibit an intriguing mix of electronic, physical and chemical characteristics which not only lead to unusual and rare phenomena but also carve a path to useful applications. There is still plenty of scope for the further manipulation of such a plentiful array of chemical and physical properties and there are many chalcogenide systems that are yet to see their remarkable features exploited. Behind such developments, the understanding of the function of chalcogenides and the underpinning chemistry responsible is critical. We hope that this introduction to some of the fundamentals of chalcogenide chemistry—including crystal chemistry, synthesis approaches and key properties and applications—might inspire the reader to read, learn further and to go on to discover still more regarding these fascinating materials.

References

1. M. Bouroushian, *Electrochemistry of Metal Chalcogenides*, Springer-Verlag, Berlin Heidelberg, 1st edn, 2010.
2. F. A. Devillanova, *Handbook of Chalcogen Chemistry: New Perspectives in Sulfur, Selenium and Tellurium*, Royal Society of Chemistry, United Kingdom, 2007.
3. U. V. Ghorpade, M. P. Suryawanshi, M. A. Green, T. Wu, X. Hao and K. M. Ryan, *Chem. Rev.*, 2023, **123**, 327.
4. L. C. Roof and J. W. Kolis, *Chem. Rev.*, 1993, **93**, 1037.
5. T. Chivers and R. S. Laitinen, *Chem. Soc. Rev.*, 2015, **44**, 1725.
6. R. Woods-Robinson, Y. B. Han, H. Y. Zhang, T. Ablekim, I. Khan, K. A. Persson and A. Zakutayev, *Chem. Rev.*, 2020, **120**, 4007.
7. G. Han, Z.-G. Chen, J. Drennan and J. Zou, *Small*, 2014, **10**, 2747.
8. X. Liu, S. Lee, J. K. Furdyna, T. Luo and Y. Zhang, *Chalcogenide: From 3D to 2D and Beyond*, Woodhead, United Kingdom, 1st edn, 2019.
9. X. Wang, R. Tang, C. Wu, C. Zhu and T. Chen, *J. Energy Chem.*, 2018, **27**, 713.
10. R. Dalven, *Infrared Phys.*, 1969, **9**, 141.
11. S. Li, X. Li, Z. Ren and Q. Zhang, *J. Mater. Chem. A*, 2018, **6**, 2432.
12. G. K. Ahluwalia, *Applications of Chalcogenides: S, Se, and Te*, Springer Nature, Switzerland, 2017.
13. Y. Cui, Z. Zhou, T. Li, K. Wang, J. Li and Z. Wei, *Adv. Funct. Mater.*, 2019, **29**, 1900040.
14. A. Kuhn, A. Chevy and R. Chevalier, *Phys. Status Solidi A*, 1975, **31**, 469.
15. H. Wang, M.-L. Chen, M. Zhu, Y. Wang, B. Dong, X. Sun, X. Zhang, S. Cao, X. Li, J. Huang, L. Zhang, W. Liu, D. Sun, Y. Ye, K. Song, J. Wang, Y. Han, T. Yang, H. Guo, C. Qin, L. Xiao, J. Zhang, J. Chen, Z. Han and Z. Zhang, *Nat. Commun.*, 2019, **10**, 2302.
16. T. Barnowsky, A. V. Krasheninnikov and R. Friedrich, *Adv. Electron. Mater.*, 2023, **9**, 2201112.
17. G. Han, Q. F. Gu, L. Yang, Z. G. Chen and J. Zou, *J. Mater. Chem. C*, 2019, **7**, 13573.
18. S. Popovic, A. Tonejc, B. Grzeta-Plenkovic and R. Trojko, *J. Appl. Crystallogr.*, 1979, **12**, 416.
19. S. Popovic, B. Celustka and D. Bidjin, *Phys. Status Solidi A*, 1971, **6**, 301.
20. K. Osamura, Y. Murakami and Y. Tomiie, *J. Phys. Soc. Jpn.*, 1966, **21**, 1848.
21. C. Manolikas, *J. Solid State Chem.*, 1988, **74**, 319.
22. J. H. C. Hogg, H. H. Sutherland and D. J. Williams, *Acta Crystallogr., Sect. B*, 1973, **29**, 1590.

23. J.-S. Rhyee, K. H. Lee, S. M. Lee, E. Cho, S. I. Kim, E. Lee, Y. S. Kwon, J. H. Shim and G. Kotliar, *Nature*, 2009, **459**, 965.
24. G. Han, Z. G. Chen, C. Sun, L. Yang, L. Cheng, Z. Li, W. Lu, Z. M. Gibbs, J. Snyder, K. Jack, J. Drennan and J. Zou, *CrystEngComm*, 2014, **16**, 393.
25. G. Han, Z.-G. Chen, L. Yang, L. Cheng, J. Drennan and J. Zou, *Cryst. Growth Des.*, 2013, **13**, 5092.
26. G. Han, Z.-G. Chen, L. Yang, L. Cheng, K. Jack, J. Drennan and J. Zou, *Appl. Phys. Lett.*, 2013, **103**, 263105.
27. X. Yin, J. Y. Liu, L. Chen and L. M. Wu, *Acc. Chem. Res.*, 2018, **51**, 240.
28. J.-S. Rhyee, K. Ahn, K. H. Lee, H. S. Ji and J.-H. Shim, *Adv. Mater.*, 2011, **23**, 2191.
29. R. B. Jacobs-Gedrim, M. Shanmugam, N. Jain, C. A. Durcan, M. T. Murphy, T. M. Murray, R. J. Matyi, R. L. Moore and B. Yu, *ACS Nano*, 2013, **8**, 514.
30. T. Zhai, X. Fang, M. Liao, X. Xu, L. Li, B. Liu, Y. Koide, Y. Ma, J. Yao, Y. Bando and D. Golberg, *ACS Nano*, 2010, **4**, 1596.
31. S. D. Lei, F. F. Wen, L. H. Ge, S. Najmaei, A. George, Y. J. Gong, W. L. Gao, Z. H. Jin, B. Li, J. Lou, J. Kono, R. Vajtai, P. Ajayan and N. J. Halas, *Nano Lett.*, 2015, **15**, 3048.
32. G. Zhang, X. Li, D. Wei, H. Yu, J. Ye, S. Chen, W. Zhang, J. Zhu and X. Duan, *Chem. Eng. J.*, 2023, 453, 139841.
33. B. Yu, S. Ju, X. Sun, G. Ng, T. D. Nguyen, M. Meyyappan and D. B. Janes, *Appl. Phys. Lett.*, 2007, **91**, 133119.
34. R. E. Abutbul, E. Segev, U. Argaman, G. Makov and Y. Golan, *Adv. Mater.*, 2018, **30**, 1706285.
35. H.-J. Yan, Z. Li, S.-C. Liu, X. Wang, X. Zhang, D.-J. Xue and J.-S. Hu, *Nano Res.*, 2022, **15**, 1013.
36. Y. Lu and J. H. Warner, *ACS Appl. Electron. Mater.*, 2020, **2**, 1777.
37. H. Zhang, Z. Zhang, Q. Zhan, D. Liu, P. Zhao and Y. Cheng, *J. Mater. Chem. C*, 2022, **10**, 7771.
38. J. P. Heremans, R. J. Cava and N. Samarth, *Nat. Rev. Mater.*, 2017, **2**, 17049.
39. Y. Du, Z. Yin, J. Zhu, X. Huang, X.-J. Wu, Z. Zeng, Q. Yan and H. Zhang, *Nat. Commun.*, 2012, **3**, 1177.
40. G. Chen, J. Zhou, J. Zuo and Q. Yang, *ACS Appl. Mater. Interfaces*, 2016, **8**, 2819.
41. T. Wu, X. Zhou, H. Zhang and X. Zhong, *Nano Res.*, 2010, **3**, 379.
42. H. Zhang, C.-X. Liu, X.-L. Qi, X. Dai, Z. Fang and S.-C. Zhang, *Nat. Phys.*, 2009, **5**, 438.
43. G. D. Moon, *Anisotropic Metal Chalcogenide Nanomaterials: Synthesis, Assembly, and Applications*, Springer, Cham, Switzerland, 1st edn, 2019.
44. A. Qurashi, *Metal Chalcogenide Nanostructures for Renewable Energy Applications*, Scrivener, United States of America, 2014.
45. V. Tsurkan, H.-A. K. Von Nidda, J. Deisenhofer, P. Lunkenheimer and A. Loidl, *Phys. Rep.-Rev. Sec. Phys. Lett.*, 2021, **926**, 1.

46. Q. Li, Y. Zhu, M. Cheng, L. Du, M. Zhao, G. Zhang, G. Wang, W. Wang, H. Liu, Y. Chen and W. Xiao, *J. Mater. Chem. A*, 2023, **12**, 59.
47. J. E. Iglesias and H. Steinfink, *J. Solid State Chem.*, 1973, **6**, 119.
48. Y. D. Tretyakov, I. V. Gordeev and Y. A. Kesler, *J. Solid State Chem.*, 1977, **20**, 345.
49. M. Liu, Y. Ding, L. Wang, Y. Wang and X. Wang, *Adv. Compos. Hybrid Mater.*, 2022, **5**, 461.
50. P. Jackson, R. Wuerz, D. Hariskos, E. Lotter, W. Witte and M. Powalla, *Phys. Status Solidi RRL*, 2016, **10**, 583.
51. H. Azimi, Y. Hou and C. J. Brabec, *Energy Environ. Sci.*, 2014, **7**, 1829.
52. X. Shi, L. Xi, J. Fan, W. Zhang and L. Chen, *Chem. Mater.*, 2010, **22**, 6029.
53. L. Hu, Y. Luo, Y.-W. Fang, F. Qin, X. Cao, H. Xie, J. Liu, J. Dong, A. Sanson, M. Giarola, X. Tan, Y. Zheng, A. Suwardi, Y. Huang, K. Hippalgaonkar, J. He, W. Zhang, J. Xu, Q. Yan and M. G. Kanatzidis, *Adv. Energy Mater.*, 2021, **11**, 2100661.
54. C. Han, Q. Sun, Z. X. Cheng, J. L. Wang, Z. Li, G. Q. Lu and S. X. Dou, *J. Am. Chem. Soc.*, 2014, **136**, 17626.
55. P. Qiu, T. Zhang, Y. Qiu, X. Shi and L. Chen, *Energy Environ. Sci.*, 2014, **7**, 4000.
56. A. J. Zhang, B. Zhang, W. Lu, D. D. Xie, H. X. Ou, X. D. Han, J. Y. Dai, X. Lu, G. Han, G. Y. Wang and X. Y. Zhou, *Adv. Funct. Mater.*, 2018, **28**, 1705117.
57. K. Doverspike, K. Dwight and A. Wold, *Chem. Mater.*, 1990, **2**, 194.
58. Q. Guo, H. W. Hillhouse and R. Agrawal, *J. Am. Chem. Soc.*, 2009, **131**, 11672.
59. M.-L. Liu, I. W. Chen, F.-Q. Huang and L.-D. Chen, *Adv. Mater.*, 2009, **21**, 3808.
60. T. Fang, X. Li, C. Hu, Q. Zhang, J. Yang, W. Zhang, X. Zhao, D. J. Singh and T. Zhu, *Adv. Funct. Mater.*, 2019, **29**, 1900677.
61. H.-S. Kim, N. A. Heinz, Z. M. Gibbs, Y. Tang, S. D. Kang and G. J. Snyder, *Mater. Today*, 2017, **20**, 452.
62. R. G. Dickinson and L. Pauling, *J. Am. Chem. Soc.*, 1923, **45**, 1466.
63. Y. Feldman, E. Wasserman, D. J. Srolovitz and R. Tenne, *Science*, 1995, **267**, 222.
64. K. S. Novoselov, A. K. Geim, S. V. Morozov, D. Jiang, Y. Zhang, S. V. Dubonos, I. V. Grigorieva and A. A. Firsov, *Science*, 2004, **306**, 666.
65. Q. H. Wang, K. Kalantar-Zadeh, A. Kis, J. N. Coleman and M. S. Strano, *Nat. Nanotechnol.*, 2012, **7**, 699.
66. F. Bonaccorso, L. Colombo, G. Yu, M. Stoller, V. Tozzini, A. C. Ferrari, R. S. Ruoff and V. Pellegrini, *Science*, 2015, **347**, 1246501.
67. G. Zhang, H. Liu, J. Qu and J. Li, *Energy Environ. Sci.*, 2016, **9**, 1190.
68. S.-L. Li, K. Tsukagoshi, E. Orgiu and P. Samori, *Chem. Soc. Rev.*, 2016, **45**, 118.
69. C. Tan, Z. Lai and H. Zhang, *Adv. Mater.*, 2017, **29**, 1701392.
70. J. H. Han, M. Kwak, Y. Kim and J. Cheon, *Chem. Rev.*, 2018, **118**, 6151.

71. J. D. Cain, E. D. Hanson, F. Shi and V. P. Dravid, *Curr. Opin. Solid State Mater. Sci.*, 2016, **20**, 374.
72. S. Manzeli, D. Ovchinnikov, D. Pasquier, O. V. Yazyev and A. Kis, *Nat. Rev. Mater.*, 2017, **2**, 17033.
73. A. V. Kolobov and J. Tominaga, *Two-Dimensional Transition-Metal Dichalcogenides*, Springer Nature, Switzerland, 2016.
74. Q. Wang, Y. Lei, D. Wang and Y. Li, *Energy Environ. Sci.*, 2019, **12**, 1730.
75. T. Zhang, J. Sun and J. Guan, *Nano Res.*, 2023, **16**, 8684.
76. Z. Yang, L. Zhu, C. Lv, R. Zhang, H. Wang, J. Wang and Q. Zhang, *Mater. Chem. Front.*, 2021, **5**, 5880.
77. Y. Jung, Y. Zhou and J. J. Cha, *Inorg. Chem. Front.*, 2016, **3**, 452.
78. Z. Guo, F. Sun and W. Yuan, *Cryst. Growth Des.*, 2017, **17**, 2238.
79. Z. Liu, S. Y. Tee, G. Guan and M.-Y. Han, *Nano-Micro Lett.*, 2024, **16**, 95.
80. G. Han, S. R. Popuri, H. F. Greer, R. Z. Zhang, L. Ferre-Llin, J. W. G. Bos, W. Z. Zhou, M. J. Reece, D. J. Paul, A. R. Knox and D. H. Gregory, *Chem. Sci.*, 2018, **9**, 3828.
81. Y. Min, E. Im, G.-T. Hwang, J.-W. Kim, C.-W. Ahn, J.-J. Choi, B.-D. Hahn, J.-H. Choi, W.-H. Yoon, D.-S. Park, D. C. Hyun and G. D. Moon, *Nano Res.*, 2019, **12**, 1750.
82. J. Jin, T. Xiao, Y.-f. Zhang, H. Zheng, H. Wang, R. Wang, Y. Gong, B. He, X. Liu and K. Zhou, *Nanoscale*, 2021, **13**, 19740.
83. C. Zhu, D. Gao, J. Ding, D. Chao and J. Wang, *Chem. Soc. Rev.*, 2018, **47**, 4332.
84. M. Chhowalla, H. S. Shin, G. Eda, L.-J. Li, K. P. Loh and H. Zhang, *Nat. Chem.*, 2013, **5**, 263.
85. S. Iijima, *Nature*, 1991, **354**, 56.
86. G. Cao, *Nanostructures and Nanomaterials: Synthesis, Properties and Applications*, World Scientific, Singapore, 2004.
87. B. Das, S. Subramaniam and M. R. Melloch, *Semicond. Sci. Technol.*, 1993, **8**, 1347.
88. C. Vieu, F. Carcenac, A. Pepin, Y. Chen, M. Mejias, A. Lebib, L. Manin-Ferlazzo, L. Couraud and H. Launois, *Appl. Surf. Sci.*, 2000, **164**, 111.
89. Y. Zhai, X. Ren, B. Wang and S. Liu, *Adv. Funct. Mater.*, 2022, **32**, 2207536.
90. S. Sha, R. Ge, Y. Li, J. M. Cairney, R. Zheng, S. Li, B. Liu, J. Zhang and W. Li, *Front. Energy*, 2023, **18**, 265.
91. S. J. Ikhmayies, *Advances in Energy Materials*, Springer Nature, Switzerland, 2020.
92. R. Tenne, L. Margulis, M. Genut and G. Hodes, *Nature*, 1992, **360**, 444.
93. Y. Feldman, G. L. Frey, M. Homyonfer, V. Lyakhovitskaya, L. Margulis, H. Cohen, G. Hodes, J. L. Hutchison and R. Tenne, *J. Am. Chem. Soc.*, 1996, **118**, 5362.
94. C. B. Murray, D. J. Norris and M. G. Bawendi, *J. Am. Chem. Soc.*, 1993, **115**, 8706.
95. C. N. R. Rao and M. Nath, *Dalton Trans.*, 2003, 1.
96. Y. Min, J. W. Roh, H. Yang, M. Park, S. I. Kim, S. Hwang, S. M. Lee, K. H. Lee and U. Jeong, *Adv. Mater.*, 2013, **25**, 1425.

97. T.-T. Zhuang, *Design, Synthesis and Applications of One-Dimensional Chalcogenide Hetero-Nanostructures: Novel Metal Sulfide Hetero-Nanorods for Enhancing Solar Energy Conversion*, Springer Nature, Singapore, 2018.
98. Q. Jin, S. Jiang, Y. Zhao, D. Wang, J. Qiu, D.-M. Tang, J. Tan, D.-M. Sun, P.-X. Hou, X.-Q. Chen, K. Tai, N. Gao, C. Liu, H.-M. Cheng and X. Jiang, *Nat. Mater.*, 2019, **18**, 62.
99. Q. Zhang, X. Ai, L. Wang, Y. Chang, W. Luo, W. Jiang and L. Chen, *Adv. Funct. Mater.*, 2015, **25**, 966.
100. B. Poudel, Q. Hao, Y. Ma, Y. C. Lan, A. Minnich, B. Yu, X. A. Yan, D. Z. Wang, A. Muto, D. Vashaee, X. Y. Chen, J. M. Liu, M. S. Dresselhaus, G. Chen and Z. F. Ren, *Science*, 2008, **320**, 634.
101. J. Tian, W. Zhang, J. Gu, T. Deng and D. Zhang, *Nano Energy*, 2015, **17**, 52.
102. M. R. Gao, J. Jiang and S. H. Yu, *Small*, 2012, **8**, 13.
103. H. Wang, Y. Liang, Y. Li and H. Dai, *Angew. Chem., Int. Ed.*, 2011, **50**, 10969.
104. T. Cao, R. Wang, R. E. Simpson and G. Li, *Prog. Quantum Electron.*, 2020, **74**, 100299.
105. J. Tomforde, W. Bensch, L. Kienle, V. Duppel, P. Merkelbach and M. Wuttig, *Chem. Mater.*, 2011, **23**, 3871.
106. H. S. Nalwa, *Nanostructured Materials and Nanotechnology*, Academic Press, The United States of America, 2000.
107. J. Zhang, Y. Chen and X. Wang, *Energy Environ. Sci.*, 2015, **8**, 3092.
108. M. R. Gao, Y. F. Xu, J. Jiang and S. H. Yu, *Chem. Soc. Rev.*, 2013, **42**, 2986.
109. J. He and T. M. Tritt, *Science*, 2017, **357**, eaak9997.
110. R. Ganatra and Q. Zhang, *ACS Nano*, 2014, **8**, 4074.
111. C. Tan and H. Zhang, *Chem. Soc. Rev.*, 2015, **44**, 2713.
112. H. Mamur, M. R. A. Bhuiyan, F. Korkmaz and M. Nil, *Renewable Sustainable Energy Rev.*, 2018, **82**, 4159.
113. P. D. Fleischauer and R. Bauer, *Tribol. Trans.*, 1988, **31**, 239.
114. M. Bernardi, M. Palummo and J. C. Grossman, *Nano Lett.*, 2013, **13**, 3664.
115. Z. Z. Wu, B. Z. Fang, Z. P. Wang, C. L. Wang, Z. H. Liu, F. Y. Liu, W. Wang, A. Alfantazi, D. Z. Wang and D. P. Wilkinson, *ACS Catal.*, 2013, **3**, 2101.
116. K. Chang and W. Chen, *Chem. Commun.*, 2011, **47**, 4252.
117. W. Zhang, P. Zhang, Z. Su and G. Wei, *Nanoscale*, 2015, **7**, 18364.
118. B. Radisavljevic, A. Radenovic, J. Brivio, V. Giacometti and A. Kis, *Nat. Nanotechnol.*, 2011, **6**, 147.
119. L. F. Mattheis, *Phys. Rev. B*, 1973, **8**, 3719.
120. R. E. Bell and R. E. Herfert, *J. Am. Chem. Soc.*, 1957, **79**, 3351.
121. B. Radisavljevic, M. B. Whitwick and A. Kis, *ACS Nano*, 2011, **5**, 9934.
122. J. C. Wildervanck and F. Jellinek, *Z. Anorg. Allg. Chem.*, 1964, **328**, 309.
123. M. R. Vazirisereshk, A. Martini, D. A. Strubbe and M. Z. Baykara, *Lubricants*, 2019, **7**, 57.
124. Y. Fang, J. Pan, J. He, R. Luo, D. Wang, X. Che, K. Bu, W. Zhao, P. Liu, G. Mu, H. Zhang, T. Lin and F. Huang, *Angew. Chem., Int. Ed.*, 2018, **57**, 1232.

125. X. Chen, Z. Chen and J. Li, *Chin. Sci. Bull.*, 2013, **58**, 1632.
126. T. Stephenson, Z. Li, B. Olsen and D. Mitlin, *Energy Environ. Sci.*, 2014, **7**, 209.
127. T. Hu, R. Li and J. M. Dong, *J. Chem. Phys.*, 2013, **139**, 174702.
128. Y. F. Yu, G. H. Nam, Q. Y. He, X. J. Wu, K. Zhang, Z. Z. Yang, J. Z. Chen, Q. L. Ma, M. T. Zhao, Z. Q. Liu, F. R. Ran, X. Z. Wang, H. Li, X. Huang, B. Li, Q. H. Xiong, Q. Zhang, Z. Liu, L. Gu, Y. H. Du, W. Huang and H. Zhang, *Nat. Chem.*, 2018, **10**, 638.
129. V. Yadav, S. Roy, P. Singh, Z. Khan and A. Jaiswal, *Small*, 2019, **15**, 1803706.
130. F. Wypych and R. Schollhorn, *J. Chem. Soc., Chem. Commun.*, 1992, 1386.
131. D. Voiry, A. Goswami, R. Kappera, C. d. C. Castro e Silva, D. Kaplan, T. Fujita, M. W. Chen, T. Asefa and M. Chhowalla, *Nat. Chem.*, 2015, **7**, 45.
132. R. Lv, J. A. Robinson, R. E. Schaak, D. Sun, Y. F. Sun, T. E. Mallouk and M. Terrones, *Acc. Chem. Res.*, 2015, **48**, 56.
133. E. R. Braithwaite and J. Haber, *Molybdenum: An Outline of its Chemistry and Uses*, Elsevier, Amsterdam, 1994.
134. K.-A. N. Duerloo, Y. Li and E. J. Reed, *Nat. Commun.*, 2014, **5**, 4214.
135. G. Eda, H. Yamaguchi, D. Voiry, T. Fujita, M. Chen and M. Chhowalla, *Nano Lett.*, 2011, **11**, 5111.
136. B. Radisavljevic and A. Kis, *Nat. Mater.*, 2013, **12**, 815.
137. X. Gan, L. Y. S. Lee, K.-y. Wong, T. W. Lo, K. H. Ho, D. Y. Lei and H. Zhao, *ACS Appl. Energy Mater.*, 2018, **1**, 4754.
138. X. Fan, P. Xu, D. Zhou, Y. Sun, Y. C. Li, M. A. T. Nguyen, M. Terrones and T. E. Mallouk, *Nano Lett.*, 2015, **15**, 5956.
139. G. Eda, T. Fujita, H. Yamaguchi, D. Voiry, M. Chen and M. Chhowalla, *ACS Nano*, 2012, **6**, 7311.
140. S. Bhattacharyya and A. K. Singh, *Phys. Rev. B:Condens. Matter Mater. Phys.*, 2012, **86**, 075454.
141. A. Kuc, N. Zibouche and T. Heine, *Phys. Rev. B:Condens. Matter Mater. Phys.*, 2011, **83**, 245213.
142. A. Splendiani, L. Sun, Y. B. Zhang, T. S. Li, J. Kim, C. Y. Chim, G. Galli and F. Wang, *Nano Lett.*, 2010, **10**, 1271.
143. K. Chang, M. Li, T. Wang, S. Ouyang, P. Li, L. Liu and J. Ye, *Adv. Energy Mater.*, 2015, **5**, 1402279.
144. H. Peelaers and C. G. Van de Walle, *J. Phys. Chem. C*, 2014, **118**, 12073.
145. W. Zhou, X. L. Zou, S. Najmaei, Z. Liu, Y. M. Shi, J. Kong, J. Lou, P. M. Ajayan, B. I. Yakobson and J. C. Idrobo, *Nano Lett.*, 2013, **13**, 2615.
146. C. Nie, M. Yin, Y. Zhao, C. Zhao, B. Zhang, X. Song, X. Yi, Y. Zhang, L. Luo and S. Wang, *J. Phys. Chem. C*, 2021, **125**, 25739.
147. Z. Wang and B. Mi, *Environ. Sci. Technol.*, 2017, **51**, 8229.
148. X. Zhang, Z. Lai, C. Tan and H. Zhang, *Angew. Chem., Int. Ed.*, 2016, **55**, 8816.
149. A. Castellanos-Gomez, M. Poot, G. A. Steele, H. S. J. van der Zant, N. Agrait and G. Rubio-Bollinger, *Adv. Mater.*, 2012, **24**, 772.

150. S. Bertolazzi, J. Brivio and A. Kis, *ACS Nano*, 2011, **5**, 9703.
151. J. Pu, Y. Yomogida, K. K. Liu, L. J. Li, Y. Iwasa and T. Takenobu, *Nano Lett.*, 2012, **12**, 4013.
152. L. Rapoport, N. Fleischer and R. Tenne, *J. Mater. Chem.*, 2005, **15**, 1782.
153. T. Lemogne, C. Donnet, J. M. Martin, A. Tonck, N. Millardpinard, S. Fayeulle and N. Moncoffre, *J. Vac. Sci. Technol., A*, 1994, **12**, 1998.
154. J. Feng, X. F. Qian, C. W. Huang and J. Li, *Nat. Photonics*, 2012, **6**, 865.
155. K.-A. N. Duerloo, M. T. Ong and E. J. Reed, *J. Phys. Chem. Lett.*, 2012, **3**, 2871.
156. J. Ma, X. Xiong, D. Wu, Y. Wang, C. Ban, Y. Feng, J. Meng, X. Gao, J.-Y. Dai, G. Han, L.-Y. Gan and X. Zhou, *Adv. Mater.*, 2023, **35**, 2300027.
157. J. Kibsgaard, Z. B. Chen, B. N. Reinecke and T. F. Jaramillo, *Nat. Mater.*, 2012, **11**, 963.
158. X. Chu, G. Yao, A. T. S. Wee and X.-S. Wang, *Nanotechnology*, 2012, **23**, 375603.
159. X. Zhong, H. D. Yang, S. J. Guo, S. W. Li, G. L. Gou, Z. Y. Niu, Z. P. Dong, Y. J. Lei, J. Jin, R. Li and J. T. Ma, *J. Mater. Chem.*, 2012, **22**, 13925.
160. H. Xu, J. J. Yi, X. J. She, Q. Liu, L. Song, S. M. Chen, Y. C. Yang, Y. H. Song, R. Vajtai, J. Lou, H. M. Li, S. Q. Yuan, J. J. Wu and P. M. Ajayan, *Appl. Catal., B*, 2018, **220**, 379.
161. K. Ai, C. Ruan, M. Shen and L. Lu, *Adv. Funct. Mater.*, 2016, **26**, 5542.
162. M. H. Francombe, *Br. J. Appl. Phys.*, 1958, **9**, 415.
163. W. M. Yim and F. D. Rosi, *Solid-State Electron.*, 1972, **15**, 1121.
164. W. S. Liu, K. C. Lukas, K. McEnaney, S. Lee, Q. Zhang, C. P. Opeil, G. Chen and Z. F. Ren, *Energy Environ. Sci.*, 2013, **6**, 552.
165. S. Wang, Y. Sun, J. Yang, B. Duan, L. Wu, W. Zhang and J. Yang, *Energy Environ. Sci.*, 2016, **9**, 3436.
166. C. A. Kagarakis, *J. Mater. Sci.*, 1978, **13**, 1594.
167. Y. Jiang, Y. Wang, J. Sagendorf, D. West, X. Kou, X. Wei, L. He, K. L. Wang, S. Zhang and Z. Zhang, *Nano Lett.*, 2013, **13**, 2851.
168. J. R. Drabble and C. H. L. Goodman, *J. Phys. Chem. Solids*, 1958, **5**, 142.
169. B. L. Huang and M. Kaviany, *Phys. Rev. B:Condens. Matter Mater. Phys.*, 2008, **77**, 125209.
170. Y. Pei, X. Shi, A. LaLonde, H. Wang, L. Chen and G. J. Snyder, *Nature*, 2011, **473**, 66.
171. Y. Pei, H. Wang and G. J. Snyder, *Adv. Mater.*, 2012, **24**, 6125.
172. G. F. Wang and T. Cagin, *Phys. Rev. B:Condens. Matter Mater. Phys.*, 2007, **76**, 075201.
173. S. K. Mishra, S. Satpathy and O. Jepsen, *J. Phys.:Condens. Matter*, 1997, **9**, 461.
174. J. Horak, K. Cermak and L. Koudelka, *J. Phys. Chem. Solids*, 1986, **47**, 805.
175. J. Jiang, L. D. Chen, Q. Yao, S. Q. Bai and Q. Wang, *Mater. Chem. Phys.*, 2005, **92**, 39.
176. D. Perrin, M. Chitroub, S. Scherrer and H. Scherrer, *J. Phys. Chem. Solids*, 2000, **61**, 1687.

177. Y. H. Wu, R. S. Zhai, T. J. Zhu and X. B. Zhao, *Mater. Today Phys.*, 2017, **2**, 62.
178. H. L. Zhuang, Y. Pan, F. H. Sun, J. F. Dong, J. Pei, Asfandiyar, B. W. Cai, H. H. Hu, H. Tang and J. F. Li, *Nano Energy*, 2019, **60**, 857.
179. K. Kim, G. Kim, H. Lee, K. H. Lee and W. Lee, *Scr. Mater.*, 2018, **145**, 41.
180. H. X. Qin, Y. Liu, Z. W. Zhang, Y. M. Wang, J. Cao, W. Cai, Q. Zhang and J. H. Sui, *Mater. Today Phys.*, 2018, **6**, 31.
181. D. M. Rowe, *Thermoelectrics Handbook: Macro to Nano*, CRC Press, Boca Raton, 1st edn, 2006.
182. R. T. Delves, A. E. Bowley, D. W. Hazelden and H. J. Goldsmid, *Proc. Phys. Soc.*, 1961, **78**, 838.
183. H. J. Goldsmid, *Proc. Phys. Soc. B*, 1956, **69**, 203.
184. J. J. Shen, L. P. Hu, T. J. Zhu and X. B. Zhao, *Appl. Phys. Lett.*, 2011, **99**, 124102.
185. Y. Xiao, W. Liu, Y. Zhang, D. Wang, H. Shi, S. Wang, Y. Jin, W. Qu, H. Wu, X. Ding, J. Sun and L.-D. Zhao, *J. Mater. Chem. A*, 2021, **9**, 23011.
186. C. B. Satterthwaite and R. W. Ure, *Phys. Rev.*, 1957, **108**, 1164.
187. F. Hao, P. Qiu, Y. Tang, S. Bai, T. Xing, H.-S. Chu, Q. Zhang, P. Lu, T. Zhang, D. Ren, J. Chen, X. Shi and L. Chen, *Energy Environ. Sci.*, 2016, **9**, 3120.
188. B. Zhu, X. Liu, Q. Wang, Y. Qiu, Z. Shu, Z. Guo, Y. Tong, J. Cui, M. Gu and J. He, *Energy Environ. Sci.*, 2020, **13**, 2106.
189. I. T. Witting, F. Ricci, T. C. Chasapis, G. Hautier and G. J. Snyder, *Research*, 2020, **2020**, 4361703.
190. T. E. Svechnikova, N. M. Maksimova and P. P. Konstantinov, *Inorg. Mater.*, 1998, **34**, 233.
191. Y. L. Chen, J. G. Analytis, J. H. Chu, Z. K. Liu, S. K. Mo, X. L. Qi, H. J. Zhang, D. H. Lu, X. Dai, Z. Fang, S. C. Zhang, I. R. Fisher, Z. Hussain and Z. X. Shen, *Science*, 2009, **325**, 178.
192. L. M. Wang, Z. M. Zhang, Y. C. Liu, B. R. Wang, L. Fang, J. J. Qiu, K. Zhang and S. R. Wang, *Nat. Commun.*, 2018, **9**, 3817.
193. I. T. Witting, T. C. Chasapis, F. Ricci, M. Peters, N. A. Heinz, G. Hautier and G. J. Snyder, *Adv. Electron. Mater.*, 2019, **5**, 1800904.
194. P. Rudolph and F. M. Kiessling, *Cryst. Res. Technol.*, 1988, **23**, 1207.
195. P. Rudolph, *Prog. Cryst. Growth Charact. Mater.*, 1994, **29**, 275.
196. K. Hoshikawa, T. Taishi, E. Ohba, C. Miyagawa, T. Kobayashi, J. Yanagisawa and M. Shinozuka, *J. Cryst. Growth*, 2014, **401**, 146.
197. M. I. Pech-Canul and N. M. Ravindra, *Semiconductors: Synthesis, Properties and Applications*, Springer Nature, Switzerland, 1st edn, 2019.
198. R. S. Rai, S. Mahajan, S. McDevitt and C. J. Johnson, *J. Vac. Sci. Technol., B*, 1991, **9**, 1892.
199. K. C. Mandal, A. Mertiri, G. W. Pabst, R. G. Roy, Y. Cui, P. Battacharya, M. Groza, A. Burger, A. M. Conway, R. J. Nikolic, A. J. Nelson and S. A. Paynec, *Proceedings of SPIE*, San Diego, 2008.
200. M. Jin, J. Jiang, R. B. Li, X. H. Wang, Y. X. Chen, R. L. Zhang and Y. Q. Chen, *Mater. Res. Bull.*, 2019, **114**, 156.
201. Y. Kim and I. H. Choi, *J. Korean Phys. Soc.*, 2018, **72**, 238.

202. A. K. Krolicka, A. Mirowska, A. Materna and M. Piersa, *Cryst. Res. Technol.*, 2017, **52**, 1600376.
203. Y. Jeon and H. S. Kang, *Opt. Rev.*, 2007, **14**, 373.
204. E. Trifonova, I. Y. Yanchev, P. Manou, K. Kambas and A. N. Anagnostopoulos, *J. Mater. Sci.*, 1996, **31**, 3647.
205. A. Hruban, A. Materna, W. Dalecki, G. Strzelecka, M. Piersa, E. Jurkiewicz-Wegner, R. Diduszko, M. Romaniec and W. Orlowski, *Acta Phys. Pol. A*, 2011, **120**, 950.
206. H. J. Wu, S. W. Chen, T. Ikeda and G. J. Snyder, *Acta Mater.*, 2012, **60**, 6144.
207. G. Marcano, L. M. de Chalbaud, C. Rincon and G. S. Perez, *Mater. Lett.*, 2002, **53**, 151.
208. Z. A. Shukri and C. H. Champness, *J. Cryst. Growth*, 1998, **191**, 97.
209. Z. Liu, J. A. Peters, H. Li, M. G. Kanatzidis and B. W. Wessels, *Semicond. Sci. Technol.*, 2013, **28**, 015022.
210. H. Bas, N. Kalkan and D. Deger, *J. Mater. Sci.:Mater. Electron.*, 2016, **27**, 7518.
211. R. H. Al Orainy, *Acta Phys. Pol. A*, 2013, **124**, 728.
212. G. E. Davidyuk, O. N. Yurchenko, O. V. Parasyuk, V. P. Sachanyuk, V. Z. Pankevich and A. P. Shavarova, *Inorg. Mater.*, 2008, **44**, 361.
213. V. F. Mikhailov, A. M. Sokolov, E. B. Sokolov and O. V. Pelevin, *Inorg. Mater.*, 1982, **18**, 1078.
214. R. J. Baughman and R. A. Lefever, *Mater. Res. Bull.*, 1969, **4**, 721.
215. T. E. Svechnikova, L. E. Shelimova, P. P. Konstantinov, M. A. Kretova, E. S. Avilov, V. S. Zemskov, C. Stiewe, A. Zuber and E. Muller, *Inorg. Mater.*, 2005, **41**, 1043.
216. L. D. Ivanova, L. I. Petrova, Y. V. Granatkina and V. S. Zemskov, *Inorg. Mater.*, 2007, **43**, 933.
217. L. D. Ivanova and Y. A. Sidorov, *Inorg. Mater.*, 1998, **34**, 222.
218. S. Hebert, T. Barbier, D. Berthebaud, O. I. Lebedev, V. Pralong and A. Maignan, *J. Phys. Chem. C*, 2021, **125**, 5386.
219. S. Roychowdhury, M. K. Jana, J. Pan, S. N. Guin, D. Sanyal, U. V. Waghmare and K. Biswas, *Angew. Chem., Int. Ed.*, 2018, **57**, 4043.
220. S. S. Wang, K. J. Bu, D. Wang, C. Zheng, X. L. Che, Z. Han and F. Q. Huang, *J. Alloys Compd.*, 2021, **853**, 157092.
221. Z. Li, X. X. Jiang, C. J. Yi, M. L. Zhou, Y. W. Guo, X. Y. Luo, Z. S. Lin, Y. C. Wu, Y. G. Shi and J. Y. Yao, *J. Mater. Chem. C*, 2018, **6**, 10042.
222. W. H. Xing, C. L. Tang, N. Z. Wang, C. X. Li, Z. Li, J. Y. Wu, Z. S. Lin, J. Y. Yao, W. L. Yin and B. Kang, *Inorg. Chem.*, 2020, **59**, 18452.
223. S. F. Li and D. Yan, *J. Solid State Chem.*, 2021, **296**, 121945.
224. M. Ohring, *Materials Science of Thin Films: Deposition and Structure*, Academic Press, San Diego, 2nd edn, 2002.
225. L. Samad, M. Caban-Acevedo, M. J. Shearer, K. Park, R. J. Hamers and S. Jin, *Chem. Mater.*, 2015, **27**, 3108.
226. J. Etzkorn, H. A. Therese, F. Rocker, N. Zink, U. Kolb and W. Tremel, *Adv. Mater.*, 2005, **17**, 2372.

227. J. Zhang, T. Zhai, F. Arifurrahman, Y. Wang, A. Hitt, Z. He, Q. Ai, Y. Liu, C.-Y. Lin, Y. Zhu, M. Tang and J. Lou, *Nano Lett.*, 2024, **24**, 2465.
228. R. Munoz and C. Gomez-Aleixandre, *Chem. Vap. Deposition*, 2013, **19**, 297.
229. M. Afzaal, K. Ellwood, N. L. Pickett, P. O'Brien, J. Raftery and J. Waters, *J. Mater. Chem.*, 2004, **14**, 1310.
230. M. Kemmler, M. Lazell, P. O'Brien, D. J. Otway, J. H. Park and J. R. Walsh, *J. Mater. Sci.:Mater. Electron.*, 2002, **13**, 531.
231. I. S. Chuprakov and K. H. Dahmen, *J. Phys. IV*, 1999, **9**, 313.
232. O. Salicio, C. Wiemer, M. Fanciulli, W. Gawelda, J. Siegel, C. N. Afonso, V. Plausinaitiene and A. Abrutis, *J. Appl. Phys.*, 2009, **105**, 033520.
233. J. S. Ritch, T. Chivers, M. Afzaal and P. O'Brien, *Chem. Soc. Rev.*, 2007, **36**, 1622.
234. Y. C. You, T. Zeng, J. S. Liu, T. S. Hu and G. A. Tai, *Prog. Chem.*, 2015, **27**, 1578.
235. X. Chen, L.-W. Wong, L. Huang, F. Zheng, R. Huang, S. P. Lau, C.-S. Lee, J. Zhao, Q. Deng and L. Thuc Hue, *Chem. Mater.*, 2021, **33**, 7039.
236. S. S. Zhou, R. Y. Wang, J. B. Han, D. L. Wang, H. Q. Li, L. Gan and T. Y. Zhai, *Adv. Funct. Mater.*, 2019, **29**, 1805880.
237. J. Waters, D. Crouch, J. Raftery and P. O'Brien, *Chem. Mater.*, 2004, **16**, 3289.
238. R. Cecchini, R. S. R. Gajjela, C. Martella, C. Wiemer, A. Lamperti, L. Nasi, L. Lazzarini, L. G. Nobili and M. Longo, *Small*, 2019, **15**, 1901743.
239. C. Pang, A. Srivastava, M. M. Lockart, T. Mewes, M. K. Bowman, N. Z. Bao, L. M. Shen and A. Gupta, *ACS Appl. Electron. Mater.*, 2019, **1**, 1424.
240. S. Selmo, R. Cecchini, S. Cecchi, C. Wiemer, M. Fanciulli, E. Rotunno, L. Lazzarini, M. Rigato, A. Pogany, A. Lugstein and M. Longo, *Appl. Phys. Lett.*, 2016, **109**, 213103.
241. C. Sanchez-Perez, C. E. Knapp, R. H. Colman, C. Sotelo-Vazquez, S. Sathasivam, R. Oilunkaniemi, R. S. Laitinen and C. J. Carmalt, *ACS Omega*, 2020, **5**, 15799.
242. Q. D. Fu, X. W. Wang, J. D. Zhou, J. Xia, Q. S. Zeng, D. H. Lv, C. Zhu, X. L. Wang, Y. Shen, X. M. Li, Y. N. Hua, F. C. Liu, Z. X. Shen, C. H. Jin and Z. Liu, *Chem. Mater.*, 2018, **30**, 4001.
243. T. Zhang, B. Jiang, Z. Xu, R. G. Mendes, Y. Xiao, L. F. Chen, L. W. Fang, T. Gemming, S. L. Chen, M. H. Ruemmeli and L. Fu, *Nat. Commun.*, 2016, **7**, 13911.
244. J. Li, Z. Wei and J. Kang, *Two-Dimensional Semiconductors: Synthesis, Physical Properties and Applications*, Wiley-VCH, Germany, 1st edn, 2020.
245. H. Wen, L. Xiong, C. B. Tan, K. M. Zhu, Y. Tang, J. B. Wang and X. L. Zhong, *Chin. Phys. B*, 2021, **30**, 057803.
246. K. S. Ahmad, A. Zafar, S. B. Jaffri, M. K. Alamgir, M. Sohail, R. F. Mehmood, M. u. Rehman and D. Ali, *Results Phys.*, 2020, **19**, 103647.
247. J. Andzane, A. Felsharuk, A. Sarakovskis, U. Malinovskis, E. Kauranens, M. Bechelany, K. A. Niherysh, I. V. Komissarov and D. Erts, *Mater. Today Energy*, 2021, **19**, 100587.

248. J. Cruz-Gomez, E. Hernandez-Cantero, D. Santos-Cruz, S. A. Mayen-Hernandez, F. DeMoure-Flores and J. Santos-Cruz, *Proceedings of IEEE*, Mexico, 2021.
249. D. D. Thornbur, *Thin Solid Films*, 1972, **11**, 219.
250. J. Anwar, K. S. Ahmad, S. B. Jaffri and M. Sohail, *Can. Metall. Q.*, 2022, **61**, 145.
251. A. Zafar, K. S. Ahmad, S. B. Jaffri and M. Sohail, *Phosphorus, Sulfur Silicon Relat. Elem.*, 2020, **196**, 36.
252. P. A. Psaras and H. D. Langford, *Advancing Materials Research: National Academy of Engineering and National Academy of Sciences*, National Academy Press, Washington, D. C., 1987.
253. Z. B. Zhuang, Q. Peng and Y. D. Li, *Chem. Soc. Rev.*, 2011, **40**, 5492.
254. M. K. Devaraju and I. Honma, *Adv. Energy Mater.*, 2012, **2**, 284.
255. M. Nath, A. Choudhury, A. Kundu and C. N. R. Rao, *Adv. Mater.*, 2003, **15**, 2098.
256. C. C. Lin, W. F. Lee, M. Y. Lu, S. Y. Chen, M. H. Hung, T. C. Chan, H. W. Tsai, Y. L. Chueh and L. J. Chen, *J. Mater. Chem.*, 2012, **22**, 7098.
257. R. C. Jin, G. Chen, J. Pei, H. M. Xu and Z. S. Lv, *RSC Adv.*, 2012, **2**, 1450.
258. W. T. Yao and S. H. Yu, *Adv. Funct. Mater.*, 2008, **18**, 3357.
259. R. C. Jin, G. Chen, J. Pei, J. X. Sun and Y. Wang, *Nanoscale*, 2011, **3**, 3893.
260. A. K. Tyagi and R. S. Ningthoujam, *Handbook on Synthesis Strategies for Advanced Materials Volume-I: Techniques and Fundamentals*, Springer, Singapore, 1st edn, 2021.
261. A. Hernández-Ramírez and I. Medina-Ramírez, *Photocatalytic Semiconductors: Synthesis, Characterization, and Environmental Applications*, Springer International, Switzerland, 2015.
262. A. Rabenau, *Angew. Chem., Int. Ed. Engl.*, 1985, **24**, 1026.
263. D. M. Rowe, *Materials, Preparation, and Characterization in Thermoelectrics*, CRC Press, Boca Raton, 2012.
264. G. Han, R. Z. Zhang, S. R. Popuri, H. F. Greer, M. J. Reece, J. W. G. Bos, W. Z. Zhou, A. R. Knox and D. H. Gregory, *Materials*, 2017, **10**, 233.
265. Z. L. Li, Y. D. Chen, J. F. Li, H. Chen, L. J. Wang, S. Q. Zheng and G. W. Lu, *Nano Energy*, 2016, **28**, 78.
266. G. Han, S. R. Popuri, H. F. Greer, L. F. Llin, J. W. G. Bos, W. Z. Zhou, D. J. Paul, H. Ménard, A. R. Knox, A. Montecucco, J. Siviter, E. A. Man, W.-g. Li, M. C. Paul, M. Gao, T. Sweet, R. Freer, F. Azough, H. Baig, T. K. Mallick and D. H. Gregory, *Adv. Energy Mater.*, 2017, **7**, 1602328.
267. M. Ibáñez, R. J. Korkosz, Z. S. Luo, P. Riba, D. Cadavid, S. Ortega, A. Cabot and M. G. Kanatzidis, *J. Am. Chem. Soc.*, 2015, **137**, 4046.
268. S. Kar and S. Chaudhuri, *Chem. Phys. Lett.*, 2004, **398**, 22.
269. B. Yuan, W. Luan and S.-t. Tu, *Dalton Trans.*, 2012, **41**, 772.
270. S. J. Lei, K. B. Tang and H. G. Zheng, *Mater. Lett.*, 2006, **60**, 1625.
271. J. Lu, P. F. Qi, Y. Y. Peng, Z. Y. Meng, Z. P. Yang, W. C. Yu and Y. T. Qian, *Chem. Mater.*, 2001, **13**, 2169.
272. L. Zhang, H. Yang, L. Li, R. Zhang, R. Liu, J. Ma, X. Xie and F. Gao, *Inorg. Chem.*, 2008, **47**, 11950.

273. L. Jiang and Y.-J. Zhu, *Eur. J. Inorg. Chem.*, 2010, 1238.
274. X. B. Zhao, X. H. Ji, Y. H. Zhang, T. J. Zhu, J. P. Tu and X. B. Zhang, *Appl. Phys. Lett.*, 2005, **86**, 062111.
275. G. a. Tai, B. Zhou and W. L. Guo, *J. Phys. Chem. C*, 2008, **112**, 11314.
276. L. Cheng, Z.-G. Chen, L. Yang, G. Han, H.-Y. Xu, G. J. Snyder, G.-Q. Lu and J. Zou, *J. Phys. Chem. C*, 2013, **117**, 12458.
277. X. F. Liu, H. Y. Wang, B. Zhang, S. K. Zheng, Y. Chen, H. Zhang, X. H. Chen, G. Y. Wang, X. Y. Zhou and G. Han, *J. Mater. Sci. Technol.*, 2022, **120**, 205.
278. J. L. Mi, N. Lock, T. Sun, M. Christensen, M. Sondergaard, P. Hald, H. H. Hng, J. Ma and B. B. Iversen, *ACS Nano*, 2010, **4**, 2523.
279. G. Han, Z. G. Chen, L. Yang, L. Cheng, J. Drennan and J. Zou, *Cryst. Growth Des.*, 2013, **13**, 5092.
280. R. B. Soriano, C. D. Malliakas, J. S. Wu and M. G. Kanatzidis, *J. Am. Chem. Soc.*, 2012, **134**, 3228.
281. R. B. Soriano, J. S. Wu and M. G. Kanatzidis, *J. Am. Chem. Soc.*, 2015, **137**, 9937.
282. C. D. Donega, P. Liljeroth and D. Vanmaekelbergh, *Small*, 2005, **1**, 1152.
283. J. Park, J. Joo, S. G. Kwon, Y. Jang and T. Hyeon, *Angew. Chem., Int. Ed.*, 2007, **46**, 4630.
284. J. H. Han, S. Lee and J. Cheon, *Chem. Soc. Rev.*, 2013, **42**, 2581.
285. M. Ibáñez, R. J. Korkosz, Z. Luo, P. Riba, D. Cadavid, S. Ortega, A. Cabot and M. G. Kanatzidis, *J. Am. Chem. Soc.*, 2015, **137**, 4046.
286. D. Cadavid, M. Ibáñez, A. Shavel, O. Juan Dura, M. A. Lopez de la Torre and A. Cabot, *J. Mater. Chem. A*, 2013, **1**, 4864.
287. Y. Liu, G. Garcia, S. Ortega, D. Cadavid, P. Palacios, J. Y. Lu, M. Ibáñez, L. L. Xi, J. De Roo, A. M. Lopez, S. Marti-Sanchez, I. Cabezas, M. de la Mata, Z. S. Luo, C. Dun, O. Dobrozhan, D. L. Carroll, W. Q. Zhang, J. Martins, M. V. Kovalenko, J. Arbiol, G. Noriega, J. M. Song, P. Wahnon and A. Cabot, *J. Mater. Chem. A*, 2017, **5**, 2592.
288. J. H. Yu, J. Joo, H. M. Park, S. I. Baik, Y. W. Kim, S. C. Kim and T. Hyeon, *J. Am. Chem. Soc.*, 2005, **127**, 5662.
289. G. Han, S. R. Popuri, H. F. Greer, J. W. G. Bos, W. Z. Zhou, A. R. Knox, A. Montecucco, J. Siviter, E. A. Man, M. Macauley, D. J. Paul, W. G. Li, M. C. Paul, M. Gao, T. Sweet, R. Freer, F. Azough, H. Baig, N. Sellami, T. K. Mallick and D. H. Gregory, *Angew. Chem., Int. Ed.*, 2016, **55**, 6433.
290. G. Han, S. R. Popuri, H. F. Greer, L. F. Llin, J. W. G. Bos, W. Z. Zhou, D. J. Paul, H. Menard, A. R. Knox, A. Montecucco, J. Siviter, E. A. Man, W. G. Li, M. C. Paul, M. Gao, T. Sweet, R. Freer, F. Azough, H. Baig, T. K. Mallick and D. H. Gregory, *Adv. Energy Mater.*, 2017, **7**, 1602328.
291. R. Lesyuk, E. Klein, I. Yaremchuk and C. Klinke, *Nanoscale*, 2018, **10**, 20640.
292. Y. Wu, C. Wadia, W. L. Ma, B. Sadtler and A. P. Alivisatos, *Nano Lett.*, 2008, **8**, 2551.
293. K. H. Park, K. Jang and S. U. Son, *Angew. Chem., Int. Ed.*, 2006, **45**, 4608.
294. F. Li, J. Fu, A. Torche, S. Kull, A. Kornowski, R. Lesyuk, G. Bester and C. Klinke, *Adv. Mater. Interfaces*, 2020, **7**, 2000410.

295. I. U. Arachchige, R. Soriano, C. D. Malliakas, S. A. Ivanov and M. C. Kanatzidis, *Adv. Funct. Mater.*, 2011, **21**, 2737.
296. H. Y. Wang, X. F. Liu, B. Zhang, L. S. Huang, M. L. Yang, X. Zhang, H. Zhang, G. Y. Wang, X. Y. Zhou and G. Han, *Chem. Eng. J.*, 2020, **393**, 124763.
297. A. H. Khan, R. Brescia, A. Polovitsyn, I. Angeloni, B. Martin-Garcia and I. Moreels, *Chem. Mater.*, 2017, **29**, 2883.
298. J. J. Choi, Y.-F. Lim, M. E. B. Santiago-Berrios, M. Oh, B.-R. Hyun, L. Sung, A. C. Bartnik, A. Goedhart, G. G. Malliaras, H. D. Abruna, F. W. Wise and T. Hanrath, *Nano Lett.*, 2009, **9**, 3749.
299. R. Malakooti, L. Cademartiri, Y. Akcakir, S. Petrov, A. Migliori and G. A. Ozin, *Adv. Mater.*, 2006, **18**, 2189.
300. J. S. Son, M. K. Choi, M. K. Han, K. Park, J. Y. Kim, S. J. Lim, M. Oh, Y. Kuk, C. Park, S. J. Kim and T. Hyeon, *Nano Lett.*, 2012, **12**, 640.
301. M. A. Malik, P. O'Brien and N. Revaprasadu, *Adv. Mater.*, 1999, **11**, 1441.
302. S. L. Castro, S. G. Bailey, R. P. Raffaelle, K. K. Banger and A. F. Hepp, *J. Phys. Chem. B*, 2004, **108**, 12429.
303. P. M. Allen and M. G. Bawendi, *J. Am. Chem. Soc.*, 2008, **130**, 9240.
304. J. Tang, S. Hinds, S. O. Kelley and E. H. Sargent, *Chem. Mater.*, 2008, **20**, 6906.
305. L. S. Huang, J. Z. Lu, D. W. Ma, C. M. Ma, B. Zhang, H. Y. Wang, G. Y. Wang, D. H. Gregory, X. Y. Zhou and G. Han, *J. Mater. Chem. A*, 2020, **8**, 1394.
306. Y. Q. Wei, Z. Z. Zhou, P. F. Jiang, S. K. Zheng, Q. H. Xiong, B. Zhang, G. Y. Wang, X. Lu, G. Han and X. Y. Zhou, *ACS Appl. Energy Mater.*, 2021, **4**, 9240.
307. F. J. Fan, Y. X. Wang, X. J. Liu, L. Wu and S. H. Yu, *Adv. Mater.*, 2012, **24**, 6158.
308. M. Ibáñez, R. Zamani, W. H. Li, A. Shavel, J. Arbiol, J. Ramon Morante and A. Cabot, *Cryst. Growth Des.*, 2012, **12**, 1085.
309. M. Ibáñez, R. Zamani, A. LaLonde, D. Cadavid, W. H. Li, A. Shavel, J. Arbiol, J. R. Morante, S. Gorsse, G. J. Snyder and A. Cabot, *J. Am. Chem. Soc.*, 2012, **134**, 4060.
310. X. Ren, L. Yuan, Q. Liang, R. Xie, Z. Geng, Y. Sun, L. Wang, K. Huang, T. Wu and S. Feng, *ChemPlusChem*, 2018, **83**, 812.
311. M. Giancaspro, T. Sibillano, F. Panzarea, C. Giannini, S. Schmitzer, F. Vischio, N. Depalo, A. Agostiano, M. L. Curri, M. Striccoli and E. Fanizza, *Mater. Chem. Front.*, 2021, **5**, 1341.
312. D. Yin, C. Dun, H. Zhang, Z. Fu, X. Gao, X. Wang, D. J. Singh, D. L. Carroll, Y. Liu and M. T. Swihart, *Small*, 2021, **17**, 2006729.
313. C. Lofton and W. Sigmund, *Adv. Funct. Mater.*, 2005, **15**, 1197.
314. S. Ithurria and B. Dubertret, *J. Am. Chem. Soc.*, 2008, **130**, 16504.
315. S. Komarneni and R. Roy, *Mater. Lett.*, 1985, **3**, 165.
316. Y. J. Zhu and F. Chen, *Chem. Rev.*, 2014, **114**, 6462.
317. T. Laird, *Org. Process Res. Dev.*, 2011, **15**, 478.
318. M. Rajamathi and R. Seshadri, *Curr. Opin. Solid State Mater. Sci.*, 2002, **6**, 337.

319. R. J. Mehta, C. Karthik, W. Jiang, B. Singh, Y. Shi, R. W. Siegel, T. Borca-Tasciuc and G. Ramanath, *Nano Lett.*, 2010, **10**, 4417.
320. A. B. Panda, G. Glaspell and M. S. El-Shall, *J. Am. Chem. Soc.*, 2006, **128**, 2790.
321. M. W. Shao, L. F. Kong, Q. Li, W. C. Yu and Y. T. Qian, *Inorg. Chem. Commun.*, 2003, **6**, 737.
322. R. Harpeness, A. Gedanken, A. M. Weiss and M. A. Slifkin, *J. Mater. Chem.*, 2003, **13**, 2603.
323. S. Park, J. Park, R. Selvaraj and Y. Kim, *J. Ind. Eng. Chem.*, 2015, **31**, 269.
324. E. Caponetti, D. C. Martino, M. Leone, L. Pedone, M. L. Saladino and V. Vetri, *J. Colloid Interface Sci.*, 2006, **304**, 413.
325. A. V. Firth, Y. Tao, D. S. Wang, J. F. Ding and F. Bensebaa, *J. Mater. Chem.*, 2005, **15**, 4367.
326. Q. Song, X. Ai, T. Topuria, P. M. Rice, F. H. Alharbi, A. Bagabas, M. Bahattab, J. D. Bass, H.-C. Kim, J. C. Scott and R. D. Miller, *Chem. Commun.*, 2010, **46**, 4971.
327. Y. Wang, W. Yang, L. Zhang, Y. Hu and X. W. Lou, *Nanoscale*, 2013, **5**, 10864.
328. D. W. Synnott, M. K. Seery, S. J. Hinder, J. Colreavy and S. C. Pillai, *Nanotechnology*, 2013, **24**, 045704.
329. H. F. Qian, X. Qiu, L. Li and J. C. Ren, *J. Phys. Chem. B*, 2006, **110**, 9034.
330. Z. M. Alvand, H. R. Rajabi, A. Mirzaei and A. Masoumiasl, *New J. Chem.*, 2019, **43**, 15126.
331. J. Q. Sheng, X. F. Liu, C. Z. Niu, Y. F. Sun, Y. Chen, H. Y. Wang, B. Zhang, G. Y. Wang, X. Y. Zhou and G. Han, *J. Mater. Chem. C*, 2020, **8**, 10333.
332. M. Maria Gonzalez-Barrios, M. Tabuyo-Martinez, V. Cascos, O. J. Dura, J. A. Alonso, D. Avila-Brande and J. Prado-Gonjal, *Ceram. Int.*, 2022, **48**, 12331.
333. L. Wang, S. Zheng and H. Chen, *J. Electron. Mater.*, 2017, **46**, 2847.
334. E. B. Hostetler, K.-J. Kim, R. P. Oleksak, R. C. Fitzmorris, D. A. Peterson, P. Chandran, C.-H. Chang, B. K. Paul, D. M. Schut and G. S. Herman, *Mater. Lett.*, 2014, **128**, 54.
335. G.-H. Dong and Y.-J. Zhu, *Chem. Eng. J.*, 2012, **193**, 227.
336. C. F. Mu, Q. Z. Yao, X. F. Qu, G. T. Zhou, M. L. Li and S. Q. Fu, *Colloids Surf., A*, 2010, **371**, 14.
337. H. Cao, X. Hu and Y. Zhu, *Mater. Lett.*, 2024, **359**, 135881.
338. X. H. Liao, N. Y. Chen, S. Xu, S. B. Yang and J. J. Zhu, *J. Cryst. Growth*, 2003, **252**, 593.
339. S. Yang, F. Ji, Z. Wang, Y. Zhu, K. Hu, Y. Ouyang, R. Wang, X. Ma and C. Cao, *Electrochim. Acta*, 2019, **324**, 134864.
340. Y. Chen, J. Chen, B. Zhang, M. Yang, X. Liu, H. Wang, L. Yang, G. Wang, G. Han and X. Zhou, *J. Mater. Sci. Technol.*, 2021, **78**, 121.
341. J. J. Zhu, O. Palchik, S. G. Chen and A. Gedanken, *J. Phys. Chem. B*, 2000, **104**, 7344.
342. Y. Chen, H. Wu, B. Zhang, J. Chen, Z. Zhou, X. Lu, Y. Zhou, G. Wang, G. Han and X. Zhou, *J. Eur. Ceram. Soc.*, 2024, **44**, 6374.

343. W. Yang, L. Zhang, Y. Hu, Y. Zhong, H. B. Wu and X. W. Lou, *Angew. Chem., Int. Ed.*, 2012, **51**, 11501.
344. X. Li, Y. Lu, K. Cai, M. Gao, Y. Li, Z. Wang, M. Wu, P. Wei, W. Zhao, Y. Du and S. Shen, *Chem. Eng. J.*, 2022, **434**, 134739.
345. J. Pei, G. Chen, D. Jia, Y. Yu, J. Sun, H. Xu and Z. Qiu, *New J. Chem.*, 2014, **38**, 59.
346. J. Kavinchan, T. Thongtem and S. Thongtem, *Mater. Lett.*, 2010, **64**, 2388.
347. Y.-Q. Liu, M. Zhang, F.-X. Wang and G.-B. Pan, *J. Mater. Chem. C*, 2014, **2**, 240.
348. S. Singh, S. Hong, W. Jeon, D. Lee, J.-Y. Hwang, S. Lim, G. D. Kwon, D. Pribat, H. Shin, S. W. Kim and S. Baik, *Chem. Mater.*, 2015, **27**, 2315.
349. X. Zhou, S. Yao, Y. Long, Z. Wang and W. Li, *Mater. Lett.*, 2015, **145**, 23.
350. J. Zheng, S. Chen, K. Cai, J. Yin and S. Shen, *Ceram. Int.*, 2017, **43**, 5920.
351. Q. Yao, Y. Zhu, L. Chen, Z. Sun and X. Chen, *J. Alloys Compd.*, 2009, **481**, 91.
352. W. J. Zhang, D. Z. Li, Z. X. Chen, M. Sun, W. J. Li, Q. Lin and X. Z. Fu, *Mater. Res. Bull.*, 2011, **46**, 975.
353. W. J. Zhang, D. Z. Li, M. Sun, Y. Shao, Z. X. Chen, G. C. Xiao and X. Z. Fu, *J. Solid State Chem.*, 2010, **183**, 2466.
354. F. Bensebaa, C. Durand, A. Aouadou, L. Scoles, X. Du, D. Wang and Y. Le Page, *J. Nanopart. Res.*, 2010, **12**, 1897.
355. S. K. Apte, S. N. Garaje, R. D. Bolade, J. D. Ambekar, M. V. Kulkarni, S. D. Naik, S. W. Gosavi, J. O. Baeg and B. B. Kale, *J. Mater. Chem.*, 2010, **20**, 6095.
356. D. D. Xie, B. Zhang, A. J. Zhang, Y. J. Chen, Y. C. Yan, H. Q. Yang, G. W. Wang, G. Y. Wang, X. D. Han, G. Han, X. Lu and X. Y. Zhou, *Nanoscale*, 2018, **10**, 14546.
357. S. W. Shin, J. H. Han, C. Y. Park, A. V. Moholkar, J. Y. Lee and J. H. Kim, *J. Alloys Compd.*, 2012, **516**, 96.
358. C. Backes, T. M. Higgins, A. Kelly, C. Boland, A. Harvey, D. Hanlon and J. N. Coleman, *Chem. Mater.*, 2017, **29**, 243.
359. Y. Y. Xu, H. Z. Cao, Y. Q. Xue, B. Li and W. H. Cai, *Nanomaterials*, 2018, **8**, 942.
360. J. N. Coleman, M. Lotya, A. O'Neill, S. D. Bergin, P. J. King, U. Khan, K. Young, A. Gaucher, S. De, R. J. Smith, I. V. Shvets, S. K. Arora, G. Stanton, H.-Y. Kim, K. Lee, G. T. Kim, G. S. Duesberg, T. Hallam, J. J. Boland, J. J. Wang, J. F. Donegan, J. C. Grunlan, G. Moriarty, A. Shmeliov, R. J. Nicholls, J. M. Perkins, E. M. Grieveson, K. Theuwissen, D. W. McComb, P. D. Nellist and V. Nicolosi, *Science*, 2011, **331**, 568.
361. M. B. Dines, *J. Chem. Educ.*, 1974, **51**, 221.
362. H. S. S. R. Matte, A. Gomathi, A. K. Manna, D. J. Late, R. Datta, S. K. Pati and C. N. R. Rao, *Angew. Chem., Int. Ed.*, 2010, **49**, 4059.
363. P. Joensen, R. F. Frindt and S. R. Morrison, *Mater. Res. Bull.*, 1986, **21**, 457.
364. B. K. Miremadi and S. R. Morrison, *J. Appl. Phys.*, 1988, **63**, 4970.
365. D. Yang and R. F. Frindt, *J. Phys. Chem. Solids*, 1996, **57**, 1113.

366. Z. Y. Lin, Y. Liu, U. Halim, M. N. Ding, Y. Y. Liu, Y. L. Wang, C. C. Jia, P. Chen, X. D. Duan, C. Wang, F. Song, M. F. Li, C. Z. Wan, Y. Huang and X. F. Duan, *Nature*, 2018, **562**, 254.
367. A. Ciesielski and P. Samori, *Chem. Soc. Rev.*, 2014, **43**, 381.
368. R. M. Tian, C. L. Wan, Y. F. Wang, Q. S. Wei, T. Ishida, A. Yamamoto, A. Tsuruta, W. S. Shin, S. Li and K. Koumoto, *J. Mater. Chem. A*, 2017, **5**, 564.
369. G. Cunningham, M. Lotya, C. S. Cucinotta, S. Sanvito, S. D. Bergin, R. Menzel, M. S. P. Shaffer and J. N. Coleman, *ACS Nano*, 2012, **6**, 3468.
370. A. K. Tyagi and R. S. Ningthoujam, *Handbook on Synthesis Strategies for Advanced Materials Volume-III: Materials Specific Synthesis Strategies*, Springer Nature, Singapore, 1st edn, 2021.
371. M. H. Enayati and F. A. Mohamed, *Int. Mater. Rev.*, 2014, **59**, 394.
372. C. Suryanarayana, *Research*, 2019, **2019**, 4219812.
373. J. Schilz, M. Riffel, K. Pixius and H. J. Meyer, *Powder Technol.*, 1999, **105**, 149.
374. C. Suryanarayana, *Prog. Mater. Sci.*, 2001, **46**, 1.
375. V. A. Volkov, I. A. El'kin, A. V. Zagainov, A. V. Protasov and E. P. Elsukov, *Phys. Met. Metallogr.*, 2014, **115**, 557.
376. F. R. Sie, H. J. Liu, C. H. Kuo, C. S. Hwang, Y. W. Chou and C. H. Yeh, *Intermetallics*, 2018, **92**, 113.
377. M. O. Melquiades, L. S. de Oliveira, Q. H. F. Rebelo, P. Chaudhuri, E. R. Leite, D. M. Triches and S. Michielon de Souza, *J. Appl. Phys.*, 2019, **126**, 135707.
378. Y.-X. Chen, Z.-H. Ge, M. Yin, D. Feng, X.-Q. Huang, W. Zhao and J. He, *Adv. Funct. Mater.*, 2016, **26**, 6836.
379. Q. Tan, L. D. Zhao, J. F. Li, C. F. Wu, T. R. Wei, Z. B. Xing and M. G. Kanatzidis, *J. Mater. Chem. A*, 2014, **2**, 17302.
380. Q. Zhang, B. L. Liao, Y. C. Lan, K. Lukas, W. S. Liu, K. Esfarjani, C. Opeil, D. Broido, G. Chen and Z. F. Ren, *Proc. Natl. Acad. Sci. U. S. A.*, 2013, **110**, 13261.
381. T. Ohtani, K. Maruyama and K. Ohshima, *Mater. Res. Bull.*, 1997, **32**, 343.
382. J. Jiang, H. T. Zhu, Y. Niu, Q. Zhu, S. W. Song, T. Zhou, C. Wang and Z. F. Ren, *J. Mater. Chem. A*, 2020, **8**, 4790.
383. Y.-X. Chen, B.-P. Zhang, Z.-H. Ge and P.-P. Shang, *J. Solid State Chem.*, 2012, **186**, 109.
384. E. Maciazek, E. Malicka, M. Karolus, J. Panek, Z. Stoklosa, T. Gron, A. Gudwanski and B. Sawicki, *IEEE Trans. Magn.*, 2017, **53**, 2400205.
385. R. Nunna, P. F. Qiu, M. J. Yin, H. Y. Chen, R. Hanus, Q. F. Song, T. S. Zhang, M. Y. Chou, M. T. Agne, J. Q. He, G. J. Snyder, X. Shi and L. D. Chen, *Energy Environ. Sci.*, 2017, **10**, 1928.
386. V. I. Tkatch, S. N. Denisenko and O. N. Beloshov, *Acta Mater.*, 1997, **45**, 2821.
387. Z. M. Zhou, W. J. Huang, J. Luo, X. P. Li, T. Zhou, B. B. Lei, C. H. Wang, H. Xia and C. Y. Peng, *Materials Science Forum*, Kunming, 2011.

388. J. S. Young and R. G. Reddy, *J. Mater. Eng. Perform.*, 2019, **28**, 5917.
389. A. L. Greer, *Science*, 1995, **267**, 1947.
390. H. Chen, Y. He, G. J. Shiflet and S. J. Poon, *Nature*, 1994, **367**, 541.
391. X. L. Su, P. Wei, H. Li, W. Liu, Y. G. Yan, P. Li, C. Q. Su, C. J. Xie, W. Y. Zhao, P. C. Zhai, Q. J. Zhang, X. F. Tang and C. Uher, *Adv. Mater.*, 2017, **29**, 1602013.
392. V. I. Tkatch, A. I. Limanovskii, S. N. Denisenko and S. G. Rassolov, *Mater. Sci. Eng., A*, 2002, **323**, 91.
393. S. Y. Wang, W. J. Xie, H. Li and X. F. Tang, *Intermetallics*, 2011, **19**, 1024.
394. X. Cai, X. a. Fan, Z. Rong, F. Yang, Z. Gan and G. Li, *J. Phys. D:Appl. Phys.*, 2014, **47**, 115101.
395. W. J. Xie, S. Y. Wang, S. Zhu, J. He, X. F. Tang, Q. J. Zhang and T. M. Tritt, *J. Mater. Sci.*, 2013, **48**, 2745.
396. Y. Chen, T. J. Zhu, S. H. Yang, S. N. Zhang, W. Miao and X. B. Zhao, *J. Electron. Mater.*, 2010, **39**, 1719.
397. B. Du, H. Li, J. Xu, X. Tang and C. Uher, *J. Solid State Chem.*, 2011, **184**, 109.
398. X. M. Yan, S. K. Zheng, Z. Z. Zhou, H. Wu, B. Zhang, Y. L. Huang, X. Lu, G. Han, G. Y. Wang and X. Y. Zhou, *Nano Energy*, 2021, **84**, 105879.
399. G. Ding, J. Si, H. Wu, S. Yang, J. Zhao and G. Wang, *J. Alloys Compd.*, 2016, **662**, 368.
400. W. Xie, J. He, H. J. Kang, X. Tang, S. Zhu, M. Laver, S. Wang, J. R. D. Copley, C. M. Brown, Q. Zhang and T. M. Tritt, *Nano Lett.*, 2010, **10**, 3283.
401. W. Xie, X. Tang, Y. Yan, Q. Zhang and T. M. Tritt, *Appl. Phys. Lett.*, 2009, **94**, 102111.
402. H. Tang, L. Guo, G. Wang, H. Wu, X. Shen, B. Zhang, X. Lu, G. Wang, X. Zhang and X. Zhou, *ACS Appl. Mater. Interfaces*, 2019, **11**, 23337.
403. J. Baxter, Z. Bian, G. Chen, D. Danielson, M. S. Dresselhaus, A. G. Fedorov, T. S. Fisher, C. W. Jones, E. Maginn, U. Kortshagen, A. Manthiram, A. Nozik, D. R. Rolison, T. Sands, L. Shi, D. Sholl and Y. Wu, *Energy Environ. Sci.*, 2009, **2**, 559.
404. L. Ren, Y. Li, N. Zhang, Z. Li, X. Lin, W. Zhu, C. Lu, W. Ding and J. Zou, *Nano-Micro Lett.*, 2023, **15**, 93.
405. G. Han, Y. Lu, H. Jia, Z. Ding, L. Wu, Y. Shi, G. Wang, Q. Luo, Y. a. Chen, J. Wang, G. Huang, X. Zhou, Q. Li and F. Pan, *J. Magnesium Alloys*, 2023, **11**, 3896.
406. M. Pumera, Z. Sofer and A. Ambrosi, *J. Mater. Chem. A*, 2014, **2**, 8981.
407. G. A. Muller, J. B. Cook, H. S. Kim, S. H. Tolbert and B. Dunn, *Nano Lett.*, 2015, **15**, 1911.
408. T. Y. Wang, S. Q. Chen, H. Pang, H. G. Xue and Y. Yu, *Adv. Sci.*, 2017, **4**, 1600289.
409. L. W. Ji, Z. Lin, M. Alcoutlabi and X. W. Zhang, *Energy Environ. Sci.*, 2011, **4**, 2682.
410. S. Qin, W. W. Lei, D. Liu and Y. Chen, *J. Mater. Chem. A*, 2016, **4**, 1440.
411. K. Tao, X. F. Wang, Y. F. Xu, J. Liu, X. F. Song, C. P. Fu, X. Q. Chen, X. Z. Qu, X. F. Zhao and L. Gao, *J. Am. Ceram. Soc.*, 2020, **103**, 4453.

412. W. Zhu, J. Zhao and X. Tao, *J. Energy Storage*, 2024, **84**, 110934.
413. S. Thomas, A. B. Gueye and R. K. Gupta, *Nanostructured Materials for Supercapacitors*, Springer Nature, Switzerland, 1st edn, 2022.
414. P. Lamba, P. Singh, P. Singh, P. Singh, Bharti, A. Kumar, M. Gupta and Y. Kumar, *J. Energy Storage*, 2022, **48**, 103871.
415. Y. Dahiya, M. Hariram, M. Kumar, A. Jain and D. Sarkar, *Coord. Chem. Rev.*, 2022, **451**, 214265.
416. J. Yang, C. Yu, X. M. Fan, S. X. Liang, S. F. Li, H. W. Huang, Z. Ling, C. Hao and J. S. Qiu, *Energy Environ. Sci.*, 2016, **9**, 1299.
417. X. L. Li, X. D. Li, J. L. Cheng, D. M. Yuan, W. Ni, Q. Guan, L. Z. Gao and B. Wang, *Nano Energy*, 2016, **21**, 228.
418. M. Acerce, D. Voiry and M. Chhowalla, *Nat. Nanotechnol.*, 2015, **10**, 313.
419. X. Hu, P. Jood, M. Ohta, M. Kunii, K. Nagase, H. Nishiate, M. G. Kanatzidis and A. Yamamoto, *Energy Environ. Sci.*, 2016, **9**, 517.
420. L. D. Zhao, C. Chang, G. J. Tan and M. G. Kanatzidis, *Energy Environ. Sci.*, 2016, **9**, 3044.
421. W.-D. Liu, L. Yang and Z.-G. Chen, *Nano Today*, 2020, **35**, 100938.
422. C. G. Fu, S. Q. Bai, Y. T. Liu, Y. S. Tang, L. D. Chen, X. B. Zhao and T. J. Zhu, *Nat. Commun.*, 2015, **6**, 8144.
423. H. Hu, Y. Ju, J. Yu, Z. Wang, J. Pei, H.-C. Thong, J.-W. Li, B. Cai, F. Liu, Z. Han, B. Su, H.-L. Zhuang, Y. Jiang, H. Li, Q. Li, H. Zhao, B.-P. Zhang, J. Zhu and J.-F. Li, *Nat. Mater.*, 2024, **23**, 527.
424. H. L. Liu, X. Shi, F. F. Xu, L. L. Zhang, W. Q. Zhang, L. D. Chen, Q. Li, C. Uher, T. Day and G. J. Snyder, *Nat. Mater.*, 2012, **11**, 422.
425. L. D. Zhao, S. H. Lo, Y. S. Zhang, H. Sun, G. J. Tan, C. Uher, C. Wolverton, V. P. Dravid and M. G. Kanatzidis, *Nature*, 2014, **508**, 373.
426. C. Li, J. Chen, J. Chen, Y. Zhang, M. Li and H. Li, *Adv. Energy Mater.*, 2024, **14**, 2303162.
427. K. Li, X. Yang, Y. Lu, J. Xue, S. Lu, J. Zheng, C. Chen and J. Tang, *Adv. Energy Mater.*, 2022, **12**, 2200725.
428. Q. Fu, J. Han, X. Wang, P. Xu, T. Yao, J. Zhong, W. Zhong, S. Liu, T. Gao, Z. Zhang, L. Xu and B. Song, *Adv. Mater.*, 2021, **33**, 1907818.
429. Y. R. Zheng, P. Wu, M. R. Gao, X. L. Zhang, F. Y. Gao, H. X. Ju, R. Wu, Q. Gao, R. You, W. X. Huang, S. J. Liu, S. W. Hu, J. F. Zhu, Z. Y. Li and S. H. Yu, *Nat. Commun.*, 2018, **9**, 2533.
430. Y. S. Wu, X. J. Liu, D. D. Han, X. Y. Song, L. Shi, Y. Song, S. W. Niu, Y. F. Xie, J. Y. Cai, S. Y. Wu, J. Kang, J. B. Zhou, Z. Y. Chen, X. S. Zheng, X. H. Xiao and G. M. Wang, *Nat. Commun.*, 2018, **9**, 1425.
431. C. Gu, S. J. Hu, X. S. Zheng, M. R. Gao, Y. R. Zheng, L. Shi, Q. Gao, X. Zheng, W. S. Chu, H. B. Yao, J. F. Zhu and S. H. Yu, *Angew. Chem., Int. Ed.*, 2018, **57**, 4020.
432. Y. Yu, S.-Y. Huang, Y. Li, S. N. Steinmann, W. Yang and L. Cao, *Nano Lett.*, 2014, **14**, 553.
433. C. Liu, F. Li, L. P. Ma and H. M. Cheng, *Adv. Mater.*, 2010, **22**, E28.
434. J. B. Goodenough and Y. Kim, *Chem. Mater.*, 2010, **22**, 587.
435. M. Armand and J. M. Tarascon, *Nature*, 2008, **451**, 652.

436. B. Chen, D. Chao, E. Liu, M. Jaroniec, N. Zhao and S.-Z. Qiao, *Energy Environ. Sci.*, 2020, **13**, 1096.
437. J. Wu, M. Ihsan-Ul-Haq, F. Ciucci, B. Huang and J.-K. Kim, *Energy Storage Mater.*, 2021, **34**, 582.
438. Z. Ye, H. Sun, H. Gao, L. Sun, J. Guo, Y. Jiang, C. Wu and S. Zheng, *Energy Storage Mater.*, 2023, **60**, 102855.
439. J. Neumann, M. Petranikova, M. Meeus, J. D. Gamarra, R. Younesi, M. Winter and S. Nowak, *Adv. Energy Mater.*, 2022, **12**, 2102917.
440. X. Zhao, Y. Zhang, Y. Hou, Z. Zhao, Y. Gong, Y. Wang, H. Wei and V. G. Pol, *Adv. Compos. Hybrid Mater.*, 2022, **5**, 2988.
441. M. Wakihara, *Mater. Sci. Eng., R*, 2001, **33**, 109.
442. D. Lin, Y. Liu and Y. Cui, *Nat. Nanotechnol.*, 2017, **12**, 194.
443. J. B. Goodenough and K.-S. Park, *J. Am. Chem. Soc.*, 2013, **135**, 1167.
444. H. Zhang, L. Qiao, H. Kuehnle, E. Figgemeier, M. Armand and G. G. Eshetu, *Energy Environ. Sci.*, 2023, **16**, 11.
445. M. B. Armand, Intercalation Electrodes, in *Materials for Advanced Batteries*, ed. D. W. Murphy, J. Broadhead and B. C. H. Steele, NATO Conference Series, Springer, Boston, MA, 1980, vol. 2. DOI: 10.1007/978-1-4684-3851-2_7.
446. Y. Nishi, *Chem. Rec.*, 2001, **1**, 406.
447. B. Dunn, H. Kamath and J.-M. Tarascon, *Science*, 2011, **334**, 928.
448. V. Etacheri, R. Marom, R. Elazari, G. Salitra and D. Aurbach, *Energy Environ. Sci.*, 2011, **4**, 3243.
449. A. G. Tamirat, Y. Lui, X. L. Dong, C. X. Wang, Y. G. Wang and Y. Y. Xia, *Small Struct.*, 2021, **2**, 2000126.
450. M. Huang, B. J. Xi, N. X. Shi, J. K. Feng, Y. T. Qian, D. F. Xue and S. L. Xiong, *Small Struct.*, 2021, **2**, 2000085.
451. S. H. Yu, S. H. Lee, D. J. Lee, Y. E. Sung and T. Hyeon, *Small*, 2016, **12**, 2146.
452. C. K. Chan, H. Peng, G. Liu, K. McIlwrath, X. F. Zhang, R. A. Huggins and Y. Cui, *Nat. Nanotechnol.*, 2008, **3**, 31.
453. H. Li, Z. X. Wang, L. Q. Chen and X. J. Huang, *Adv. Mater.*, 2009, **21**, 4593.
454. M. H. Park, Y. Cho, K. Kim, J. Kim, M. L. Liu and J. Cho, *Angew. Chem., Int. Ed.*, 2011, **50**, 9647.
455. Z. Y. Wang, L. Zhou and X. W. Lou, *Adv. Mater.*, 2012, **24**, 1903.
456. J. Cabana, L. Monconduit, D. Larcher and M. Rosa Palacin, *Adv. Mater.*, 2010, **22**, E170.
457. G. C. Jia, D. L. Chao, N. H. Tiep, Z. Zhang and H. J. Fan, *Energy Storage Mater.*, 2018, **14**, 136.
458. Y. H. Li, K. Chang, E. Shangguan, D. L. Guo, W. Zhou, Y. Hou, H. W. Tang, B. Li and Z. R. Chang, *Nanoscale*, 2019, **11**, 1887.
459. A. H. Biby, B. A. Ali and N. K. Allam, *Mater. Today Energy*, 2021, **20**, 100677.
460. G. Du, Z. Guo, S. Wang, R. Zeng, Z. Chen and H. Liu, *Chem. Commun.*, 2010, **46**, 1106.
461. S.-H. Yu, M. J. Zachman, K. Kang, H. Gao, X. Huang, F. J. DiSalvo, J. Park, L. F. Kourkoutis and H. D. Abruna, *Adv. Energy Mater.*, 2019, **9**, 1902773.

462. K. Yao, Z. W. Xu, J. F. Huang, M. Ma, L. C. Fu, X. T. Shen, J. Li and M. S. Fu, *Small*, 2019, **15**, 1805405.
463. H. Liu, D. Su, R. F. Zhou, B. Sun, G. X. Wang and S. Z. Qiao, *Adv. Energy Mater.*, 2012, **2**, 970.
464. C.-L. Zhang, Z.-H. Jiang, B.-R. Lu, J.-T. Liu, F.-H. Cao, H. Li, Z.-L. Yu and S.-H. Yu, *Nano Energy*, 2019, **61**, 104.
465. K. Chang and W. Chen, *ACS Nano*, 2011, **5**, 4720.
466. Y. M. Chen, X. Y. Yu, Z. Li, U. Paik and X. W. Lou, *Sci. Adv.*, 2016, **2**, e1600021.
467. Q. Zeng, S. Tian, G. Liu, H. Yang, X. Sun, D. Wang, J. Huang, D. Yan and S. Peng, *ACS Appl. Mater. Interfaces*, 2022, **14**, 6958.
468. J. Lee, C. Choi, H. Lee, S. Ma, J. Tan, G. Jang, S. G. Shim, Y. S. Park, J. Yun, D.-W. Kim and J. Moon, *Adv. Energy Mater.*, 2022, **12**, 2103138.
469. C. An, Y. Ni, Z. Wang, X. Li and X. Liu, *Inorg. Chem. Front.*, 2018, **5**, 1045.
470. C. He, Y. Wei, Z. Wang, J. Wang and Y. Kwok Kit Richard, *J. Colloid Interface Sci.*, 2024, **664**, 156.
471. J. Wu, W.-M. Lau and D.-S. Geng, *Rare Met.*, 2017, **36**, 307.
472. A. Wang, X. Chen, G. Yu and Y. Wang, *Ionics*, 2021, **27**, 541.
473. L. Fei, Q. Lin, B. Yuan, G. Chen, P. Xie, Y. Li, Y. Xu, S. Deng, S. Smirnov and H. Luo, *ACS Appl. Mater. Interfaces*, 2013, **5**, 5330.
474. L. Yu, L. Qin, X. Xu, K. Kim, J. Liu, J. Kang and K. H. Kim, *Chem. Eng. J.*, 2022, **431**, 133463.
475. X. Hu, X. Liu, K. Chen, G. Wang and H. Wang, *J. Mater. Chem. A*, 2019, **7**, 11016.
476. X. Lv, F. Li, J. Gong, J. Gu, S. Lin and Z. Chen, *Phys. Chem. Chem. Phys.*, 2020, **22**, 8902.
477. A.-R. Park and C.-M. Park, *ACS Nano*, 2017, **11**, 6074.
478. F. Ersan, H. D. Ozaydin, G. Gokoglu and E. Akturk, *Appl. Surf. Sci.*, 2017, **425**, 301.
479. S. Yin, Y. Wang, X. Zhang, C. Wei, X. Huang and G. Wen, *Appl. Surf. Sci.*, 2023, **609**, 155435.
480. S. U. Rehman, A. Samad, M. Saeed, B. Amin, M. Hafeez, I. A. Mir and L. Zhu, *Appl. Surf. Sci.*, 2021, **551**, 149304.
481. V. Vega-Mayoral, R. Tian, A. G. Kelly, A. Griffin, A. Harvey, M. Borrelli, K. Nisi, C. Backes and J. N. Coleman, *Nanoscale*, 2019, **11**, 6206.
482. X. Dong, F. Chen, G. Chen, B. Wang, X. Tian, X. Yan, Y.-X. Yin, C. Deng, D. Wang, J. Mao, S. Xu and S. Zhang, *J. Colloid Interface Sci.*, 2022, **619**, 359.
483. D. Wu, C. Wang, M. Wu, Y. Chao, P. He and J. Ma, *J. Energy Chem.*, 2020, **43**, 24.
484. L. Kong, Y. Liu, H. Huang, M. Liu, W. Xu, B. Li and X.-H. Bu, *Sci. China Mater.*, 2021, **64**, 820.
485. Y. Wang, X. Qian, W. Zhou, H. Liao and S. Cheng, *RSC Adv.*, 2014, **4**, 36597.
486. H. Fan, H. Yu, X. Wu, Y. Zhang, Z. Luo, H. Wang, Y. Guo, S. Madhavi and Q. Yan, *ACS Appl. Mater. Interfaces*, 2016, **8**, 25261.
487. T. Liang, H. Wang, R. Wang, B. He, Y. Gong and C. Yan, *Electrochim. Acta*, 2021, **389**, 138686.

488. Y. Chen, J. Wang, Y. Hong, Y. Yang, L. Tan, N. Li, C. Ma, J. Wang, X. Fan and Y. Zhu, *J. Mater. Chem. A*, 2023, **11**, 6168.
489. Y. Yuan, M. Yang, L. Liu, J. Xia, H. Yan, J. Liu, J. Wen, Y. Zhang and X. Wang, *Nanoscale*, 2020, **12**, 20337.
490. S. Yao, J. Cui, Y. Deng, W. G. Chong, J. Wu, M. Ihsan-Ul-Haq, Y.-W. Mai and J.-K. Kim, *Energy Storage Mater.*, 2019, **20**, 36.
491. W. Luo, A. Calas, C. Tang, F. Li, L. Zhou and L. Mai, *ACS Appl. Mater. Interfaces*, 2016, **8**, 35219.
492. X. Sun, B. Zhang, M. Chen, L. Wang, D. Wang, R. Man, S. Iqbal, F. Tian, Y. Qian and L. Xu, *Nano Today*, 2022, **43**, 101408.
493. P. Kumari, P. Pal, K. Shinzato, K. Awasthi, T. Ichikawa, A. Jain and M. Kumar, *Int. J. Hydrogen Energy*, 2020, **45**, 16992.
494. Y. Wei, J. Chen, S. Wang, X. Zhong, R. Xiong, L. Gan, Y. Ma, T. Zhai and H. Li, *ACS Appl. Mater. Interfaces*, 2020, **12**, 16264.
495. G. Han, Z. G. Chen, D. L. Ye, L. Yang, L. Z. Wang, J. Drennan and J. Zou, *J. Mater. Chem. A*, 2014, **2**, 7109.
496. L. Zhang, Q. Wei, D. Sun, N. Li, H. Ju, J. Feng, J. Zhu, L. Mai, E. J. Cairns and J. Guo, *Nano Energy*, 2018, **51**, 391.
497. G. Han, Z.-G. Chen, D. Ye, B. Wang, L. Yang, Y. Zou, L. Wang, J. Drennan and J. Zou, *J. Mater. Chem. A*, 2015, **3**, 7560.
498. M. Imran, T. H. Bokhari, Y. Wu, Z. Rana, E. Gul, G. Rahman, A. Zafar, T. Ali, S. Javaid, S. Hussain, K. Maaz, S. Karim, M. Ahmad, G. Xiang and A. Nisar, *New J. Chem.*, 2024, **48**, 2755.
499. L. Chen, W. Yang, X. Li, L. Han and M. Wei, *J. Mater. Chem. A*, 2019, **7**, 10331.
500. K. Thiagarajan, J. Theerthagiri, R. A. Senthil and J. Madhavan, *J. Mater. Sci.:Mater. Electron.*, 2017, **28**, 17354.
501. C. Zhong, Y. D. Deng, W. B. Hu, J. L. Qiao, L. Zhang and J. J. Zhang, *Chem. Soc. Rev.*, 2015, **44**, 7484.
502. K. Karuppasamy, K. Prasanna, D. Kim, Y. H. Kang and H. W. Rhee, *RSC Adv.*, 2017, **7**, 19211.
503. K. Karuppasamy, H. W. Rhee, P. A. Reddy, D. Gupta, L. Mitu, A. R. Polu and X. S. Shajan, *J. Ind. Eng. Chem.*, 2016, **40**, 168.
504. K. Karuppasamy, H. S. Kim, D. Kim, D. Vikraman, K. Prasanna, A. Kathalingam, R. Sharma and H. W. Rhee, *Sci. Rep.*, 2017, **7**, 11103.
505. S. Karthikprabhu, K. Karuppasamy, D. Vikraman, K. Prasanna, T. Maiyalagan, A. Nichelson, A. Kathalingam and H. S. Kim, *Appl. Surf. Sci.*, 2018, **449**, 435.
506. P. Simon, Y. Gogotsi and B. Dunn, *Science*, 2014, **343**, 1210.
507. M.-x. Wang, D. He, M. Zhu, L. Wu, Z. Wang, Z.-H. Huang and H. Yang, *J. Power Sources*, 2023, **560**, 232703.
508. M. Pershaanaa, S. Bashir, S. Ramesh and K. Ramesh, *J. Energy Storage*, 2022, **50**, 104599.
509. E. G. da Silveira Firmiano, A. C. Rabelo, C. J. Dalmaschio, A. N. Pinheiro, E. C. Pereira, W. H. Schreiner and E. R. Leite, *Adv. Energy Mater.*, 2014, **4**, 1301380.

510. Y.-j. Gao, W. Song, J. Louise Liu and S. Bashir, *Advances in Sustainable Energy: Policy, Materials and Devices*, Springer Nature, Switzerland, 1st edn, 2021.
511. M. Z. Iqbal and U. Aziz, *J. Energy Storage*, 2022, **46**, 103823.
512. J. T. Zhang and X. S. Zhao, *ChemSusChem*, 2012, **5**, 818.
513. A. Laheaar, A. Arenillas and F. Beguin, *J. Power Sources*, 2018, **396**, 220.
514. A. Borenstein, O. Hanna, R. Attias, S. Luski, T. Brousse and D. Aurbach, *J. Mater. Chem. A*, 2017, **5**, 12653.
515. V. Augustyn, P. Simon and B. Dunn, *Energy Environ. Sci.*, 2014, **7**, 1597.
516. Y. X. Zhang, S. Yu, G. B. Lou, Y. L. Shen, H. Chen, Z. H. Shen, S. Y. Zhao, J. Z. Zhang, S. G. Chai and Q. C. Zou, *J. Mater. Sci.*, 2017, **52**, 11201.
517. D. P. Chatterjee and A. K. Nandi, *J. Mater. Chem. A*, 2021, **9**, 15880.
518. S. S. Nardekar, K. Krishnamoorthy, P. Pazhamalai, S. Sahoo, V. K. Mariappan and S.-J. Kim, *J. Mater. Chem. A*, 2020, **8**, 13121.
519. M. Mohan, N. P. Shetti and T. M. Aminabhavi, *J. Energy Storage*, 2023, **58**, 106321.
520. L. Q. Fan, G. J. Liu, C. Y. Zhang, J. H. Wu and Y. L. Wei, *Int. J. Hydrogen Energy*, 2015, **40**, 10150.
521. C. Liu, Y. Bai, Y. Zhao, H. Yao and H. Pang, *Energy Storage Mater.*, 2020, **33**, 470.
522. H. Wang, D. Tran, J. Qian, F. Ding and D. Losic, *Adv. Mater. Interfaces*, 2019, **6**, 1900915.
523. Y. Ouyang, H. T. Ye, X. F. Xia, X. Y. Jiao, G. M. Li, S. Mutahir, L. Wang, D. Mandler, W. Lei and Q. L. Hao, *J. Mater. Chem. A*, 2019, **7**, 3228.
524. H. Chen, J. Jiang, L. Zhang, H. Wan, T. Qi and D. Xia, *Nanoscale*, 2013, **5**, 8879.
525. J. Pu, F. L. Cui, S. B. Chu, T. L. Wang, E. H. Sheng and Z. H. Wang, *ACS Sustainable Chem. Eng.*, 2014, **2**, 809.
526. H. Chen, J. Jiang, Y. Zhao, L. Zhang, D. Guo and D. Xia, *J. Mater. Chem. A*, 2015, **3**, 428.
527. L. E. Bell, *Science*, 2008, **321**, 1457.
528. Z. Zhou, G. Han, X. Lu, G. Wang and X. Zhou, *J. Magnesium Alloys*, 2022, **10**, 1719.
529. F. J. DiSalvo, *Science*, 1999, **285**, 703.
530. J. R. Sootsman, D. Y. Chung and M. G. Kanatzidis, *Angew. Chem., Int. Ed.*, 2009, **48**, 8616.
531. T. J. Seebeck, *Abh. Dtsch. Akad. Wiss. Berlin*, 1822, **265**, 1822.
532. J. C. A. Peltier, *Ann. Chim. Phys.*, 1834, **56**, 371.
533. W. Thomson, *Proc. R. Soc. Edinburgh*, 1857, **3**, 91.
534. B. C. Sales, *Science*, 2002, **295**, 1248.
535. M. S. Dresselhaus, G. Chen, M. Y. Tang, R. Yang, H. Lee, D. Wang, Z. Ren, J.-P. Fleurial and P. Gogna, *Adv. Mater.*, 2007, **19**, 1043.
536. G. J. Snyder and E. S. Toberer, *Nat. Mater.*, 2008, **7**, 105.
537. J. H. Yang and T. Caillat, *MRS Bull.*, 2006, **31**, 224.
538. A. Khajepour and F. Rahmani, *Appl. Radiat. Isot.*, 2017, **119**, 51.

539. T. C. Holgate, R. Bennett, T. Hammel, T. Caillat, S. Keyser and B. Sievers, *J. Electron. Mater.*, 2015, **44**, 1814.
540. B. Orr, A. Akbarzadeh, M. Mochizuki and R. Singh, *Appl. Therm. Eng.*, 2016, **101**, 490.
541. J. H. Yang and F. R. Stabler, *J. Electron. Mater.*, 2009, **38**, 1245.
542. L. Rebenklau, P. Gierth, A. Paproth, K. Irrgang, L. Lippmann, A. Wodtke, L. Niedermeyer, K. Augsburg and F. Bechtold, *European Microelectronics Packaging Conference*, Friedrichshafen, 2015.
543. T. M. Tritt and M. A. Subramanian, *MRS Bull.*, 2006, **31**, 188.
544. V. Leonov, *J. Manuf. Syst.*, 2011, **21**, 125013.
545. V. Leonov and R. J. M. Vullers, *J. Renewable Sustainable Energy*, 2009, **1**, 062701.
546. Y. Yang, X. J. Wei and J. Liu, *J. Phys. D:Appl. Phys.*, 2007, **40**, 5790.
547. H. J. Goldsmid, *Introduction to Thermoelectricity*, Springer Verlag, Berlin, 2nd edn, 2016.
548. M. Cutler, J. F. Leavy and R. L. Fitzpatrick, *Phys. Rev.*, 1964, **133**, 1143.
549. G. D. Mahan and M. Bartkowiak, *Appl. Phys. Lett.*, 1999, **74**, 953.
550. X.-L. Shi, J. Zou and Z.-G. Chen, *Chem. Rev.*, 2020, **120**, 7399.
551. K. Peng, X. Lu, H. Zhan, S. Hui, X. Tang, G. Wang, J. Dai, C. Uher, G. Wang and X. Zhou, *Energy Environ. Sci.*, 2016, **9**, 454.
552. J. P. Heremans, V. Jovovic, E. S. Toberer, A. Saramat, K. Kurosaki, A. Charoenphakdee, S. Yamanaka and G. J. Snyder, *Science*, 2008, **321**, 554.
553. J. P. Heremans, B. Wiendlocha and A. M. Chamoire, *Energy Environ. Sci.*, 2012, **5**, 5510.
554. Y. Lin, M. Wood, K. Imasato, J. J. Kuo, D. Lam, A. N. Mortazavi, T. J. Slade, S. A. Hodge, K. Xi, M. G. Kanatzidis, D. R. Clarke, M. C. Hersam and G. J. Snyder, *Energy Environ. Sci.*, 2020, **13**, 4114.
555. L. Yang, Z.-G. Chen, M. Hong, L. Wang, D. Kong, L. Huang, G. Han, Y. Zou, M. Dargusch and J. Zou, *Nano Energy*, 2017, **31**, 105.
556. Y. Pei, L. Zheng, W. Li, S. Lin, Z. Chen, Y. Wang, X. Xu, H. Yu, Y. Chen and B. Ge, *Adv. Electron. Mater.*, 2016, **2**, 1600019.
557. Y.-K. Zhu, Y. Sun, J. Zhu, K. Song, Z. Liu, M. Liu, M. Guo, X. Dong, F. Guo, X. Tan, B. Yu, W. Cai, J. Jiang and J. Sui, *Small*, 2022, **18**, 2201352.
558. K. Biswas, J. Q. He, I. D. Blum, C.-I. Wu, T. P. Hogan, D. N. Seidman, V. P. Dravid and M. G. Kanatzidis, *Nature*, 2012, **489**, 414.
559. B. Yu, W. S. Liu, S. Chen, H. Wang, H. Z. Wang, G. Chen and Z. F. Ren, *Nano Energy*, 2012, **1**, 472.
560. G. A. Slack and D. M. Rowe, *CRC Handbook of Thermoelectrics*, CRC press, Boca Raton, 1995.
561. L. P. Hu, H. J. Wu, T. J. Zhu, C. G. Fu, J. Q. He, P. J. Ying and X. B. Zhao, *Adv. Energy Mater.*, 2015, **5**, 1500411.
562. J. S. Rhyee, K. H. Lee, S. M. Lee, E. Cho, S. I. Kim, E. Lee, Y. S. Kwon, J. H. Shim and G. Kotliar, *Nature*, 2009, **459**, 965.
563. Y. Z. Pei, A. LaLonde, S. Iwanaga and G. J. Snyder, *Energy Environ. Sci.*, 2011, **4**, 2085.

564. L. Yang, Z. G. Chen, G. Han, M. Hong, Y. C. Zou and J. Zou, *Nano Energy*, 2015, **16**, 367.
565. M. Hong, Z. G. Chen, Y. Z. Pei, L. Yang and J. Zou, *Phys. Rev. B:Condens. Matter Mater. Phys.*, 2016, **94**, 161201.
566. W. Li, L. L. Zheng, B. H. Ge, S. Q. Lin, X. Y. Zhang, Z. W. Chen, Y. J. Chang and Y. Z. Pei, *Adv. Mater.*, 2017, **29**, 1605887.
567. X. Shi, J. Yang, J. R. Salvador, M. F. Chi, J. Y. Cho, H. Wang, S. Q. Bai, J. H. Yang, W. Q. Zhang and L. D. Chen, *J. Am. Chem. Soc.*, 2011, **133**, 7837.
568. X. F. Meng, Z. H. Liu, B. Cui, D. D. Qin, H. Y. Geng, W. Cai, L. W. Fu, J. Q. He, Z. F. Ren and J. H. Sui, *Adv. Energy Mater.*, 2017, **7**, 1602582.
569. B. C. Sales, D. Mandrus and R. K. Williams, *Science*, 1996, **272**, 1325.
570. H. Kleinke, *Chem. Mater.*, 2010, **22**, 604.
571. M. Christensen, S. Johnsen and B. B. Iversen, *Dalton Trans.*, 2010, **39**, 978.
572. Y. Liu, L. M. Wu, L. H. Li, S. W. Du, J. D. Corbett and L. Chen, *Angew. Chem., Int. Ed.*, 2009, **48**, 5305.
573. E. S. Toberer, A. F. May and G. J. Snyder, *Chem. Mater.*, 2010, **22**, 624.
574. L. Wu, Z. Zhou, G. Han, B. Zhang, J. Yu, H. Wang, Y. Chen, X. Lu, G. Wang and X. Zhou, *Chem. Eng. J.*, 2023, **475**, 145988.
575. S. Xiao, K. Peng, Z. Zhou, H. Wang, S. Zheng, X. Lu, G. Han, G. Wang and X. Zhou, *J. Magnesium Alloys*, 2023, **11**, 2486.
576. X. Tang, B. Zhang, X. Zhang, S. Wang, X. Lu, G. Han, G. Wang and X. Zhou, *ACS Appl. Mater. Interfaces*, 2020, **12**, 8359.
577. S. R. Brown, S. M. Kauzlarich, F. Gascoin and G. J. Snyder, *Chem. Mater.*, 2006, **18**, 1873.
578. G. J. Snyder, M. Christensen, E. Nishibori, T. Caillat and B. B. Iversen, *Nat. Mater.*, 2004, **3**, 458.
579. C. K. Mai, R. A. Schlitz, G. M. Su, D. Spitzer, X. J. Wang, S. L. Fronk, D. G. Cahill, M. L. Chabinyc and G. C. Bazan, *J. Am. Chem. Soc.*, 2014, **136**, 13478.
580. O. Bubnova, Z. U. Khan, A. Malti, S. Braun, M. Fahlman, M. Berggren and X. Crispin, *Nat. Mater.*, 2011, **10**, 429.
581. M. He, F. Qiu and Z. Q. Lin, *Energy Environ. Sci.*, 2013, **6**, 1352.
582. S. Maensiri and W. Nuansing, *Mater. Chem. Phys.*, 2006, **99**, 104.
583. R. Robert, S. Romer, A. Reller and A. Weidenkaff, *Adv. Eng. Mater.*, 2005, **7**, 303.
584. Y. Y. Wang, N. S. Rogado, R. J. Cava and N. P. Ong, *Nature*, 2003, **423**, 425.
585. Y. Xiao and L.-D. Zhao, *npj Quantum Mater.*, 2018, **3**, 55.
586. B. Jia, Y. Huang, Y. Wang, Y. Zhou, X. Zhao, S. Ning, X. Xu, P. Lin, Z. Chen, B. Jiang and J. He, *Energy Environ. Sci.*, 2022, **15**, 1920.
587. L. D. Zhao, H. J. Wu, S. Q. Hao, C. I. Wu, X. Y. Zhou, K. Biswas, J. Q. He, T. P. Hogan, C. Uher, C. Wolverton, V. P. Dravid and M. G. Kanatzidis, *Energy Environ. Sci.*, 2013, **6**, 3346.
588. Z.-G. Chen, X. Shi, L.-D. Zhao and J. Zou, *Prog. Mater. Sci.*, 2018, **97**, 283.

589. L.-D. Zhao, G. Tan, S. Hao, J. He, Y. Pei, H. Chi, H. Wang, S. Gong, H. Xu, V. P. Dravid, C. Uher, G. J. Snyder, C. Wolverton and M. G. Kanatzidis, *Science*, 2016, **351**, 141.
590. H. Liu, X. Yuan, P. Lu, X. Shi, F. Xu, Y. He, Y. Tang, S. Bai, W. Zhang, L. Chen, Y. Lin, L. Shi, H. Lin, X. Gao, X. Zhang, H. Chi and C. Uher, *Adv. Mater.*, 2013, **25**, 6607.
591. L. P. Hu, T. J. Zhu, X. H. Liu and X. B. Zhao, *Adv. Funct. Mater.*, 2014, **24**, 5211.
592. Y. B. Luo, J. Y. Yang, Q. H. Jiang, W. X. Li, Y. Xiao, L. W. Fu, D. Zhang, Z. W. Zhou and Y. D. Cheng, *Nano Energy*, 2015, **18**, 37.
593. H. Kitagawa, A. Kurata, H. Araki, S. Morito and E. Tanabe, *J. Electron. Mater.*, 2010, **39**, 1692.
594. X. A. Yan, B. Poudel, Y. Ma, W. S. Liu, G. Joshi, H. Wang, Y. C. Lan, D. Z. Wang, G. Chen and Z. F. Ren, *Nano Lett.*, 2010, **10**, 3373.
595. L. P. Hu, T. J. Zhu, Y. G. Wang, H. H. Xie, Z. J. Xu and X. B. Zhao, *NPG Asia Mater.*, 2014, **6**, e88.
596. D. Bao, J. Chen, Y. Yu, W. Liu, L. Huang, G. Han, J. Tang, D. Zhou, L. Yang and Z.-G. Chen, *Chem. Eng. J.*, 2020, **388**, 124295.
597. Y. Wang, W.-D. Liu, X.-L. Shi, M. Hong, L.-J. Wang, M. Li, H. Wang, J. Zou and Z.-G. Chen, *Chem. Eng. J.*, 2020, **391**, 123513.
598. W. J. Xie, X. F. Tang, Y. G. Yan, Q. J. Zhang and T. M. Tritt, *J. Appl. Phys.*, 2009, **105**, 113713.
599. J. H. Li, Q. Tan, J. F. Li, D. W. Liu, F. Li, Z. Y. Li, M. M. Zou and K. Wang, *Adv. Funct. Mater.*, 2013, **23**, 4317.
600. S. I. Kim, K. H. Lee, H. A. Mun, H. S. Kim, S. W. Hwang, J. W. Roh, D. J. Yang, W. H. Shin, X. S. Li, Y. H. Lee, G. J. Snyder and S. W. Kim, *Science*, 2015, **348**, 109.
601. R. Deng, X. Su, S. Hao, Z. Zheng, M. Zhang, H. Xie, W. Liu, Y. Yan, C. Wolverton, C. Uher, M. G. Kanatzidis and X. Tang, *Energy Environ. Sci.*, 2018, **11**, 1520.
602. Z. Chen, Z. Jian, W. Li, Y. Chang, B. Ge, R. Hanus, J. Yang, Y. Chen, M. Huang, G. J. Snyder and Y. Pei, *Adv. Mater.*, 2017, **29**, 1606768.
603. Z.-Z. Luo, S. Cai, S. Hao, T. P. Bailey, Y. Luo, W. Luo, Y. Yu, C. Uher, C. Wolverton, V. P. Dravid, Z. Zou, Q. Yan and M. G. Kanatzidis, *Energy Environ. Sci.*, 2022, **15**, 368.
604. J. Q. He, L. D. Zhao, J. C. Zheng, J. W. Doak, H. J. Wu, H. Q. Wang, Y. Lee, C. Wolverton, M. G. Kanatzidis and V. P. Dravid, *J. Am. Chem. Soc.*, 2013, **135**, 4624.
605. Q. Y. Zhang, H. Wang, W. S. Liu, H. Z. Wang, B. Yu, Q. Zhang, Z. T. Tian, G. Ni, S. Lee, K. Esfarjani, G. Chen and Z. F. Ren, *Energy Environ. Sci.*, 2012, **5**, 5246.
606. A. D. LaLonde, Y. Z. Pei and G. J. Snyder, *Energy Environ. Sci.*, 2011, **4**, 2090.
607. J. Zhang, D. Wu, D. He, D. Feng, M. Yin, X. Qin and J. He, *Adv. Mater.*, 2017, **29**, 1703148.

608. R. Hanus, M. T. Agne, A. J. E. Rettie, Z. Chen, G. Tan, D. Y. Chung, M. G. Kanatzidis, Y. Pei, P. W. Voorhees and G. J. Snyder, *Adv. Mater.*, 2019, **31**, 1900108.
609. S. N. Girard, J. He, C. Li, S. Moses, G. Wang, C. Uher, V. P. Dravid and M. G. Kanatzidis, *Nano Lett.*, 2010, **10**, 2825.
610. D. Wu, L. D. Zhao, X. Tong, W. Li, L. J. Wu, Q. Tan, Y. L. Pei, L. Huang, J. F. Li, Y. M. Zhu, M. G. Kanatzidis and J. Q. He, *Energy Environ. Sci.*, 2015, **8**, 2056.
611. B. B. Jiang, Y. Yu, J. Cui, X. X. Liu, L. Xie, J. C. Liao, Q. H. Zhang, Y. Huang, S. C. Ning, B. H. Jia, B. Zhu, S. Q. Bai, L. D. Chen, S. J. Pennycook and J. Q. He, *Science*, 2021, **371**, 830.
612. C. Chang, M. Wu, D. He, Y. Pei, C.-F. Wu, X. Wu, H. Yu, F. Zhu, K. Wang, Y. Chen, L. Huang, J.-F. Li, J. He and L.-D. Zhao, *Science*, 2018, **360**, 778.
613. Y. M. Zhou and L. D. Zhao, *Adv. Mater.*, 2017, **29**, 1702676.
614. J. Liu, P. Wang, M. Y. Wang, R. Xu, J. Zhang, J. Z. Liu, D. Li, N. N. Liang, Y. W. Du, G. Chen and G. D. Tang, *Nano Energy*, 2018, **53**, 683.
615. X. L. Shi, A. Wu, T. L. Feng, K. Zheng, W. D. Liu, Q. Sun, M. Hong, S. T. Pantelides, Z. G. Chen and J. Zou, *Adv. Energy Mater.*, 2019, **9**, 1803242.
616. X. Shi, K. Zheng, M. Hong, W. Liu, R. Moshwan, Y. Wang, X. Qu, Z.-G. Chen and J. Zou, *Chem. Sci.*, 2018, **9**, 7376.
617. Q. Zhang, E. K. Chere, J. Y. Sun, F. Cao, K. Dahal, S. Chen, G. Chen and Z. F. Ren, *Adv. Energy Mater.*, 2015, **5**, 1500360.
618. Y. M. Han, J. Zhao, M. Zhou, X. X. Jiang, H. Q. Leng and L. F. Li, *J. Mater. Chem. A*, 2015, **3**, 4555.
619. C. C. Lin, R. Lydia, J. H. Yun, H. S. Lee and J. S. Rhyee, *Chem. Mater.*, 2017, **29**, 5344.
620. J. Guo, J. K. Jian, J. Liu, B. L. Cao, R. B. Lei, Z. H. Zhang, B. Song and H. Z. Zhao, *Nano Energy*, 2017, **38**, 569.
621. M. Zhou, G. J. Snyder, L. F. Li and L. D. Zhao, *Inorg. Chem. Front.*, 2016, **3**, 1449.
622. Y. K. Lee, Z. Luo, S. P. Cho, M. G. Kanatzidis and I. Chung, *Joule*, 2019, **3**, 719.
623. Y. B. Luo, S. T. Cai, X. Hua, H. J. Chen, Q. H. Liang, C. F. Du, Y. Zheng, J. H. Shen, J. W. Xu, C. Wolverton, V. P. Dravid, Q. Y. Yan and M. G. Kanatzidis, *Adv. Energy Mater.*, 2019, **9**, 1803072.
624. H. Zhang, Y. Chen, X. Liu, H. Wang, C. Niu, S. Zheng, B. Zhang, X. Lu, G. Wang, G. Han and X. Zhou, *Mater. Today Energy*, 2022, **30**, 101137.
625. X. Liu, H. Wang, Y. Chen, B. Zhang, H. Zhang, S. Zheng, X. Chen, X. Lu, G. Wang, X. Zhou and G. Han, *Scr. Mater.*, 2022, **218**, 114846.
626. X. Liu, Y. Chen, H. Wang, S. Liu, B. Zhang, X. Lu, G. Wang, G. Han, X. Chen and X. Zhou, *ACS Appl. Mater. Interfaces*, 2024, **16**, 2240.
627. A. A. Olvera, N. A. Moroz, P. Sahoo, P. Ren, T. P. Bailey, A. A. Page, C. Uher and P. F. P. Poudeu, *Energy Environ. Sci.*, 2017, **10**, 1668.
628. K. P. Zhao, P. F. Qiu, Q. F. Song, A. B. Blichfeld, E. Eikeland, D. D. Ren, B. H. Ge, B. B. Iversen, X. Shi and L. D. Chen, *Mater. Today Phys.*, 2017, **1**, 14.

629. Y. He, T. Day, T. S. Zhang, H. L. Liu, X. Shi, L. D. Chen and G. J. Snyder, *Adv. Mater.*, 2014, **26**, 3974.
630. H. Wang, S. K. Zheng, H. Wu, X. Xiong, Q. H. Xiong, H. Y. Wang, Y. Wang, B. Zhang, X. Lu, G. Han, G. Y. Wang and X. Y. Zhou, *Small*, 2022, **18**, 2104592.
631. S. Zheng, X. Yang, H. Wang, B. Zhang, X. Xiong, X. Lu, G. Han, G. Wang and X. Zhou, *Adv. Energy Mater.*, 2022, **12**, 2202575.
632. A. J. Hong, T. Li, H. X. Zhu, X. H. Zhou, Q. Y. He, W. S. Liu, Z. B. Yan, J. M. Liu and Z. F. Ren, *Solid State Ionics*, 2014, **261**, 21.
633. X. Shi, H. Y. Chen, F. Hao, R. H. Liu, T. Wang, P. F. Qiu, U. Burkhardt, Y. Grin and L. D. Chen, *Nat. Mater.*, 2018, **17**, 421.
634. S. Yang, Z. Gao, P. Qiu, J. Liang, T.-R. Wei, T. Deng, J. Xiao, X. Shi and L. Chen, *Adv. Mater.*, 2021, **33**, 2007681.
635. S. He, Y. Li, L. Liu, Y. Jiang, J. Feng, W. Zhu, J. Zhang, Z. Dong, Y. Deng, J. Luo, W. Zhang and G. Chen, *Sci. Adv.*, 2020, **6**, eaaz8423.
636. H. Wang, X. Liu, Z. Zhou, H. Wu, Y. Chen, B. Zhang, G. Wang, X. Zhou and G. Han, *Acta Mater.*, 2022, **223**, 117502.
637. H. Wang, G. Han, B. Zhang, Y. Chen, X. Liu, K. Zhang, X. Lu, G. Wang and X. Zhou, *Acta Mater.*, 2023, **248**, 118753.
638. J. Li, X. Zhang, Z. Chen, S. Lin, W. Li, J. Shen, I. T. Witting, A. Faghaninia, Y. Chen, A. Jain, L. Chen, G. J. Snyder and Y. Pei, *Joule*, 2018, **2**, 976.
639. B. Jiang, W. Wang, S. Liu, Y. Wang, C. Wang, Y. Chen, L. Xie, M. Huang and J. He, *Science*, 2022, **377**, 208.
640. N. S. Lewis, *Science*, 2007, **315**, 798.
641. Z. H. Zhang and P. Wang, *Energy Environ. Sci.*, 2012, **5**, 6506.
642. M. Garín, T. P. Pasanen, G. López, V. Vähänissi, K. Chen, I. Martín and H. Savin, *Small*, 2023, **19**, 2302250.
643. F. Rehman, I. H. Syed, S. Khanam, S. Ijaz, H. Mehmood, M. Zubair, Y. Massoud and M. Q. Mehmood, *Energy Adv.*, 2023, **2**, 1239.
644. M. Ren, L. Fang, Y. Zhang, F. T. Eickemeyer, Y. Yuan, S. M. Zakeeruddin, M. Grätzel and P. Wang, *Adv. Mater.*, 2024, 2403403.
645. R. Zhou, J. Xu, P. Luo, L. Hu, X. Pan, J. Xu, Y. Jiang and L. Wang, *Adv. Energy Mater.*, 2021, **11**, 2101923.
646. B. Kim, S.-W. Baek, C. Kim, J. Kim and J.-Y. Lee, *Adv. Energy Mater.*, 2022, **12**, 2102689.
647. X. Huang, L. Zhang, Y. Cheng, J. Oh, C. Li, B. Huang, L. Zhao, J. Deng, Y. Zhang, Z. Liu, F. Wu, X. Hu, C. Yang, L. Chen and Y. Chen, *Adv. Funct. Mater.*, 2022, **32**, 2108634.
648. L. Kujovic, X. Liu, A. Abbas, L. O. Jones, A. M. Law, M. Togay, K. M. Curson, K. L. Barth, J. W. Bowers, J. M. Walls, O. Oklobia, D. A. Lamb, S. J. C. Irvine, W. Zhang, C. Lee, T. Nagle, D. Lu and G. Xiong, *Adv. Funct. Mater.*, 2024, **34**, 2312528.
649. G. Q. Zhang, S. Finefrock, D. X. Liang, G. G. Yadav, H. R. Yang, H. Y. Fang and Y. Wu, *Nanoscale*, 2011, **3**, 2430.
650. G. Makrides, B. Zinsser, M. Norton, G. E. Georghiou, M. Schubert and J. H. Werner, *Renewable Sustainable Energy Rev.*, 2010, **14**, 754.

651. C.-Z. Ning, L. Dou and P. Yang, *Nat. Rev. Mater.*, 2017, **2**, 17070.
652. A. S. Al-Ezzi and M. N. M. Ansari, *Appl. Syst. Innov.*, 2022, **5**, 67.
653. C. Lin, G. Liu, X. Xi, L. Wang, Q. Wang, Q. Sun, M. Li, B. Zhu, D. P. de Lara and H. Zai, *Materials*, 2022, **15**, 8142.
654. B. W. H. Baugher, H. O. H. Churchill, Y. Yang and P. Jarillo-Herrero, *Nat. Nanotechnol.*, 2014, **9**, 262.
655. A. Pospischil, M. M. Furchi and T. Mueller, *Nat. Nanotechnol.*, 2014, **9**, 257.
656. R. W. Birkmire and B. E. McCandless, *Curr. Opin. Solid State Mater. Sci.*, 2010, **14**, 139.
657. C. S. Ferekides, U. Balasubramanian, R. Mamazza, V. Viswanathan, H. Zhao and D. L. Morel, *Sol. Energy*, 2004, **77**, 823.
658. D. Lincot, J. F. Guillemoles, S. Taunier, D. Guimard, J. Sicx-Kurdi, A. Chaumont, O. Roussel, O. Ramdani, C. Hubert, J. P. Fauvarque, N. Bodereau, L. Parissi, P. Panheleux, P. Fanouillere, N. Naghavi, P. P. Grand, M. Benfarah, P. Mogensen and O. Kerrec, *Sol. Energy*, 2004, **77**, 725.
659. E. W. Jones, V. Barrioz, S. J. C. Irvine and D. Lamb, *Thin Solid Films*, 2009, **517**, 2226.
660. J. D. Major, Y. Y. Proskuryakov and K. Durose, *Prog. Photovoltaics*, 2013, **21**, 436.
661. A. Rios-Flores, J. L. Pena, V. Castro-Pena, O. Ares, R. Castro-Rodriguez and A. Bosio, *Sol. Energy*, 2010, **84**, 1020.
662. M. Afzaal and P. O'Brien, *J. Mater. Chem.*, 2006, **16**, 1597.
663. C. S. Ferekides, D. Marinskiy, V. Viswanathan, B. Tetali, V. Palekis, P. Selvaraj and D. L. Morel, *Thin Solid Films*, 2000, **361**, 520.
664. M. A. Green, K. Emery, Y. Hishikawa, W. Warta and E. D. Dunlop, *Prog. Photovoltaics*, 2013, **21**, 827.
665. S. M. Pawar, B. S. Pawar, J. H. Kim, O. S. Joo and C. D. Lokhande, *Curr. Appl. Phys.*, 2011, **11**, 117.
666. I. Repins, M. A. Contreras, B. Egaas, C. DeHart, J. Scharf, C. L. Perkins, B. To and R. Noufi, *Prog. Photovoltaics*, 2008, **16**, 235.
667. H. Aboulfadl, K. V. Sopiha, J. Keller, J. K. Larsen, J. J. S. Scragg, C. Persson, M. Thuvander and M. Edoff, *ACS Appl. Mater. Interfaces*, 2021, **13**, 7188.
668. A. Allouhi, S. Rehman, M. S. Buker and Z. Said, *J. Cleaner Prod.*, 2022, **362**, 132339.
669. I. Massiot, A. Cattoni and S. Collin, *Nat. Energy*, 2020, **5**, 959.
670. J. R. Tuttle, J. S. Ward, A. Duda, T. A. Berens, M. A. Contreras, K. R. Ramanathan, A. L. Tennant, J. Keane, E. D. Cole, K. Emery and R. Noufi, *Materials Research Society Symposium Proceedings*, San Francisco, 1996.
671. L. Stolt, J. Hedstrom, J. Kessler, M. Ruckh, K. O. Velthaus and H. W. Schock, *Appl. Phys. Lett.*, 1993, **62**, 597.
672. T. Negami, M. Nishitani, N. Kohara, Y. Hashimoto and T. Wada, *Materials Research Society Symposium Proceedings*, San Francisco, 1996.
673. J. Yin, J. Jin, H. Lin, Z. Yin, J. Li, M. Lu, L. Guo, P. Xi, Y. Tang and C.-H. Yan, *Adv. Sci.*, 2020, **7**, 1903070.

674. M. Wang, L. Zhang, Y. He and H. Zhu, *J. Mater. Chem. A*, 2021, **9**, 5320.
675. Y. Guo, T. Park, J. W. Yi, J. Henzie, J. Kim, Z. Wang, B. Jiang, Y. Bando, Y. Sugahara, J. Tang and Y. Yamauchi, *Adv. Mater.*, 2019, **31**, 1807134.
676. D.-R. Chen, J. Muthu, X.-Y. Guo, H.-T. Chin, Y.-C. Lin, G. Haider, C.-C. Ting, M. Kalbac, M. Hofmann and Y.-P. Hsieh, *J. Mater. Chem. A*, 2023, **11**, 15802.
677. Y. Jiao, Y. Zheng, M. Jaroniec and S. Z. Qiao, *Chem. Soc. Rev.*, 2015, **44**, 2060.
678. J.-X. Feng, J.-Q. Wu, Y.-X. Tong and G.-R. Li, *J. Am. Chem. Soc.*, 2018, **140**, 610.
679. N. Zhang and Y. Chai, *Energy Environ. Sci.*, 2021, **14**, 4647.
680. Y. Zheng, Y. Jiao, M. Jaroniec and S. Z. Qiao, *Angew. Chem., Int. Ed.*, 2015, **54**, 52.
681. W. C. Sheng, H. A. Gasteiger and Y. Shao-Horn, *J. Electrochem. Soc.*, 2010, **157**, B1529.
682. H. C. Tao, Y. N. Gao, N. Talreja, F. Guo, J. Texter, C. Yan and Z. Y. Sun, *J. Mater. Chem. A*, 2017, **5**, 7257.
683. S. Jayabal, G. Saranya, J. Wu, Y. Q. Liu, D. S. Geng and X. B. Meng, *J. Mater. Chem. A*, 2017, **5**, 24540.
684. S. Marini, P. Salvi, P. Nelli, R. Pesenti, M. Villa, M. Berrettoni, G. Zangari and Y. Kiros, *Electrochim. Acta*, 2012, **82**, 384.
685. C. G. Morales-Guio and X. L. Hu, *Acc. Chem. Res.*, 2014, **47**, 2671.
686. J. Joo, T. Kim, J. Lee, S.-I. Choi and K. Lee, *Adv. Mater.*, 2019, **31**, 1806682.
687. J. Wang, X. Yue, Y. Yang, S. Sirisomboonchai, P. Wang, X. Ma, A. Abudula and G. Guan, *J. Alloys Compd.*, 2020, **819**, 153346.
688. T. Sun, Z. Tang, W. Zang, Z. Li, J. Li, Z. Li, L. Cao, J. S. D. Rodriguez, C. O. M. Mariano, H. Xu, P. Lyu, X. Hai, H. Lin, X. Sheng, J. Shi, Y. Zheng, Y.-R. Lu, Q. He, J. Chen, K. S. Novoselov, C.-H. Chuang, S. Xi, X. Luo and J. Lu, *Nat. Nanotechnol.*, 2023, **18**, 763.
689. A. P. Tiwari, K. Kim and S. Jeon, *Curr. Opin. Electrochem.*, 2022, **34**, 100982.
690. Z. W. Seh, J. Kibsgaard, C. F. Dickens, I. B. Chorkendorff, J. K. Norskov and T. F. Jaramillo, *Science*, 2017, **355**, eaad4998.
691. Y. T. Meng, W. Q. Song, H. Huang, Z. Ren, S. Y. Chen and S. L. Suib, *J. Am. Chem. Soc.*, 2014, **136**, 11452.
692. P. D. Tran, T. V. Tran, M. Orio, S. Torelli, Q. D. Truong, K. Nayuki, Y. Sasaki, S. Y. Chiam, R. Yi, I. Honma, J. Barber and V. Artero, *Nat. Mater.*, 2016, **15**, 640.
693. Y. Yin, J. C. Han, Y. M. Zhang, X. H. Zhang, P. Xu, Q. Yuan, L. Samad, X. J. Wang, Y. Wang, Z. H. Zhang, P. Zhang, X. Z. Cao, B. Song and S. Jin, *J. Am. Chem. Soc.*, 2016, **138**, 7965.
694. D. F. Yan, Y. X. Li, J. Huo, R. Chen, L. M. Dai and S. Y. Wang, *Adv. Mater.*, 2017, **29**, 1606459.
695. J. Di, C. Chen, S.-Z. Yang, S. Chen, M. Duan, J. Xiong, C. Zhu, R. Long, W. Hao, Z. Chi, H. Chen, Y.-X. Weng, J. Xia, L. Song, S. Li, H. Li and Z. Liu, *Nat. Commun.*, 2019, **10**, 2840.

696. H. Wu, T. Yang, Y. H. Du, L. Shen and G. W. Ho, *Adv. Mater.*, 2018, **30**, 1804341.
697. J. S. Kim, B. Kim, H. Kim and K. Kang, *Adv. Energy Mater.*, 2018, **8**, 1702774.
698. B. Zhang, X. L. Zheng, O. Voznyy, R. Comin, M. Bajdich, M. Garcia-Melchor, L. L. Han, J. X. Xu, M. Liu, L. R. Zheng, F. P. G. de Arquer, C. T. Dinh, F. J. Fan, M. J. Yuan, E. Yassitepe, N. Chen, T. Regier, P. F. Liu, Y. H. Li, P. De Luna, A. Janmohamed, H. L. L. Xin, H. G. Yang, A. Vojvodic and E. H. Sargent, *Science*, 2016, **352**, 333.
699. M. S. Faber, R. Dziedzic, M. A. Lukowski, N. S. Kaiser, Q. Ding and S. Jin, *J. Am. Chem. Soc.*, 2014, **136**, 10053.
700. J. Wang, F. Xu, H. Y. Jin, Y. Q. Chen and Y. Wang, *Adv. Mater.*, 2017, **29**, 1605838.
701. Y. Lei, S. Pakhira, K. Fujisawa, X. Wang, O. O. Iyiola, N. P. Lopez, A. L. Elias, L. P. Rajukumar, C. Zhou, B. Kabius, N. Alem, M. Endo, R. Lv, J. L. Mendoza-Cortes and M. Terrones, *ACS Nano*, 2017, **11**, 5103.
702. Y. Zhao, N. Dongfang, C. A. Triana, C. Huang, R. Erni, W. Wan, J. Li, D. Stoian, L. Pan, P. Zhang, J. Lan, M. Iannuzzi and G. R. Patzke, *Energy Environ. Sci.*, 2022, **15**, 727.
703. Y. Zhao, S. Wei, K. Pan, Z. Dong, B. Zhang, H.-H. Wu, Q. Zhang, J. Lin and H. Pang, *Chem. Eng. J.*, 2021, **421**, 129645.
704. X. Shan, J. Liu, H. Mu, Y. Xiao, B. Mei, W. Liu, G. Lin, Z. Jiang, L. Wen and L. Jiang, *Angew. Chem., Int. Ed.*, 2020, **59**, 1659.
705. N. Zhang, Y. Hu, L. An, Q. Li, J. Yin, J. Li, R. Yang, M. Lu, S. Zhang, P. Xi and C.-H. Yan, *Angew. Chem., Int. Ed.*, 2022, **61**, e202207217.
706. Z. Xue, X. Li, Q. Liu, M. Cai, K. Liu, M. Liu, Z. Ke, X. Liu and G. Li, *Adv. Mater.*, 2019, **31**, 1900430.
707. W. J. Zhou, X. J. Wu, X. H. Cao, X. Huang, C. L. Tan, J. Tian, H. Liu, J. Y. Wang and H. Zhang, *Energy Environ. Sci.*, 2013, **6**, 2921.
708. Q. Gao, C. Q. Huang, Y. M. Ju, M. R. Gao, J. W. Liu, D. An, C. H. Cui, Y. R. Zheng, W. X. Li and S. H. Yu, *Angew. Chem., Int. Ed.*, 2017, **56**, 7769.
709. M. R. Gao, W. T. Yao, H. B. Yao and S. H. Yu, *J. Am. Chem. Soc.*, 2009, **131**, 7486.
710. M. R. Gao, Y. F. Xu, J. Jiang, Y. R. Zheng and S. H. Yu, *J. Am. Chem. Soc.*, 2012, **134**, 2930.
711. M. R. Gao, X. Cao, Q. Gao, Y. F. Xu, Y. R. Zheng, J. Jiang and S. H. Yu, *ACS Nano*, 2014, **8**, 3970.
712. Y. R. Zheng, M. R. Gao, Q. Gao, H. H. Li, J. Xu, Z. Y. Wu and S. H. Yu, *Small*, 2015, **11**, 182.
713. B. Hinnemann, P. G. Moses, J. Bonde, K. P. Jorgensen, J. H. Nielsen, S. Horch, I. Chorkendorff and J. K. Norskov, *J. Am. Chem. Soc.*, 2005, **127**, 5308.
714. T. F. Jaramillo, K. P. Jorgensen, J. Bonde, J. H. Nielsen, S. Horch and I. Chorkendorff, *Science*, 2007, **317**, 100.
715. H. I. Karunadasa, E. Montalvo, Y. J. Sun, M. Majda, J. R. Long and C. J. Chang, *Science*, 2012, **335**, 698.

716. J. F. Xie, J. J. Zhang, S. Li, F. Grote, X. D. Zhang, H. Zhang, R. X. Wang, Y. Lei, B. C. Pan and Y. Xie, *J. Am. Chem. Soc.*, 2013, **135**, 17881.
717. M. R. Gao, M. K. Y. Chan and Y. G. Sun, *Nat. Commun.*, 2015, **6**, 7493.
718. J. Zhang, T. Wang, P. Liu, S. H. Liu, R. H. Dong, X. D. Zhuang, M. W. Chen and X. L. Feng, *Energy Environ. Sci.*, 2016, **9**, 2789.
719. J. Staszak-Jirkovsky, C. D. Malliakas, P. P. Lopes, N. Danilovic, S. S. Kota, K. C. Chang, B. Genorio, D. Strmcnik, V. R. Stamenkovic, M. G. Kanatzidis and N. M. Markovic, *Nat. Mater.*, 2016, **15**, 197.
720. Q. Z. Xiong, Y. Wang, P. F. Liu, L. R. Zheng, G. Z. Wang, H. G. Yang, P. K. Wong, H. M. Zhang and H. J. Zhao, *Adv. Mater.*, 2018, **30**, 1801450.
721. R. Miao, B. Dutta, S. Sahoo, J. K. He, W. Zhong, S. A. Cetegen, T. Jiang, S. P. Alpay and S. L. Suib, *J. Am. Chem. Soc.*, 2017, **139**, 13604.
722. J. Yin, J. Jin, H. Zhang, M. Lu, Y. Peng, B. L. Huang, P. X. Xi and C. H. Yan, *Angew. Chem., Int. Ed.*, 2019, **58**, 18676.
723. A. Paracchino, V. Laporte, K. Sivula, M. Graetzel and E. Thimsen, *Nat. Mater.*, 2011, **10**, 456.
724. A. Fujishima and K. Honda, *Nature*, 1972, **238**, 37.
725. M. Ni, M. K. H. Leung, D. Y. C. Leung and K. Sumathy, *Renewable Sustainable Energy Rev.*, 2007, **11**, 401.
726. S. U. M. Khan, M. Al-Shahry and W. B. Ingler, *Science*, 2002, **297**, 2243.
727. J. H. Park, S. Kim and A. J. Bard, *Nano Lett.*, 2006, **6**, 24.
728. G. L. Chiarello, E. Selli and L. Forni, *Appl. Catal., B*, 2008, **84**, 332.
729. H. J. Wu and Z. H. Zhang, *Int. J. Hydrogen Energy*, 2011, **36**, 13481.
730. H. J. Wu and Z. H. Zhang, *J. Solid State Chem.*, 2011, **184**, 3202.
731. K. Maeda, T. Takata, M. Hara, N. Saito, Y. Inoue, H. Kobayashi and K. Domen, *J. Am. Chem. Soc.*, 2005, **127**, 8286.
732. K. Maeda and K. Domen, *Chem. Mater.*, 2010, **22**, 612.
733. X. Y. Yang, A. Wolcott, G. M. Wang, A. Sobo, R. C. Fitzmorris, F. Qian, J. Z. Zhang and Y. Li, *Nano Lett.*, 2009, **9**, 2331.
734. A. Wolcott, W. A. Smith, T. R. Kuykendall, Y. P. Zhao and J. Z. Zhang, *Adv. Funct. Mater.*, 2009, **19**, 1849.
735. K. G. Kanade, B. B. Kale, J. O. Baeg, S. M. Lee, C. W. Lee, S. J. Moon and H. J. Chang, *Mater. Chem. Phys.*, 2007, **102**, 98.
736. H. M. Chen, C. K. Chen, Y.-C. Chang, C.-W. Tsai, R.-S. Liu, S.-F. Hu, W.-S. Chang and K.-H. Chen, *Angew. Chem., Int. Ed.*, 2010, **49**, 5966.
737. Z. H. Zhang and P. Wang, *J. Mater. Chem.*, 2012, **22**, 2456.
738. A. Paracchino, J. C. Brauer, J. E. Moser, E. Thimsen and M. Graetzel, *J. Phys. Chem. C*, 2012, **116**, 7341.
739. K. Sivula, F. Le Formal and M. Graetzel, *Chem. Mater.*, 2009, **21**, 2862.
740. D. W. Hwang, J. Kim, T. J. Park and J. S. Lee, *Catal. Lett.*, 2002, **80**, 53.
741. H. L. Wang, T. Lindgren, J. J. He, A. Hagfeldt and S. E. Lindquist, *J. Phys. Chem. B*, 2000, **104**, 5686.
742. A. Hameed, M. A. Gondal and Z. H. Yamani, *Catal. Commun.*, 2004, **5**, 715.
743. S. J. Hong, H. Jun, P. H. Borse and J. S. Lee, *Int. J. Hydrogen Energy*, 2009, **34**, 3234.

744. Q. Wang and K. Domen, *Chem. Rev.*, 2020, **120**, 919.
745. J. H. Wu, Y. Huang, W. Ye and Y. G. Li, *Adv. Sci.*, 2017, **4**, 1700194.
746. Q. P. Lu, Y. F. Yu, Q. L. Ma, B. Chen and H. Zhang, *Adv. Mater.*, 2016, **28**, 1917.
747. B. Mei, K. Han and G. Mul, *ACS Catal.*, 2018, **8**, 9154.
748. M. Gratzel, *Nature*, 2001, **414**, 338.
749. R. Asahi, T. Morikawa, T. Ohwaki, K. Aoki and Y. Taga, *Science*, 2001, **293**, 269.
750. S. Chen, J. J. M. Vequizo, Z. Pan, T. Hisatomi, M. Nakabayashi, L. Lin, Z. Wang, K. Kato, A. Yamakata, N. Shibata, T. Takata, T. Yamada and K. Domen, *J. Am. Chem. Soc.*, 2021, **143**, 10633.
751. C. M. Wolff, P. D. Frischmann, M. Schulze, B. J. Bohn, R. Wein, P. Livadas, M. T. Carlson, F. Jaeckel, J. Feldmann, F. Wuerthner and J. K. Stolarczyk, *Nat. Energy*, 2018, **3**, 862.
752. M. Sathish, B. Viswanathan and R. P. Viswanath, *Int. J. Hydrogen Energy*, 2006, **31**, 891.
753. N. Bao, L. Shen, T. Takata and K. Domen, *Chem. Mater.*, 2008, **20**, 110.
754. N. Zhao, J. Peng, G. Liu, Y. Zhang, W. Lei, Z. Yin, J. Li and M. Zhai, *J. Mater. Chem. A*, 2018, **6**, 18458.
755. J. Chen, X.-J. Wu, L. Yin, B. Li, X. Hong, Z. Fan, B. Chen, C. Xue and H. Zhang, *Angew. Chem., Int. Ed.*, 2015, **54**, 1210.
756. Z. Li, X. Meng and Z. Zhang, *J. Photochem. Photobiol., C*, 2018, **35**, 39.
757. W.-Z. Xiao, L. Xu, Q.-Y. Rong, X.-Y. Dai, C.-P. Cheng and L.-L. Wang, *Appl. Surf. Sci.*, 2020, **504**, 144425.

CHAPTER 2

Pnictides: An Overview of Crystal Structures, Properties and Applications

ALEXANDER OVCHINNIKOV[a,b†], SVIATOSLAV BARANETS[c,d†] AND SVILEN BOBEV[c*]

[a]Department of Materials and Environmental Chemistry, Stockholm University, Stockholm 10691, Sweden; [b]Faculty of Chemistry and Food Chemistry, Technische Universität Dresden, 01062 Dresden, Germany; [c]Department of Chemistry and Biochemistry, University of Delaware, Newark, Delaware, 19716, USA; [d]Department of Chemistry, Louisiana State University, Baton Rouge, Louisiana 70803, USA
*E-mail: bobev@udel.edu

2.1 Introduction

Group 15 in the Periodic Table includes six elements, commonly referred to as pnictogens (Pn). These are nitrogen (N), phosphorus (P), arsenic (As), antimony (Sb), bismuth (Bi), and the synthetic element moscovium (Mc).[1] Pnictogens, excluding N, which exists in nature as the chemically stable molecular species N_2, are chemically reactive and form compounds with many other elements. In this chapter, we will focus on pnictides—compounds with anionic Pn species. Since moscovium is an unstable element with poorly

[†]These authors contributed equally.

Inorganic Materials Series
Non-oxide and Heteroanionic Materials
Edited by Duncan H. Gregory
A catalogue record for this book is available from the British Library

known chemical properties, we will exclude it from consideration. In addition, nitrides will not be covered here due to their unique crystal chemistry, which is very different from that of other pnictides, and requires a separate discussion.

Pnictides with Pn = P, As, Sb, and Bi are known to adopt a plethora of crystal structure types, sometimes with substantial structural complexity (Figure 2.1). Despite the great variability of atomic arrangements found in pnictides, some general traits can be identified. Unlike nitrides, pnictides with heavier Group 15 congeners often exhibit homoatomic polyanionic species $[Pn_x]^{y-}$ of varying dimensionalities. While three-dimensional (3D) $[Pn_x]^{y-}$ units, such as frameworks or cages, are not so common in pnictides (in contrast to, *e.g.*, 3D polyanions $[Tt_x]^{y-}$ in compounds of Group 14 elements, tetrels, Tt), zero-, one-, and two-dimensional polyanionic building blocks are frequently observed.[2]

Chemical bonding in such moieties, as well as in more complex heteroatomic polyanions containing Pn and other main-group elements, can often be rationalised using the Zintl concept, a method widely used to explain chemical interactions in solids.[3] Within this approach, it is assumed that the

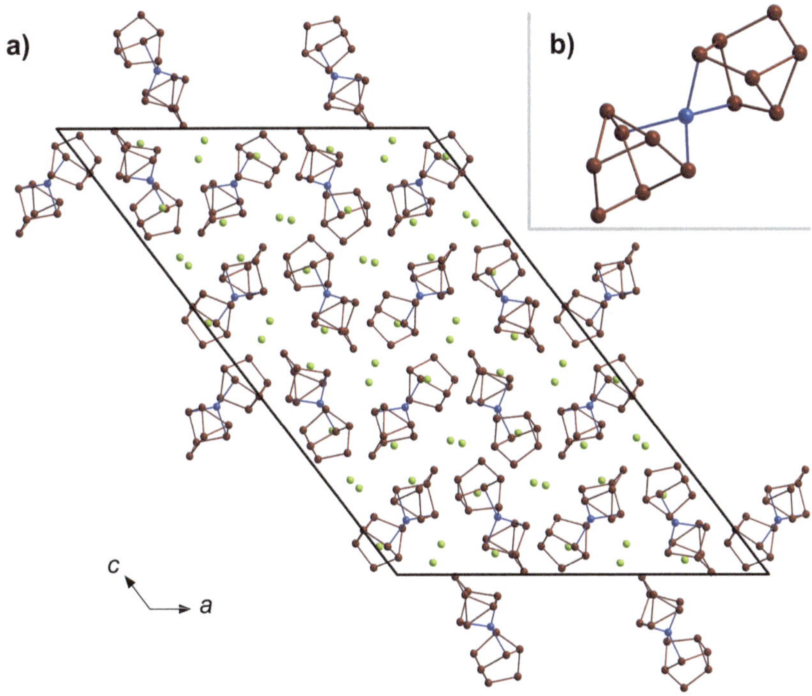

Figure 2.1 Crystal structure of the complex pnictide $Cs_4Zn(As_7)_2$ (a), space group $P2_1/c$, Pearson code (*mP*304) and an enlarged view of a single $[Zn(As_7)_2]^{4-}$ polyanion (b). Reproduced from ref. 2, https://doi.org/10.1088/1361-648X/ab6b87, under the terms of the CC-BY 3.0 License, https://creativecommons.org/licenses/by/3.0/.

charge transfer from the electropositive species present in the structure onto the Pn-bearing anions is complete, and the chemical bonding is optimised when the total number of valence electrons in the anion provides the realisation of an octet electronic state for all the atoms.

To illustrate the application of the Zintl concept, we may first consider the composition Na_3P. With one electron donated from each Na atom—three in total per formula unit—and five of its own valence electrons, the P atom adopts an octet state without any extra P–P bonds, according to the charge partitioning $(Na^+)_3(P^{3-})$. However, in the phosphorus-richer phase NaP, each P atom has a total of six electrons. To increase its electron count and satisfy the octet rule, each P atom forms homoatomic covalent bonds with two neighbors, yielding helical (P^-) chains, so that the charge-balanced $(Na^+)(P^-)$ composition results.

Compounds satisfying the Zintl electron count are called Zintl phases. Since the bonding picture described within the Zintl formalism perfectly optimises the chemical interactions due to the stable octet configuration, the bonding and antibonding electronic states are expected to be separated by an energy gap; therefore, a semiconducting behaviour should be anticipated (Figure 2.2(a)). While many Zintl phases are indeed semiconducting, the ionicity of the chemical bonding is often exaggerated in this simple electron-counting approach, with many formally charge-balanced compounds being bad metals due to significant hybridisation of the electronic states (Figure 2.2(b)). Despite the somewhat naïve treatment of the

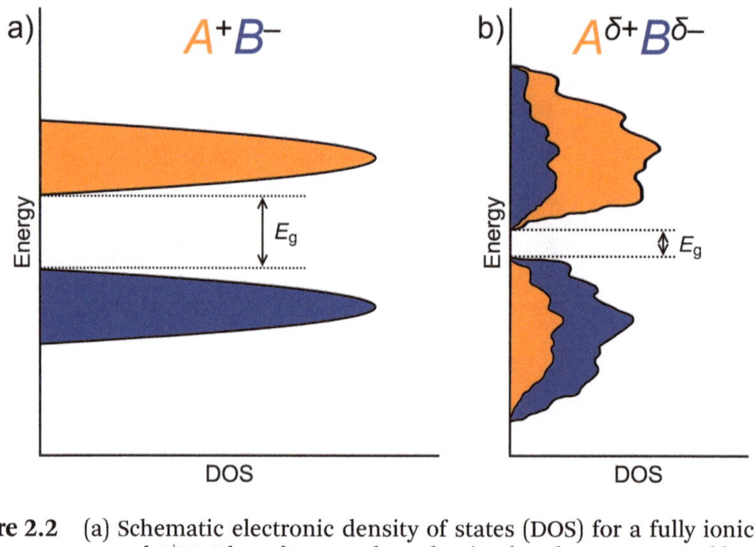

Figure 2.2 (a) Schematic electronic density of states (DOS) for a fully ionic compound A^+B^-. The valence and conduction bands are separated by a sizeable energy gap. (b) In the presence of covalent bonding, the charge transfer is incomplete, and the hybridisation of electronic states reduces the size of the gap. Reproduced with from ref. 4 with permission Elsevier, Copyright 2019.

valence electrons, the Zintl concept is a powerful tool for rationalising the crystal structures of pnictides.

Several factors may lead to deviations from the simple bonding picture assumed by the Zintl formalism, making its application challenging. One such factor is the presence of transition metals in the structure. Having partially filled d- or f-states, these elements naturally disobey the octet rule. Nevertheless, the Zintl approach can still be applied to certain compounds of transition elements, in particular when the stable d^5 or d^{10} electron configuration is achieved (*e.g.*, in Mn^{2+} and Cu^{+}, respectively), or when the partially occupied states do not participate in chemical bonding, as is usually the case with f orbitals, *e.g.*, in compounds with rare-earth metals.[4]

For some structural families of pnictides containing transition metals, special electron counting rules were developed, often based on the similarities of the electronic structures within a single family. Thus, a large group of intermetallic compounds, which also includes some pnictide representatives, is the Heusler family. Materials belonging to the full-Heusler (Figure 2.3(a), Cu_2MnAl type structure, space group $Fm\bar{3}m$, Pearson code *cF*16) and half-Heusler types (Figure 2.3(b), MgAgAs type structure, space group $F\bar{4}3m$, Pearson code *cF*12), as well as their structurally distorted derivatives, are often referred to by the generic name Heusler phases. The full-Heusler and half-Heusler compounds have optimal valence electron counts (VEC) of 24 and 18 electrons, respectively. If the optimal VEC is achieved, the Heusler phase is expected to be semiconducting.[5]

Another factor that complicates analysis of chemical bonding in pnictides is the deviation of the interatomic interactions from simple two-center two-electron bonds. Formation of multicenter metal–metal bonding often occurs in intermetallic compounds. In those cases, proper accounting of valence electrons becomes challenging.

The bonding in the pnictide polyanions may also deviate from a two-centre two-electron model, even without formation of metal–metal bonds. An excellent example is the so-called hypervalent bonding, realised in some electron-rich species, *e.g.*, in the linear polyanion $[Sb_3]^{7-}$. The Sb–Sb contacts in

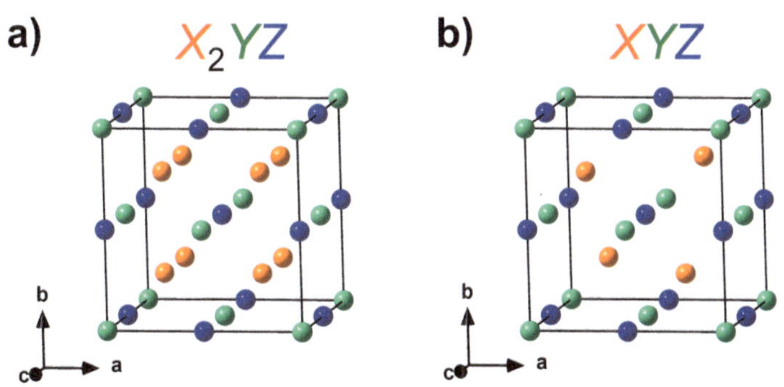

Figure 2.3 Crystal structures of a full-Heusler (a) and a half-Heusler (b) phase.

this species are usually described as one-electron bonds, which allows realisation of an octet configuration and brings the description of the chemical bonding back to the realm of the Zintl approach. However, the application of the Zintl concept may be more problematic for polyanionic hypervalent species of higher dimensionalities, such as ribbons and sheets, which are typical for the structures of pnictides with Pn = Sb and Bi. Calculations show that one- and two-dimensional homoatomic hypervalent polyanions get progressively destabilised upon going up the pnictogen group in the Periodic System.[6]

The vast diversity of pnictide crystal structures and availability of simple predictive tools based on the valence electron accounting, open up great opportunities for design of materials with exciting properties. In the following discussion, we will present some common synthetic approaches used to produce pnictides and outline the most common applications and emerging technologies in the field of pnictide materials.

2.2 Synthesis

Numerous synthetic approaches have been employed for the production of pnictides. The most commonly used method in solid-state chemistry is direct reaction of the elements, usually at elevated temperatures. The thermal energy for such reactions is habitually provided by resistance or induction furnaces, laser sources, spark plasma sintering, *etc.* The nature of the starting materials is a crucial factor for the selection of the most appropriate synthesis technique. In many cases, when highly electropositive metals, such as alkali, alkaline-earth, or rare-earth metals are present alongside pnictogens, many types of crucibles cannot be used due to potential side reactions. Examples of such incompatible 'reaction vessel' materials include ceramics, fused silica, graphite, and glassy carbon, all of which are prone to be reduced by electropositive metals. In such cases, the preference should be given to containers made of inert metals, such as niobium, tantalum, molybdenum, or platinum. Metal containers provide yet another advantage: They often can be sealed using, *e.g.*, arc welding, which helps reduce losses of volatile components at elevated temperatures over the course of the reaction. However, one must be mindful of the fact that many of these metal containers may react with pnictogens, especially with the lighter congeners, P and As, resulting in the formation of side phases. In fact, many new pnictides were originally discovered as impurities resulting from adverse interactions with the 'reaction vessel' material (Figure 2.4).[7,8]

The proper choice of starting materials can improve reaction kinetics and help avoid side reactions. A good illustration of this approach is the synthesis of LaOFeAs (a parent compound for iron pnictide superconductors) by a solid-state metathesis reaction of LaOCl and NaFeAs, which results in higher sample purity in comparison with the conventional approach involving the binary precursors LaAs, La_2O_3, and FeAs.[9] Another notable example is the synthesis of the ternary phosphides $RESiP_3$ from the binary precursor RESi

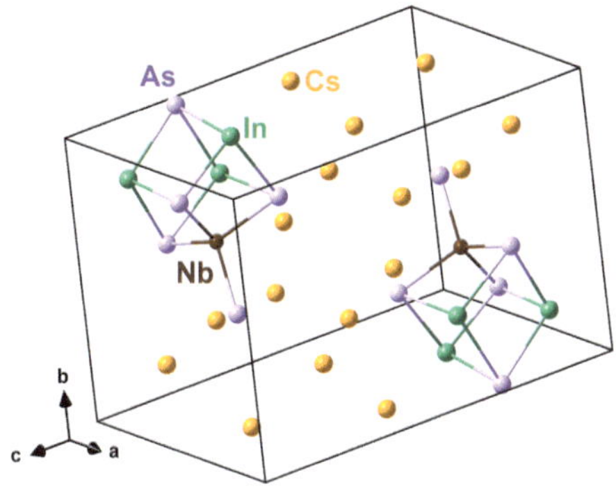

Figure 2.4 Crystal structure of $Cs_7NbIn_3As_5$ (space group $P\bar{1}$, Pearson code *aP*32), comprising As-decorated cubane-like In_3As_4Nb–As clusters. This compound was originally produced by an accidental reaction of a Cs–In–As elemental mixture with a Nb container.

and elemental P (RE denotes rare-earth metals, the abbreviation is also used hereafter). This method enables preparation of pure materials, while synthesis from the elements results in the formation of the highly stable and unreactive binary phase SiP.[10]

A number of pnictides were produced by high-pressure high-temperature synthesis, usually in multi-anvil apparatuses or diamond anvil cells. Such experiments are typically aimed at obtaining metastable materials not accessible by other techniques. Examples include bismuthides $Cu_{11}Bi_7$, CuBi, and $FeBi_2$ prepared at pressures of up to 30 GPa. The corresponding binary systems, Cu–Bi and Fe–Bi, at ambient pressure, lack any phases of compositions located between the end members.[11–13]

The methods described above usually produce polycrystalline materials, although, occasionally, single crystals can be obtained in such reactions. However, high-quality single crystals, suitable for, *e.g.*, characterisation of physical properties, are more often manufactured employing various dedicated crystal growth techniques. One of the commonly used methods is flux-assisted crystal growth. In this approach, suitable liquid media, usually molten metals, salts, or other individual compounds and mixtures, are employed as solvents for the starting materials. The dissolution, either partial or complete, enables efficient diffusion and allows precipitation of single crystals upon cooling the flux. The crystals can be separated from the flux mechanically or by using chemical etching, centrifugation, evaporation of the flux material in high vacuum, and other techniques.[14,15] Due to its simplicity, the flux method has been actively employed to grow single crystals of pnictides. In particular, detailed studies on iron pnictide superconductors were possible owing to the availability of single crystals grown from liquid Sn or molten media containing the same elements as the targeted crystals, the technique referred to as the self-flux approach (Figure 2.5).[16]

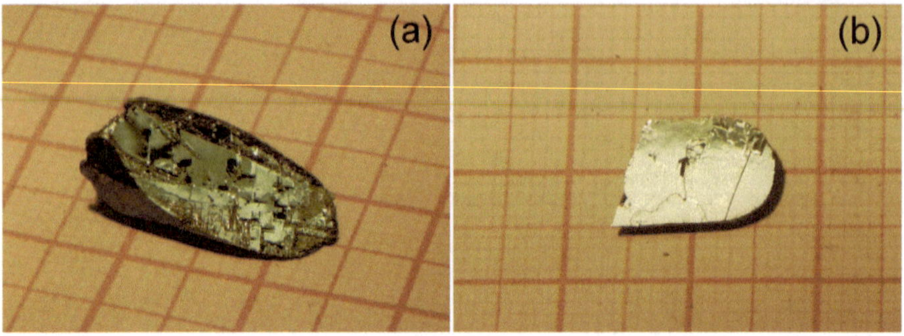

Figure 2.5 Single crystals of the iron arsenide superconductors $NaFe_{1-x}Co_xAs$ (a) and $NaFe_{1-x}Rh_xAs$ (b) grown using the self-flux approach. Reproduced from ref. 17 with permission from American Physical Society, Copyright 2015.

Other crystal growth techniques widely utilised for pnictides are the Bridgman and Czochralski methods, floating zone methods, and gas transport reactions. The latter approach provided access to high-quality single crystals of numerous pnictides, owing to the ability of pnictogens to form volatile intermediates upon interaction with a suitable chemical transport agent, such as halogens or halogen-containing compounds.[18]

Besides high-temperature routes, soft chemistry methods can be used for the production of some pnictides. Such techniques are especially useful when manufacturing metastable compounds, nanomaterials, or composites. Thus, nanostructured nickel phosphide Ni_2P, a well-known electrocatalyst for the hydrogen evolution reaction (HER), can be produced by reacting nickel(II) acetylacetonate with oleylamine and tri-*n*-octylphosphine (TOP) in 1-octadecene at a moderately high temperature of 320 °C.[19] Similar procedures employing TOP as the phosphorus source have been used to synthesise various binary phosphides.[20] Other binary pnictides are also accessible by solution chemistry. The binary nanocrystalline $FeSb_2$, a material with a colossal Seebeck coefficient, is formed under solvothermal conditions in ethanol from Fe(II) and Sb(III) acetates and $NaBH_4$ as a reducing agent.[21]

Examples of ternary and other multinary pnictides synthesised by soft chemistry approaches have been reported as well. For instance, one such material is the semiconducting LiZnP, with promising thermoelectric performance. Nanocrystals of this phosphide can be prepared from LiH, diethyl zinc, and TOP.[22]

An important class of multicomponent materials that is habitually synthesised by solution chemistry is pnictides with complex ionic (usually anionic) clusters.[23,24] Many of such compounds are stabilised by solvent molecules or organic ligands/ions when isolated in crystalline form. An illustrative example is the salt $[Bu_4P]_3[As@Ni_{12}@As_{20}]$ (Bu_4P = tetra-*n*-butylphosphonium), which can be precipitated from an ethylenediamine solution of K_3As_7 and dicyclooctadiene-nickel in the presence of Bu_4PBr. This

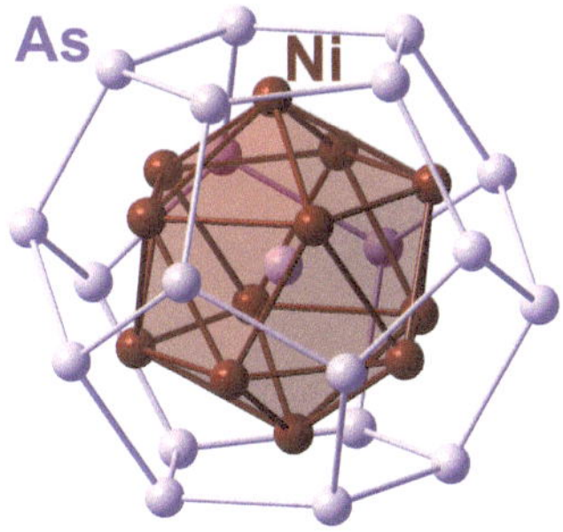

Figure 2.6 Structure of the multi-shell anion $[As@Ni_{12}@As_{20}]^{3-}$, produced by solution chemistry.

salt is characterised by the presence of a complex multi-shell anion $[As@Ni_{12}@As_{20}]^{3-}$ in the crystal structure (Figure 2.6).[25]

Finally, it is worth mentioning that special tasks may require synthetic methods other than those described above. For instance, molecular beam epitaxy (MBE) is widely used for thin-film deposition in semiconducting technologies. In particular, this technique has been employed for the fabrication of the diluted magnetic semiconductor (Ga,Mn)As.[26]

2.3 Pnictide Semiconductors

There are two main types of binary pnictide semiconducting materials, designated as II–V type and III–V type semiconductors, formed by elements of Groups 12/15 and 13/15, respectively.[27] Tetragonal $P4_2/nmc$ M_3Pn_2 (M = Zn, Cd; Pn = P, As, Sb) phases crystallise in the Zn_3P_2 structure type (Figure 2.7(a)).[28] The band gaps of these formally charge-balanced compounds $(M^{2+})_3(Pn^{3-})_2$ vary depending on the composition and can be tuned by isovalent substitutions due to the existence of continuous solid solutions, such as the one in the Zn–Cd–P–As system.[29] M_3Pn_2 phases possess interesting structural, optical, and transport properties. For instance, due to strong optical absorption and a suitable band gap, Zn_3P_2 and Cd_3P_2 were reported to be promising materials for thin-film photovoltaic applications.[30,31]

More well-known are semiconducting pnictides belonging to the III–V type, with the composition of MPn (M = B, Al, Ga, In; Pn = P, As, Sb). Hardly any other binary pnictide has received so much attention as the sphalerite-type GaAs, a structurally simple compound (Figure 2.7(b)), which offers a wide range of applications. The properties of this direct-band gap semiconductor can be modified considerably by chemical doping, nanostructuring, or application of external stimuli, such as mechanical strains.[32]

Controlled alloying of different MPn binaries allows the creation of ternary and quaternary compounds with a range of band gaps (Figure 2.8). Multiple applications for MPn-based semiconductors and related multinary heterostructures include ultrathin transparent electrode layers in solar

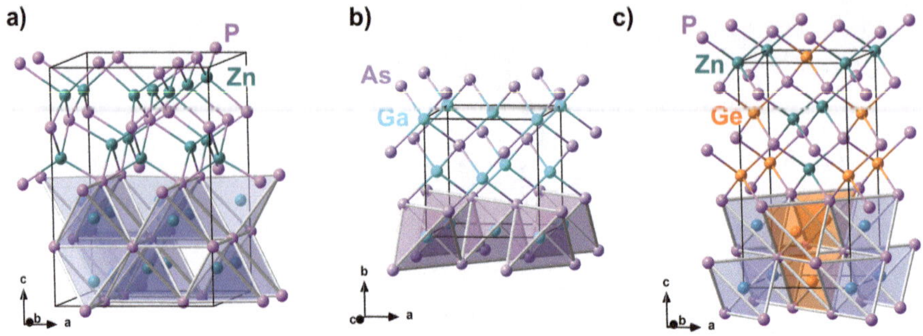

Figure 2.7 Crystal structures of some common pnictide semiconductors: (a) Zn_3P_2 (space group $P4_2/nmc$, Pearson code $tP40$), (b) GaAs (space group $F\bar{4}3m$, Pearson code $cF8$), (c) $ZnGeP_2$ (space group $I\bar{4}2d$, Pearson code $tI16$). The frameworks of principal tetrahedral building blocks are emphasised. The unit cells are outlined in black.

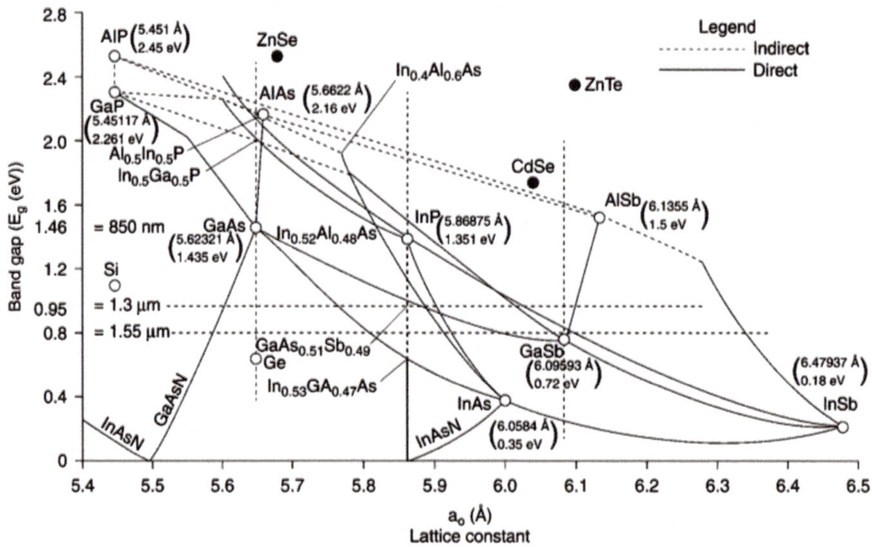

Figure 2.8 Lattice constants and the electronic band gap (E_g) for selected binary (node points), ternary (curves connecting the nodes) and quaternary (areas defined by these curves) sphalerite-type semiconductors. Solid and dashed curves indicate direct and indirect band gaps, respectively. Reproduced from ref. 40 with permission from Elsevier, Copyright 2013.

cells,[33,34] materials for manufacturing of optoelectronic devices, such as light-emitting diodes,[35–37] laser technologies,[38–40] high-frequency electronics transistors and circuit components,[41,42] X-ray detectors,[43] and infrared photodiode detectors.[44]

A family of ternary chalcopyrite-type $ABPn_2$ phases (A = Mg, Zn, Cd; B = Si, Ge, Sn; Pn = P, As) with wide band gaps can be considered as electronic analogs of MPn phases, according to the charge-balanced notation (A^{2+})

$(B^{4+})(Pn^{3-})_2$ (Figure 2.7(c)). II–IV–V_2 materials are known for their luminescent properties, non-linear optical properties, and potential for the development of new optoelectronic photovoltaic devices that are briefly recapped in Section 2.10.[45–48] The electronic band gaps of these materials can be tuned in wide regions, similar to the II–V type and III–V type semiconductors.

While applications in such fields as solar cells call for semiconductors with wide electronic band gaps, narrow-gap pnictide semiconductors are of technological importance too. The main proposed area of use for such materials is thermoelectric technologies. Materials for thermoelectric applications will be discussed in detail in the next section.

2.4 Thermoelectric Materials

The development of environmentally friendly and renewable energy sources is the main driving force behind the progress of solid-state thermoelectric (TE) technology, which enables direct conversion between heat and electricity.[49–51] The performance of thermoelectric materials is characterised by the dimensionless figure of merit zT, defined as

$$zT = \sigma S^2 T/(\kappa_e + \kappa_L), \tag{2.1}$$

where σ, S, T, κ_e, κ_L are the electrical conductivity, Seebeck coefficient (also called thermopower), absolute temperature, electronic and lattice contribution to the thermal conductivity κ, respectively. An excellent thermoelectric material should possess high σ and S and low κ, although these parameters are strongly correlated, which makes effective optimisation challenging.[52,53] The Phonon-Glass Electron-Crystal (PGEC) concept proposed by Slack is used as a guide for maximising zT by developing materials with low 'glass-like' thermal conductivity and high 'electron-crystal' electrical conductivity.[54] Usually, narrow-gap semiconducting properties are desirable for a good thermoelectric material, since in this case, high values of the Seebeck coefficient and low electrical resistivity (ρ) can be attained.

Pnictide-bearing compounds often satisfy necessary criteria for excellent thermoelectric materials due to the diversified bonding interactions, structural complexity, and amenability to crystallise in large unit cells with a noticeable degree of disorder while preserving high electrical conductivities. Large unit cells are of particular interest in designing novel thermoelectrics due to the reduced lattice thermal conductivity.[2] The presence of the heavy pnictogens Sb and Bi is often especially favourable for the efficiency of pnictide thermoelectric materials due to the typically lower thermal conductivities and better electrical transport properties of antimonides and bismuthides in comparison with compounds with the lighter Group 15 elements. Pnictide semiconducting materials have been an excellent playground for developing thermoelectric technologies since their transport properties can be easily optimised by controlled doping and alloying.

Most known pnictide thermoelectric materials can be rationalised using the valence rules. Among them, half-Heusler compounds, filled skutterudites, pnictide clathrates, complex Zintl pnictides, and some binary compounds are considered the most promising thermoelectric materials and are the subject of the discussion in the following subsections.

2.4.1 Half-Heusler Compounds

The Heusler phases, described in Section 2.1 are known for the wide range of physical properties they show. Especially interesting for thermoelectric applications are half-Heusler compounds with 18 valence electrons. This valence electron count (VEC) is compatible with narrow-gap semiconducting properties, while electron- or hole-doping can be used to optimise the electric transport properties (Figure 2.9(a)).[5,55–57]

Two well-studied pnictide half-Heusler systems are the semiconducting MFeSb (M = V, Nb, Ta) and MCoSb (M = Ti, Zr, Hf) phases. The former system is used to design p-type materials, where conductivity is dominated by holes as charge carriers, while the latter system can be tuned to be n- or p-type, *i.e.*, with electrons or holes as principal charge carriers, respectively. Both

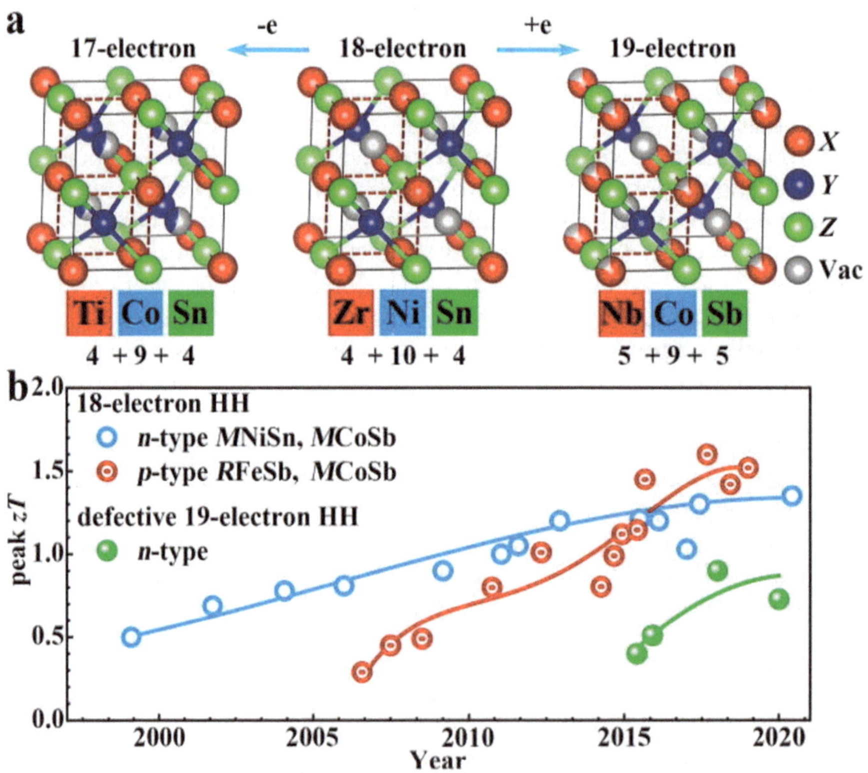

Figure 2.9 (a) Examples of 17-, 18-, and 19-electron half-Heusler phases. (b) The evolution of the reported peak zT values for half-Heusler thermoelectrics. Reproduced from ref. 56 with permission from AIP Publishing, Copyright 2021.

antimonide subfamilies are characterised by high values of power factors (σS^2), but the main challenge in improving the TE performance lies in their high lattice thermal conductivities, κ_L (10–20 W m^{-1} K^{-1} at 300 K). Studies show that κ_L can be considerably reduced by grain refinement, phase separation, and alloying, which ultimately results in higher thermoelectric figures of merit (Figure 2.9(b)). For instance, p-type (Nb,Ta)FeSb solid solutions exhibit peak zT values of 1.6 at 1200 K,[58] whereas n-type (Zr,Hf)CoSb alloy possesses a zT value of 1.0 at 1173 K.[59] Since different atomic sites in a half-Heusler compound can be doped simultaneously, there are many mechanisms for optimisation of the thermoelectric properties.

2.4.2 Filled Skutterudites

Filled skutterudite pnictides are promising thermoelectric materials with a wide range of properties depending on their composition, generally expressed as AM_4Pn_{12} (A = alkali, alkaline-earth, rare-earth metal, Th, U; M = transition elements of Group 8, 9; Pn = P, As, Sb).[4,60–62] A prototype for this family is the mineral skutterudite with the formula $CoAs_3$, which is an electron-balanced phase according to the formal notation $(Co^{3+})_4([As_4]^{4-})_3$. The square $[As_4]^{4-}$ species can be viewed as a typical Zintl polyanion. The Co atoms in the structure are octahedrally coordinated by As, with the corner-sharing $[CoAs_6]$ units forming a cage-like three-dimensional framework. The latter accommodates voids of considerable size, which can be filled by cations of electropositive metals (Figure 2.10).[4] The introduction of a filler cation increases the overall electron count. Therefore, simultaneous hole doping is required to maintain charge balance, which can be achieved by replacing Co^{3+} with a lower

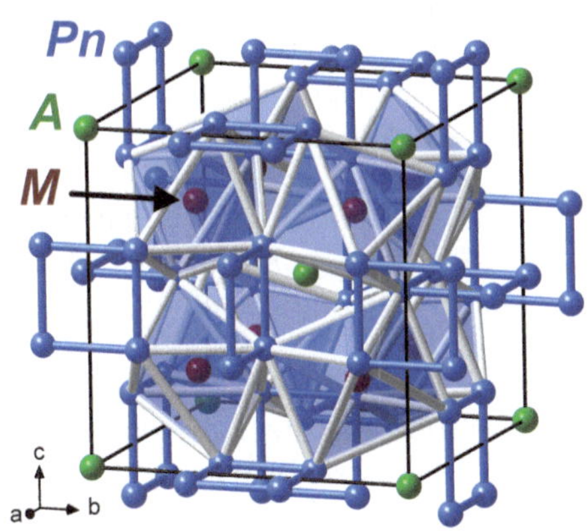

Figure 2.10 Crystal structure of a filled skutterudite, AM_4Pn_{12} (space group $Im\bar{3}$, Pearson code $cI34$, A = alkali, alkaline-earth, or rare-earth metal, Th, U; M = transition elements of Group 8, 9; Pn = P, As, Sb). The framework of corner-sharing $[MPn_6]$ octahedra and $[Pn_4]$ squares are emphasised. The unit cell is outlined in black.

formal charge species, such as Fe^{2+}. The tolerance of the structure to this kind of substitution creates premises for doping and further modification of the charge transport properties. An excellent example is the narrow-gap semiconducting $LaFe_3CoSb_4$ phase, which despite an intense compositional modification, exemplifies a charge-balanced composition $(La^{3+})(Fe^{2+})_3(Co^{3+})([Sb_4]^{4-})_3$ with decent thermoelectric performance.[63–65]

Contrary to the half-Heusler phases, the increased thermoelectric efficiency of filled skutterudites originates from the significantly lower lattice thermal conductivity compared to the unfilled structure, in line with the PGEC concept. Recent advances in thermoelectric performance of skutterudites resulted in the synthesis of highly efficient p-type TE materials, with reported zT values of 1.2 at 700 K for $La_{0.8}Ti_{0.1}Ga_{0.1}Fe_3CoSb_{12}$,[66] 1.0 at 750 K for $Ce_{0.5}Yb_{0.5}Fe_{3.25}Co_{0.75}Sb_{12}$,[67] and 1.1 at 700 K for $Mm_{0.9}Fe_{3.1}Co_{0.9}Sb_{12}$ (Mm = Mischmetal, *i.e.*, a mixture of rare-earth metals),[68] to name a few. At the same time, n-type filled skutterudites exhibit even better thermoelectric performance, with high reported zTs of 1.9 at 835 K for $(Sr,Ba,Yb)_yCo_4Sb_{12}$,[69] 1.8 at 800 K for its In-doped analog,[70] 1.5 at 850 K for $Y_{0.3}Co_4Sb_{12}$,[71] and 1.7 at 850 K for $Ba_xLa_yYb_zCo_4Sb_{14}$.[72]

Although high zT values are observed for the filled Sb-based bulk skutterudites, incorporation of homogeneously dispersed nanoparticles into the thermoelectric matrix was shown to be a successful route to increase the figure of merit even further. This approach noticeably reduces the lattice thermal conductivity keeping high values for electrical conductivity and Seebeck coefficient.[73]

2.4.3 Inorganic Clathrates

Similar to the skutterudites, inorganic clathrates are cage-like compounds with potential for thermoelectric applications. The chemical bonding in clathrates can often be treated within the Zintl concept, which enables rational optimisation of the transport properties based on electron count. The crystal structures of most clathrates feature a tetrahedral framework of covalently bonded atoms, with large polyhedral voids of various sizes accommodated within the framework. To stabilise such an open structure, guest atoms have to occupy the voids. The weak bonding between the guest species and the framework may result in a rattling motion of the former, which provides efficient phonon scattering.[74,75]

While most clathrates are formed by tetrel elements, Si, Ge, or Sn,[76,77] multiple pnictogen-based phases have also been discovered and characterised.[78–86] Pnictogens, such as P, As, or Sb, usually occupy framework sites along with late d-elements, *e.g.*, Zn, Cd, Cu, and Au, or Group 13 elements, such as Ga or In. The atomic distribution in the framework may be short-range or long-range ordered, which sometimes gives rise to very complex clathrate superstructures.[80–82] Thermal conductivities for such materials are exceptionally low,[80,81] often falling below 0.5 W m^{-1} K^{-1} with reported zT values of about 0.6 for both $Ba_8Cu_{14}Ge_6P_{26}$ (Figure 2.11) at 812 K, and $Ba_{6.4}La_{1.6}Cu_{16}P_{30}$ at 900 K.[83,86]

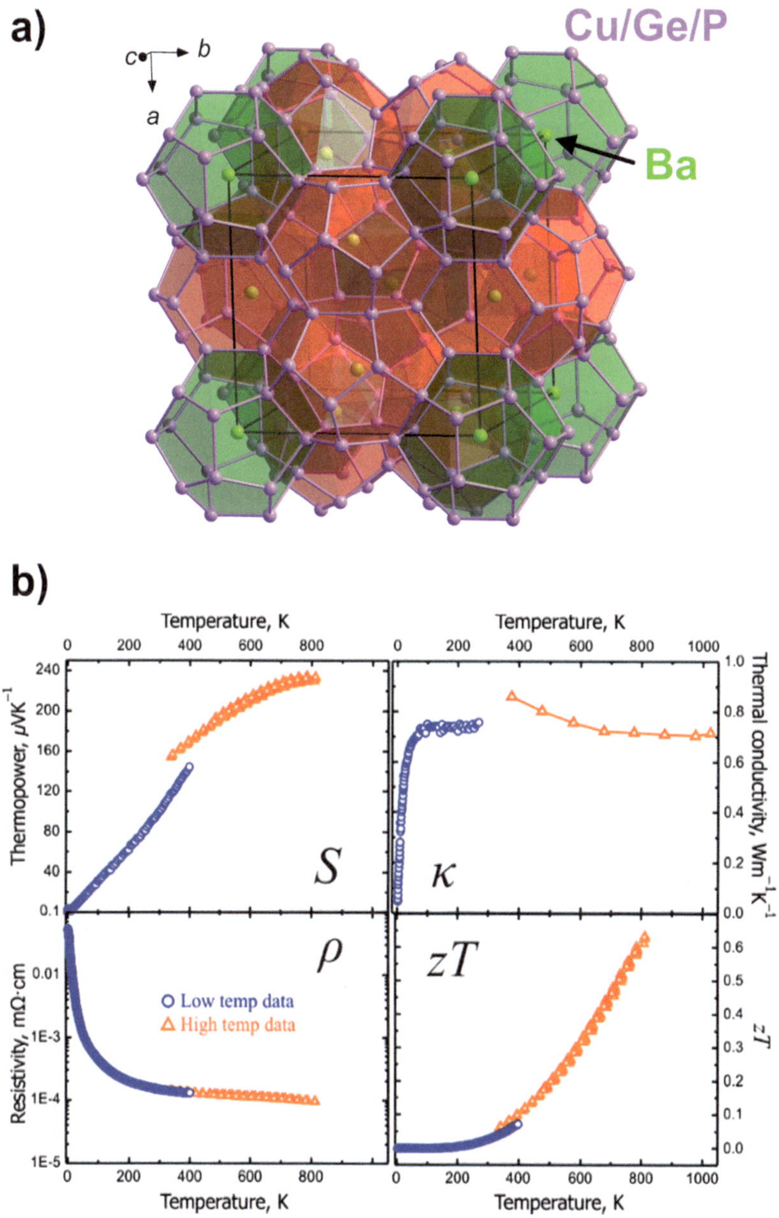

Figure 2.11 (a) Crystal structure of the type-I clathrate $Ba_8Cu_{14}Ge_6P_{26}$ (space group $Pm\bar{3}n$, Pearson code *cP*54). Dodecahedral and tetrakaidecahedral cages formed by the framework atoms are shown in green and red, respectively. The framework is composed of statistically distributed Cu, Ge, and P atoms, with no local ordering. The unit cell is outlined in black. (b) Low-temperature (blue) and high-temperature (orange) transport properties of a $Ba_8Cu_{14}Ge_6P_{26}$ single-crystalline sample. Reproduced from ref. 83, https://doi.org/10.1039/C7SC03482B, under the terms of the CC-BY 3.0 Unported License, https://creativecommons.org/licenses/by/3.0/.

2.4.4 Complex Zintl Pnictides

As described earlier, the Zintl concept can be applied to explain the chemical bonding patterns in skutterudites and clathrates. However, a large number of Zintl pnictides exhibit complex and diverse crystal structures, which precludes their placement in a single structural family.[3]

Within the field of thermoelectrics, a particular interest in the Zintl pnictides emerged after the discovery of promising thermoelectric properties in certain representatives of this family in the early 2000s.[87,88] The design of thermoelectric Zintl pnictides has become an actively developing topic since then.[89–93]

$Yb_{14}MnSb_{11}$ is one of the most well-known Zintl pnictides, suitable for thermoelectric applications (Figure 2.12).[94] The crystal structure of this compound hosts three kinds of anions: isolated $[MnSb_4]^{10-}$ tetrahedra, linear $[Sb_3]^{7-}$ units with hypervalent bonding, and simple Sb^{3-} species. Since Yb is found in the structure as a cation with charge 2+ (recall that Yb^{2+} is very stable and has a closed shell electronic configuration $[Xe]4f^{14}$), a simple electron counting suggests that this material displays a shortage of electrons, according to $(Yb^{2+})_{14}[MnSb_4]^{10-}[Sb_3]^{7-}(Sb^{3-})_4(h^+)$, where h^+ denotes a hole, located on the tetrahedral $[MnSb_4]$ unit and able to propagate by hopping between such units. In accordance with this assignment, $Yb_{14}MnSb_{11}$ exhibits metallic behaviour. The thermoelectric figure of merit, zT, reaches values greater than 1.0 at 1200 K. The charge carrier concentration and mobility can be tuned by aliovalent doping and defect design, which results in further improvement of zT, leading to peak zT values of up to 1.3.[95–98]

Being the most studied $Ca_{14}AlSb_{11}$-type thermoelectric Zintl pnictide, $Yb_{14}MnSb_{11}$ is just a single representative of a big family of compounds, referred to as 14-1-11 phases, many of which demonstrate outstanding thermoelectric properties. Thus, the magnesium analog, $Yb_{14}MgSb_{11}$, has a zT of *ca.* 1.26 at 1200 K for a sample prepared by hydride-assisted synthesis.[99] Due to the complex crystal structure, 14-1-11 phases exhibit remarkably low thermal conductivity of 0.6–0.9 W m^{-1} K^{-1}, an essential contributing factor for the overall thermoelectric performance.

Zintl phases with 1-2-2 compositions (AB_2X_2) adopt several structure types, among which pnictides with the $ThCr_2Si_2$ type structure are known for their magnetic and superconducting properties, as discussed in Sections 2.5 and 2.6. Another common structure type for 1-2-2 pnictides is $CaAl_2Si_2$, which can be viewed as a ternary derivative of the *anti*-La_2O_3-type. The structure of $CaAl_2Si_2$ features $[Al_2Si_2]^{2-}$ layers made by edge-sharing $[AlSi_4]$ tetrahedra and Ca atoms sandwiched between the layers and located in an octahedral surrounding of the Si atoms (Figure 2.13). Pnictide 1-2-2 phases often exhibit excellent thermoelectric performance with both n- and p-type transport properties.[100] The hallmark of the $CaAl_2Si_2$-type Zintl pnictides is high carrier mobility originating from the anionic layers, which results in high reported figures of merit. For instance, the maximum zT value for a Zn-doped $YbCd_2Sb_2$ phase is 1.2,[101] while it is around 1.0 for $Ca_{1-x}Yb_xMg_2Bi_2$.[102]

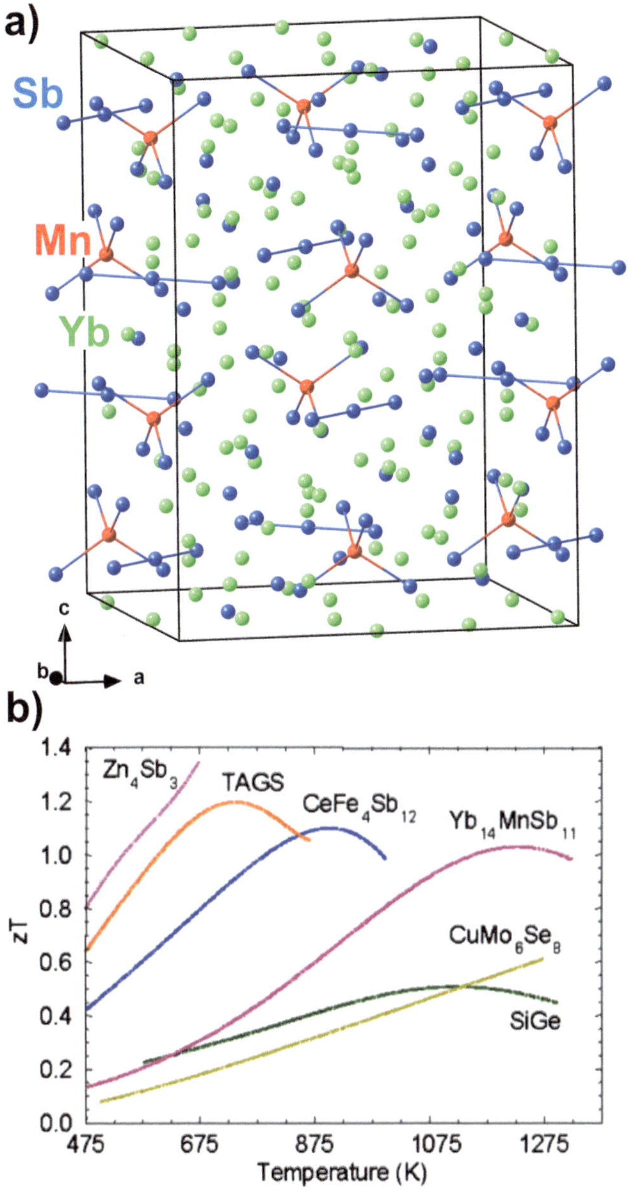

Figure 2.12 (a) Crystal structure of $Yb_{14}MnSb_{11}$ (space group $I4_1/acd$, Pearson code *tI*208). The unit cell is outlined in black. (b) Thermoelectric figure of merit zT of $Yb_{14}MnSb_{11}$ (pink) compared to the values in other p-type thermoelectric materials. Reproduced from ref. 94 with permission from American Chemical Society, Copyright 2006.

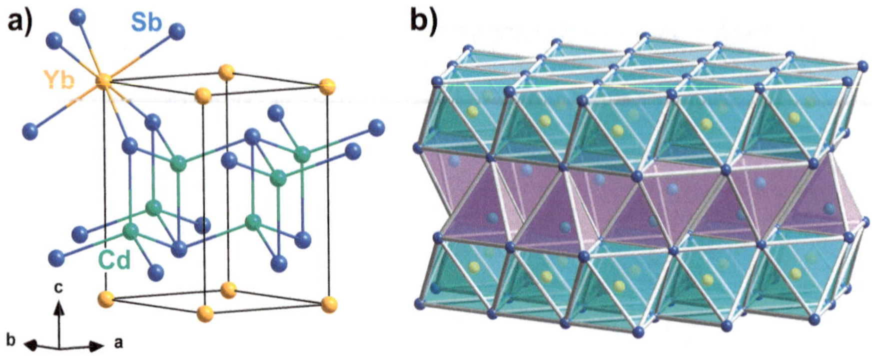

Figure 2.13 Crystal structure of $YbCd_2Sb_2$, adopting the $CaAl_2Si_2$ type (a, space group $P\bar{3}m1$, Pearson code *hP*5). Local coordination of selected atoms is shown. The unit cell is outlined in black. Polyhedral representation of the structure (b), showing the interconnection of the $[YbSb_6]$ octahedra (blue) and $[CdSb_4]$ tetrahedra (purple).

The *anti*-La_2O_3-type phases with the composition of Mg_3Pn_2 (Pn = Sb, Bi) formally belong to the $CaAl_2Si_2$ structure type. Their crystal structure can be broken down to $[Mg_2Pn_2]^{2-}$ layers, similar to the $[Al_2Si_2]^{2-}$ species in $CaAl_2Si_2$, with Mg^{2+} cations occupying the octahedral voids built up of Pn atoms. These binary compounds are reported to be intrinsically p-type materials[103] with poor thermoelectric performance. However, electrical transport properties can be optimised by doping, which can change the type of principal charge carriers and increase carrier concentration and mobility. For instance, Te-doped $Mg_3Sb_{1.5}Bi_{0.5}$ is a low-cost n-type material with very high peak zT values, exceeding 1.5,[104] and even approaching 1.65.[105]

Several other families of Zintl pnictides have also been studied with respect to their thermoelectric properties. Thus, pnictides with the general formula $AE_9M_{4+x}Pn_9$ (AE = Ca, Sr, Eu, Yb; M = Mn, Zn, Cd; Pn = As, Sb, Bi) exhibit potential for thermoelectric applications.[4,106] The complex crystal structures of these compounds are made of $[M_4Sb_9]^{19-}$ ribbons connected through interstitial M^{2+} atoms. Because of the low thermal conductivity, the zT can reach as high as 1.1 at 875 K, as has been shown for $Ca_9Zn_{4+x}Sb_9$.[107]

Other promising compounds with quasi-one-dimensional anionic substructures include representatives of the $AE_5M_2Pn_6$ and AE_3MPn_3 classes (AE = Ca, Sr, Eu, Yb; M = Al, Ga, In; Pn = P, As, Sb, Bi). The former family crystallises in three different structure types, depending on the composition, although the charge partitioning for these phases is identical and can be expressed by the electron-precise formula $(AE^{2+})_5(Tr^{3+})_2(Pn^{3-})_4[Pn_2]^{4-}$.[106,108] All $AE_5M_2Pn_6$ structures accommodate $[M_2Pn_6]$ chains of different shapes, based on interconnected $[MPn_4]$ tetrahedra. Members of the $AE_5M_2Pn_6$ family display low thermal conductivity and semiconducting behaviour. Thus, $Ca_5Al_2Sb_6$ exhibits κ_L values of less than 0.6 W m^{-1} K^{-1} at 850 K and optimised zT of up to 0.6 upon, *e.g.*, Na doping.[108,109]

The AE_3MPn_3 classes can also adopt different structures. Of particular interest are compounds crystallising in the Ca_3InP_3 and Sr_3GaSb_3 types, with

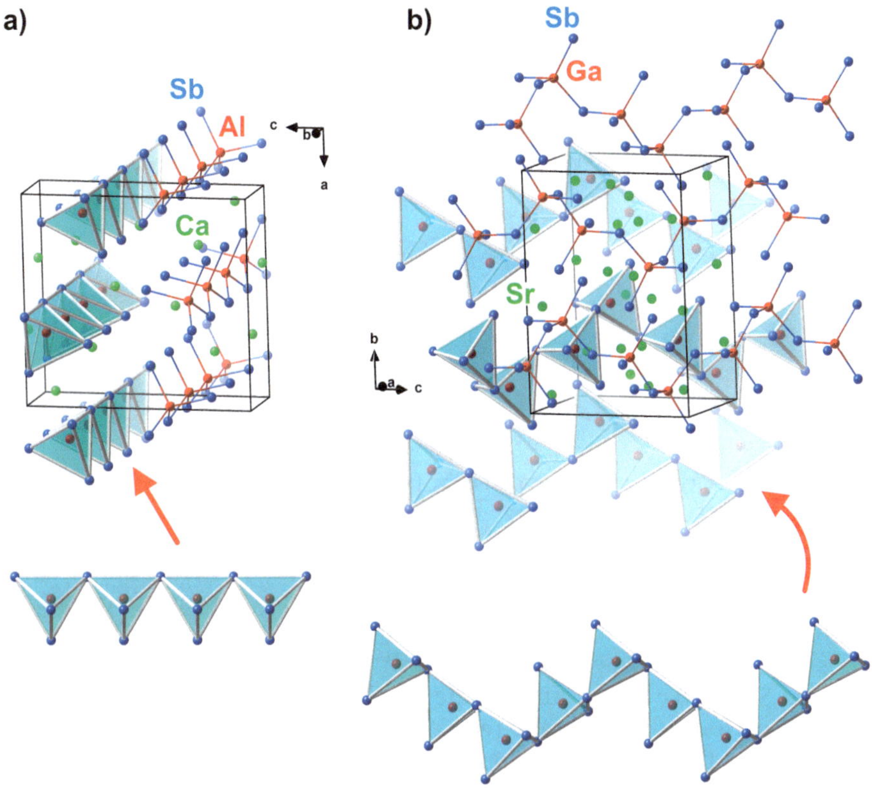

Figure 2.14 Crystal structures (top) of Ca_3AlSb_3 (a, space group *Pnma*, Pearson code *oP*28) and Sr_3GaSb_3 (b, space group $P2_1/n$, Pearson code *mP*56). The chain-line anionic substructures based on corner-sharing [MSb_4] tetrahedra (M = Al, Ga) are emphasised. The respective individual chains are depicted at the bottom. The M atoms in Ca_3AlSb_3 are arranged in linear chains, whereas they form zigzag chains in Sr_3GaSb_3. The unit cells are outlined in black.

crystal structures accommodating chains of corner-sharing [MPn_4] tetrahedra (Figure 2.14). Thus, a Na-doped Ca_3AlSb_3, adopting the former type, demonstrates a *zT* value of 0.8 at *T* = 1050 K, whereas for a Zn-doped Sr_3GaSb_3, *zT* reaches values above 0.9 in the same temperature range.[110,111]

The Eu_2ZnSb_2 phase possesses ultralow thermal conductivity, which results in the reported maximum *zT* of *ca.* 1.0 at 823 K,[112] while the silver-doped $Eu_2Zn_{0.95}Ag_{0.06}Sb_2$ phase exhibits a peak value of 1.1 at 823 K.[113] Heavily doped $CaZn_{1-x}Ag_{1-y}Sb$ phases with the TiNiSi structure type are also good p-type thermoelectric materials with reported *zT* of 1.3 at 873 K.[114]

2.4.5 Binary Phases

Some binary pnictides have been proposed as promising thermoelectric materials due to their low cost, low toxicity, and abundance. The actively studied Mg_3Pn_2 phases (Pn = Sb, Bi) were described earlier. Other binary pnictides

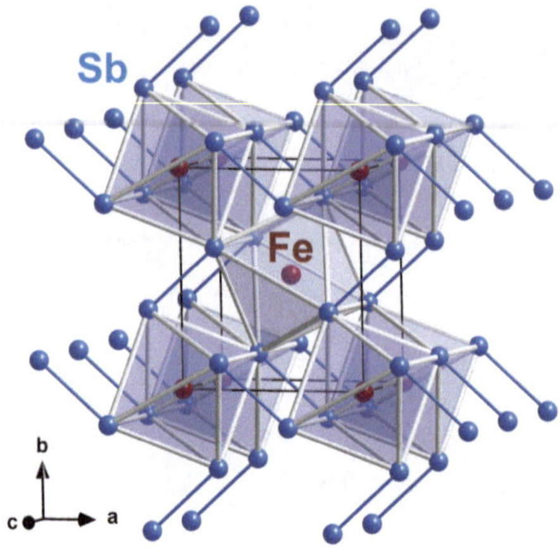

Figure 2.15 Crystal structure of $FeSb_2$ (space group *Pnnm*, Pearson code *oP*6). $[FeSb_6]$ octahedra are linked by corner-sharing and *via* Sb–Sb bonds. The unit cell is outlined in black.

of particular interest are zinc antimonides, exhibiting decent thermoelectric performance in the mid-temperature range (300–700 K). Thermoelectric properties of inexpensive binaries, ZnSb,[115] Zn_4Sb_3,[116–119] Zn_3Sb_2,[120] Zn_8Sb_7[121] and $Zn_{13}Sb_{10}$[122] have been studied intensively in recent years. Reported zT values vary for these materials and reach high values of 0.8 at 573 K for the ZnSb phase[123] and 1.4 at 700 K for a doped Zn_4Sb_3.[124] Due to their complex and disordered crystal structures, some zinc antimonides exhibit low glass-like thermal conductivity, which lends these compounds their good thermoelectric properties. However, this crystallographic complexity poses challenges for detailed structural investigations. For these reasons, the crystal structures of some phases in the Zn–Sb system remain a matter of debate.[117,125]

Another notable binary pnictide compound made of earth-abundant elements is the marcasite-type $FeSb_2$ (Figure 2.15). This correlated narrow-gap semiconductor displays colossal negative thermopower values of about −45 mV K^{-1} at *ca.* 10 K, resulting in excellent power factor values.[126,127] However, its large thermal conductivity leads to low zT values.[128] Nevertheless, $FeSb_2$ has been proposed as a candidate system for thermoelectric cooling at cryogenic temperatures.

2.5 Pnictide Superconductors

Superconductors containing pnictogens have been known for many decades. Thus, the binary compound Au_2Bi, crystallising in the cubic Laves structure (Figure 2.16), was one of the first metal alloys discovered to be superconducting.[129,130] The observation of superconductivity in this material in the early 1930s was particularly exciting due to the fact that the constituent elements lack superconductivity above 0.5 mK, while Au_2Bi enters the state of zero resistance below 1.9 K.

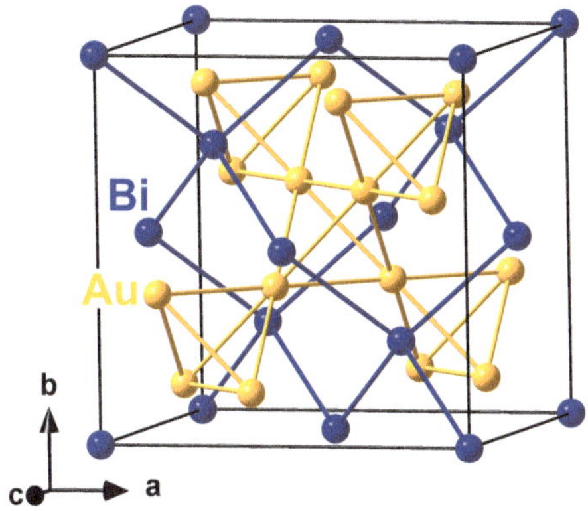

Figure 2.16 Crystal structure of Au_2Bi (space group $Fd\bar{3}m$, Pearson code *cF*24). The diamond-like framework of Bi atoms and the Au-based linked tetrahedra are emphasised. The unit cell is outlined in black.

Most of the superconductors discovered in the first half of the twentieth century can be well described by the Bardeen–Cooper–Schrieffer (BCS) theory of superconductivity. The superconducting ground state in BCS superconductors is induced by electron–phonon coupling, and the properties of such superconductors are generally well understood.

Many more superconducting pnictides were discovered in the second half of the twentieth century and the early 2000s. Among them, filled skutterudites, already described in Section 2.4.2, form a large subfamily and remain an important topic in solid-state research. Thus, superconductivity was found in a number of La-containing members, *e.g.*, LaM_4Pn_{12} (M = Ru, Os; Pn = P, As, Sb),[131–136] as well as in some compositions with Pr. The Pr representatives, such as $PrOs_4Sb_{12}$,[133,137] belong to the family of heavy-fermion superconductors, which are materials with unconventional superconductivity exhibiting quasiparticles with large effective masses. Such superconductors cannot be described by the classic BCS theory and thus represent fascinating systems for condensed matter science.

A turning point in the field of pnictide superconductivity was the discovery of iron-based superconductors in 2008. Since then, the term 'pnictide superconductor' has been mainly associated with iron pnictides. The first reported iron-based superconductor was $LaO_{1-x}F_xFeAs$, crystallising in the widespread ZrCuSiAs structure type (Figure 2.17).[138] (The mixed anion superconducting pnictides (principally oxypnictides) are covered in more detail in Chapter 5.) The maximum critical temperature (T_c) in this material was found to be around 26 K for $x \approx 0.1$. A peculiar feature of the parent phase, LaOFeAs, is the presence of quasi-two-dimensional [FeAs] layers in the structure with magnetic Fe atoms. Usually, local magnetic fields destroy superconductivity, and for these reasons, compounds exhibiting coexistence of superconductivity and magnetism are rare. The well-known class of superconducting high T_c cuprates offers an example of such materials. Just like cuprates,

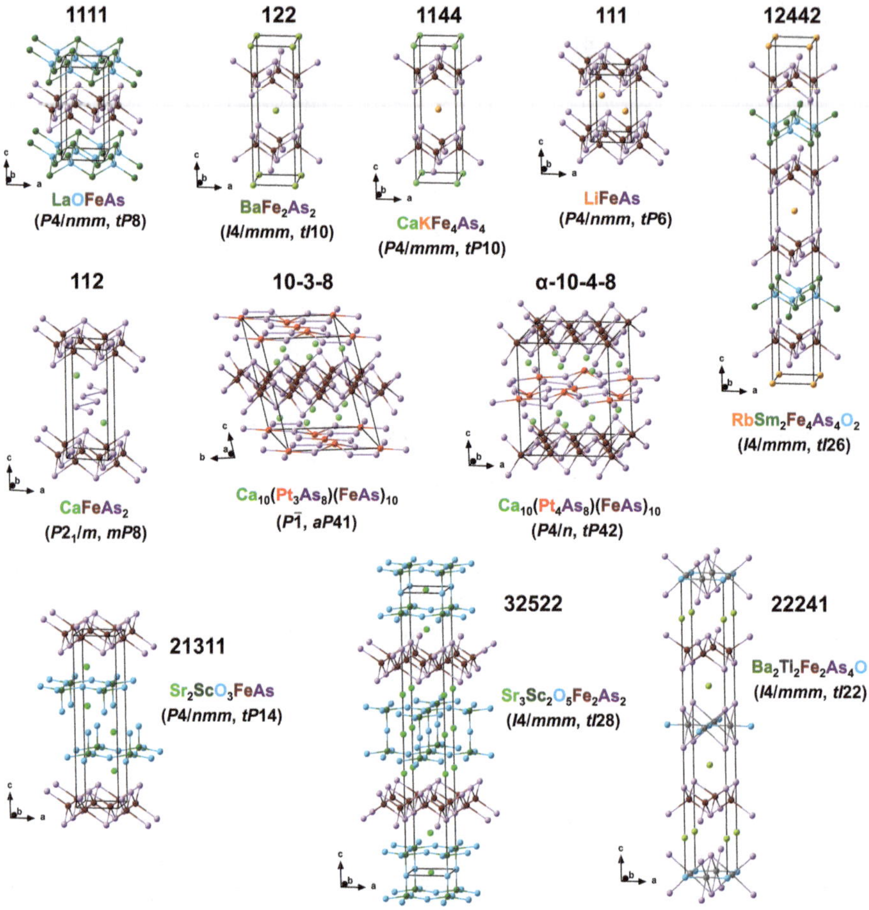

Figure 2.17 Most common families of iron pnictide superconductors exemplified by the crystal structures of selected representatives. The atom colour codes are explained in the formula labels. Given in parentheses are the space groups and Pearson codes, respectively. The unit cells are outlined in black.

superconducting iron arsenides turned out to be a large structural family, encompassing various compositions and atomic arrangements but always comprising the mentioned [FeAs] layers in their structures. The magnetism in these phases reveals itself in a spin density wave (SDW) ordering. Upon suppression of the SDW instability by, *e.g.*, chemical doping, unconventional superconductivity arises.

Several factors have contributed to the interest of the scientific community in the iron-based superconductors. Besides the mentioned coexistence of magnetism and superconductivity, iron pnictides display a wide spectrum of exciting features. The parent phases of these materials are metallic, in contrast to the insulating ones in cuprates. The multiband structure of iron pnictides at the Fermi level is considered one of the key features that determines their superconducting properties, although the exact origin of superconductivity has been widely disputed and remains not fully understood. From the application perspective, iron-based superconductors are appealing for various electronic and magnetic applications due to high isotropic critical fields.[16,139,140]

In iron pnictides, the highest superconducting transition temperature was achieved for several compositions, crystallising, just like $LaO_{1-x}F_xFeAs$, in the ZrCuSiAs structure type (the so-called 1111 phases). Examples of such compounds include $Gd_{0.8}Th_{0.2}OFeAs$,[141] $SmO_{0.9}F_{0.1}FeAs$,[142] $Sr_{0.5}Sm_{0.5}FFeAs$[143] and $Ca_{0.4}Nd_{0.6}FFeAs$,[144] all exhibiting T_c of about 56 K. Although the critical temperatures are still relatively low in comparison with, *e.g.*, the values observed in some cuprates, iron pnictides are usually classified as high-temperature superconductors.

The crystal structure of all 1111 phases can be described as consisting of two kinds of layers stacked in an alternating manner (Figure 2.17). The two types of layers have different chemical compositions but are isostructural and can be viewed as fluorite-type slabs. The [FeAs] blocks, responsible for superconductivity, are built of edge-sharing $FeAs_4$ tetrahedra. In these units, the Fe atoms form square nets with As atoms residing above and below the nets in a checkerboard order. The layers of the second type are usually described as a charge reservoir since they compensate the negative charge of the anionic $[FeAs]^{n-}$ layers. The charge reservoir blocks are based on edge-sharing metal tetrahedra built around atoms of non-metallic elements: O, as in LaOFeAs,[138] or F, as in CaFFeAs.[145] Nitride layers are present in the structure of ThNFeAs.[146] In all these parent compounds, the [FeAs] layers can be assigned a formal charge of 1−, corresponding to a charge of 2+ on Fe. Although this charge assignment yields electron-balanced compositions, such as $(La^{3+})(O^{2-})(Fe^{2+})(As^{3-})$, in the spirit of the Zintl concept, the parent phases of iron-based superconductors are multi-band metals, as was already mentioned above, and thus do not fall in the group of Zintl phases.

The tetragonal structure of the parent 1111 phases undergoes an orthorhombic distortion upon cooling, associated with the presence of nematic magnetic fluctuations, *i.e.*, magnetic instabilities that break the four-fold rotation symmetry. At lower temperatures, antiferromagnetic ordering occurs. Both instabilities get suppressed in the doped superconducting samples.[147] As alluded to above, this behaviour is prevalent for iron pnictides.

Although the crystal structural description in terms of stacked independent layers does not grasp all peculiarities of the chemical bonding in the 1111 phases, it allows a straightforward identification and design of new superconducting phases. Not least because of such a modular approach to the crystal chemistry of iron pnictides, many structural families of iron-based superconductors have been discovered.

Besides the 1111 phases, one of the most thoroughly studied groups is represented by the compounds with the general formula AFe_2As_2 (A = alkali or alkaline-earth metal), referred to as 122 iron pnictides. They crystallise in the tetragonal $ThCr_2Si_2$ type (Figure 2.17), which comprises, besides iron pnictides, a large number of other non-oxide compositions, including chalcogenides, tetrelides (compounds of Group 14 elements), trielides (compounds of Group 13 elements) and other intermetallic compounds. Some examples of magnetic pnictides crystallizing in this structure type will be covered in Section 2.6. In the crystal structures of the 122 iron pnictides, the role of the charge reservoir is played by flat nets of the monovalent or divalent A^{n+} cations accommodated between the $[FeAs]^{n/2-}$ layers.

The first 122 phase in the iron pnictide family identified as a superconductor was $Ba_{1-x}K_xFe_2As_2$.[148] While the pristine $BaFe_2As_2$ does not enter the superconducting state and displays an orthorhombic structural distortion and SDW ordering at low temperatures, hole-doping *via* K substitution for Ba results in emergence of superconductivity with T_c up to 38 K for $x \approx 0.4$. Interestingly, the end member on the K-rich side, the ternary KFe_2As_2, is also a superconductor with T_c = 3.4 K.[149] Since the formal charge of Fe in this material is 2.5+, the [FeAs] layers can be considered heavily intrinsically hole-doped in comparison with the pristine $BaFe_2As_2$ or LaOFeAs with Fe^{2+}.

Superconductivity in $BaFe_2As_2$ can also be induced by electron-doping, as in, *e.g.*, $Ba(Fe_{1-x}Co_x)_2As_2$, although in this case, the maximum T_c is only about 22 K.[150] Interestingly, the doping behaviour of superconductivity follows similar trends for hole- and electron-doped samples. The x–T phase diagrams exhibit a characteristic dome-shaped superconducting region (Figure 2.18).[150,151] Such behaviour appears to be typical for many iron pnictides.[16,139,140]

In addition to chemical doping, the application of external pressure proves to be an efficient way of suppressing the SDW instability and inducing

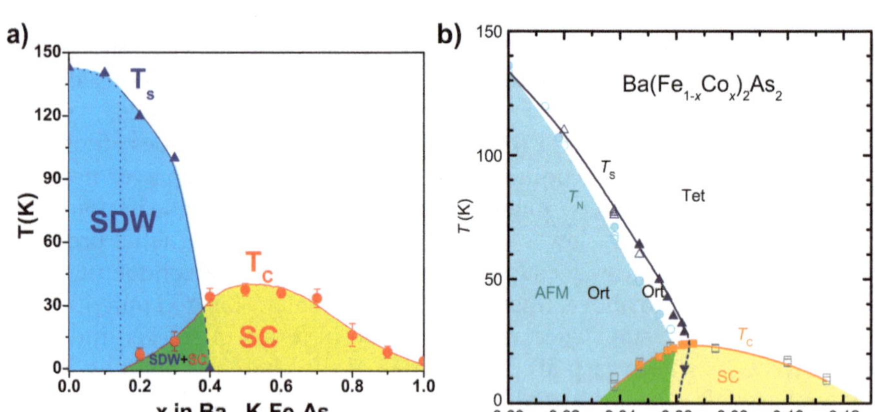

Figure 2.18 x–T phase diagrams for the hole-doped $Ba_{1-x}K_xFe_2As_2$ (a) (reproduced from ref. 152 with permission from Institute of Physics, Copyright 2009) and electron-doped $Ba(Fe_{1-x}Co_x)_2As_2$ (b) (reproduced from ref. 150 with permission from American Physical Society, Copyright 2010).

superconductivity in 122 phases. Thus, $SrFe_2As_2$ and $BaFe_2As_2$ were reported to exhibit superconductivity with T_c up to about 30 K at 4 GPa.[153] Another representative of the 122 family, $CaFe_2As_2$, is notable for undergoing a drastic unit cell volume collapse by about 5% at pressures above 1.5 GPa (at room temperature).[154] The shrinkage of the unit cell is most pronounced in the tetragonal *c* direction. From the chemistry perspective, the formation of the collapsed tetragonal phase is accompanied by the development of As–As chemical bonds between the adjacent [FeAs] layers, which affects the electronic properties. In particular, the orthorhombic structural distortion and magnetic ordering observed for the ambient pressure $CaFe_2As_2$, are suppressed in the collapsed phase. Studies suggest that superconductivity does not develop in $CaFe_2As_2$ under hydrostatic pressure conditions.[155]

Upon partial substitution of the AE metal (AE = alkaline-earth metal or Eu) by an alkali metal (A) in a 122 phase, a complete crystallographic ordering of the AE and A species in charge reservoir layers can develop at the ratio AE : A = 1 : 1, resulting in a so-called 1144 iron pnictide, such as $CaKFe_4As_4$ (Figure 2.17).[156] This ordering is accompanied by the space group symmetry change from the parent *I*4/*mmm* in the $ThCr_2Si_2$-type 122 structure to *P*4/*mmm* in the 1144 derivative. The 1144 structure can also be stabilised for iron pnictides containing trivalent rare-earth metals, for instance, (La,Na) $RbFe_4As_4$.[157] An exciting combination of magnetic and superconducting properties is offered by the widely studied Eu-containing 1144 member, $RbEuFe_4As_4$. Upon cooling, this material undergoes a transition to a superconducting state at T_c = 37 K, followed by a helical antiferromagnetic ordering in the Eu^{2+} sublattice at T_N = 15 K (Figure 2.19).[158,159]

The 1144 phases can formally be viewed as crystallographically ordered amalgamations of two different 122 iron pnictides, *e.g.*, $RbEuFe_4As_4$ = $RbFe_2As_2$ + $EuFe_2As_2$. Using the same logic, it is possible to imagine even more complex structures resulting from a hypothetical synthesis of various iron pnictide types. Some of these structures have been experimentally obtained. For example, the so-called 12442 phases, such as $RbSm_2Fe_4As_4O_2$ (= $RbFe_2As_2$ + 2SmFeAsO) and $RbCa_2Fe_4As_4F_2$ (= $RbFe_2As_2$ + 2CaFeAsF) represent compounds with alternating blocks of the 122 and 1111 types in their crystal structures (Figure 2.17).[161,162] A noteworthy feature of these materials is their superconductivity without extrinsic doping. Since the average formal charge of Fe in these phases is 2.25+, they are intrinsically hole-doped and demonstrate critical temperatures around 30 K.

While heavier alkali metals K, Rb, and Cs form the 122 type iron arsenides, in the A–Fe–As systems with A = Li and Na, *anti*-PbFCl-type AFeAs compounds can be produced, referred to as 111 phases (Figure 2.17). LiFeAs become superconducting without extrinsic doping below 18 K.[163] The superconductivity in this phase appears to originate from a mechanism different from that in other iron pnictides. LiFeAs does not display a spin density wave (SDW) order.[164] Furthermore, both electron and hole doping cause suppression of superconductivity.[165,166] In contrast, NaFeAs is reminiscent of such 122 phases as $BaFe_2As_2$.[167] An orthorhombic distortion and SDW order are

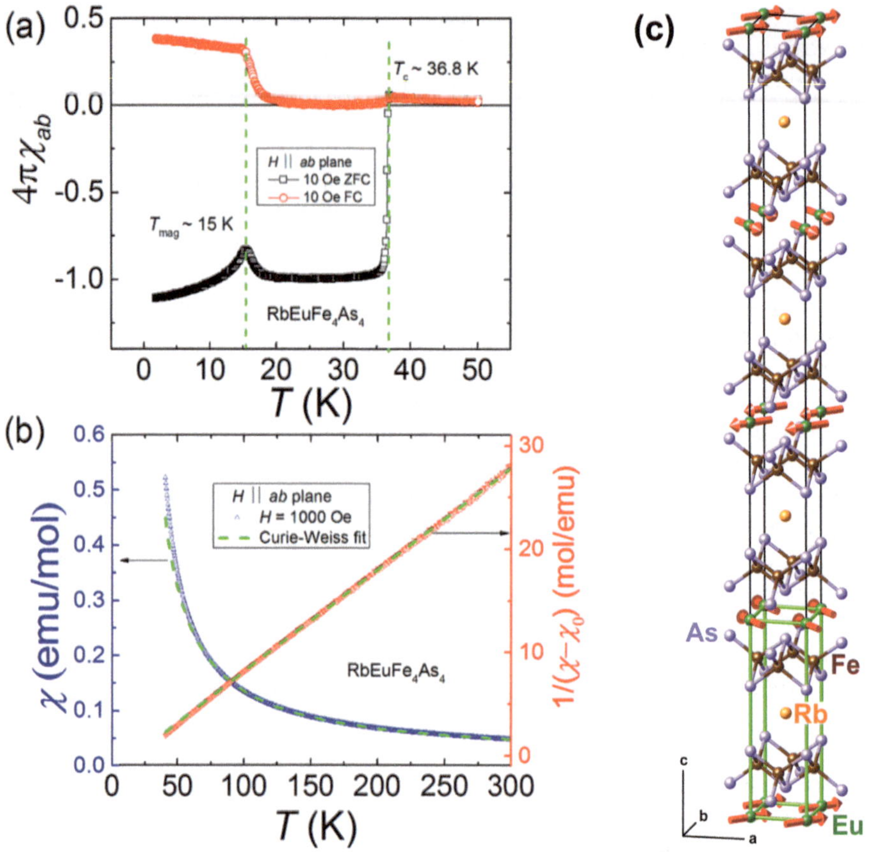

Figure 2.19 Temperature dependence of magnetic susceptibility for $RbEuFe_4As_4$ under low and high fields, respectively, (a) and (b). The superconducting (T_c) and magnetic (T_{mag}) transitions are indicated. Reproduced from ref. 160 with permission from American Chemical Society, Copyright 2018. Magnetic structure of $RbEuFe_4As_4$ based on the data from ref. 159 (c). The crystallographic and magnetic unit cells are outlined in green and black, respectively.

observed for NaFeAs and can be suppressed by doping with electrons or holes, leading to superconductivity.[168,169] Due to high moisture sensitivity, samples of NaFeAs can develop superconductivity upon exposure to air by transforming to the hole-doped Na-deficient $Na_{1-\delta}FeAs$ with T_c up to 25 K.[170] Similar critical temperatures can be achieved by electron doping, as in, *e.g.*, $NaFe_{1-x}Co_xAs$.[168]

The 112 family of iron pnictides, represented by such solid solutions as $Ca_{1-x}La_xFeAs_2$, adopts a crystal structure that features the typical [FeAs] layers and zig-zag chains of As, the latter located in the charge reservoir layers along with the Ca atoms (Figure 2.17).[171–173] The zig-zag chains can be understood as the Zintl polyanions $^{1}_{\infty}[As]^{1-}$. The hypothetical ternary phase

$CaFeAs_2$, which has not been reported so far, can then be expressed with the notation $(Ca^{2+})(Fe^{2+})(As^{3-})(As^{-})$ and is expected to be non-superconducting based on the formal charge of Fe. In fact, small amounts of rare-earth metal appear to be necessary for stabilisation of this structure. The $Ca_{1-x}RE_xFeAs_2$ compositions (RE = rare-earth metal) are electron-doped and show critical temperatures of up to 47 K.[173]

Related building blocks are present in the crystal structures of the noble-metal-containing arsenides $Ca_{10}(M_3As_8)(FeAs)_{10}$ and $Ca_{10}(M_4As_8)(FeAs)_{10}$ (M = Pd, Ir, Pt) with even higher structural complexity. These two families are also known as 10-3-8 and 10-4-8 phases, respectively (Figure 2.17).[174–177] The layered structures of these materials contain three types of quasi-two-dimensional building blocks: the typical [FeAs] slabs, nets of Ca atoms and $[M_3As_8]$ or $[M_4As_8]$ layers built up of corner-sharing MAs_4 squares additionally linked by As–As bonds. The $[M_3As_8]$ or $[M_4As_8]$ layers can be regarded as slices of a skutterudite structure. The $[M_3As_8]$ nets accommodate ordered vacancies in the square sites and are essentially flat, while $[M_4As_8]$ are vacancy-free but corrugated.[174] The corrugation has been explained by repulsive interactions with the Ca species in the adjacent layers. Due to the steric factors and weak interactions between the principal building blocks, the structures of the 10-3-8 and 10-4-8 phases display stacking disorder and associated polymorphism.[175,176] The highest superconducting transitions temperatures up to about 38 K can be realised for the Pt families by, *e.g.*, rare-earth doping on the Ca site (10-3-8) or by self-doping *via* partial substitution of Pt for Fe in the [FeAs] slabs (10-4-8).[174,178]

Several iron arsenides with complex oxide slabs in the crystal structures have been produced. The discovery of these phases serves as another example of the structural variability in iron pnictides. Thus, the crystal structures of the so-called 21311 (also referred to as 42622) and 32522 phases, exemplified by Sr_2ScO_3FeAs ($=Sr_4Sc_2O_6Fe_2As_2$) and $Sr_3Sc_2O_5Fe_2As_2$, respectively, accommodate perovskite-type $[Sr_{n+1}Sc_nO_{2n+1}]$ blocks sandwiched between the [FeAs] layers (Figure 2.17).[179–182] The successful synthesis of the related homologous series $Ca_{n+1}(Sc,Ti)_nO_yFe_2As_2$ ($n = 3$–5) indicates that the thickness of the perovskite slabs can be tuned in wide ranges.[183] The critical temperatures in these phases can exceed 40 K by proper chemical doping.

The crystal structure of another oxygen-containing arsenide, $Ba_2Ti_2Fe_2As_4O$ (the 22241 phase, Figure 2.17), is hallmarked by the presence of the oxyarsenide $[Ti_2As_2O]$ slabs, isostructural to the $[Ti_2Pn_2O]$ units (Pn = Sb, Bi) in the Ti-based superconductors $BaTi_2Pn_2O$, described later in this chapter. $Ba_2Ti_2Fe_2As_4O$ superconducts without extrinsic doping below 21 K. The superconductivity is believed to originate from the [FeAs] layers only, while the presence of the $[Ti_2As_2O]$ layers gives rise to an anomaly at 125 K, likely associated with a charge density wave (CDW) or spin density wave (SDW) order (Figure 2.20).[184]

With such a great variety of crystal structure types for iron-based superconductors, attempts have been made to find correlations between the observed critical temperatures and crystal–chemical parameters. The degree of distortion in the $FeAs_4$ tetrahedra building the [FeAs] layers was proposed

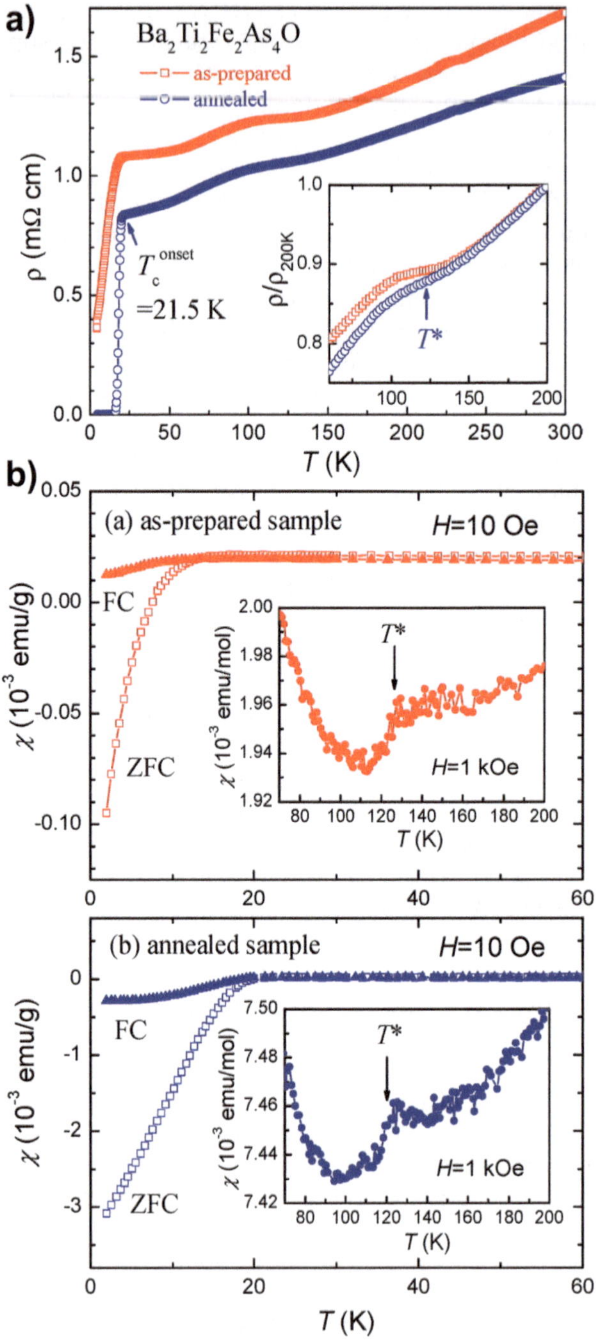

Figure 2.20 Temperature dependence of electrical resistivity (a) and magnetic susceptibility (b) for different samples of $Ba_2Ti_2Fe_2As_4O$ (as-prepared and annealed at 773 K for 40 h). Insets show the presence of the CDW/SDW anomaly at $T = 125$ K. Reproduced from ref. 184 with permission from American Chemical Society, Copyright 2012.

as one of the possible indicators, since iron pnictides with higher T_c values usually exhibit more regular tetrahedral coordination of Fe in their crystal structures.[185] Another structural feature that appears to be favorable for high critical temperatures is the location of the As atoms at an optimal distance from the Fe plane, which was found to be around 1.38 Å.[186] Although these correlations are helpful for the identification of promising iron-based superconductors, a predictive theory is yet to be developed.

All iron pnictides described above belong to the subgroup of arsenides. Superconductivity has also been observed in iron phosphides crystallising in some of the structures shown in Figure 2.17, although in most cases, the critical temperatures are lower than for the arsenide counterparts, which may be related to weaker magnetic fluctuations in the representatives of the phosphide subfamily.[187,188]

After the discovery of superconductivity in iron-based pnictides, research on pnictide systems with electronic or magnetic instabilities yielded some new superconducting compounds. Most of them demonstrate relatively low critical temperatures. Nevertheless, the unconventional nature of superconductivity in these materials has gained considerable attention.

The multinary titanium pnictides-oxides $BaTi_2Pn_2O$ (Pn = Sb, Bi, Figure 2.21(a)) display superconductivity below 1.2 K (Pn = Sb) and 4.6 K (Pn = Bi).[189–191] $BaTi_2Sb_2O$ has been shown to display an anomaly consistent with a charge density wave (CDW) or spin density wave (SDW) transition.[192] This anomaly coexists with superconductivity in the pristine phase. However, much like in the case of iron pnictides, suppression of the CDW/SDW transition enhances superconductivity, reflected in a higher transition temperature. Such suppression can be accomplished by chemical doping, like in the example of the hole-doped $Ba_{0.8}Na_{0.2}Ti_2Sb_2O$ with $T_c \approx 6$ K.[193]

As noted in the discussion of iron pnictides, electronic and magnetic instabilities can be tuned not only by chemical doping but also by applying

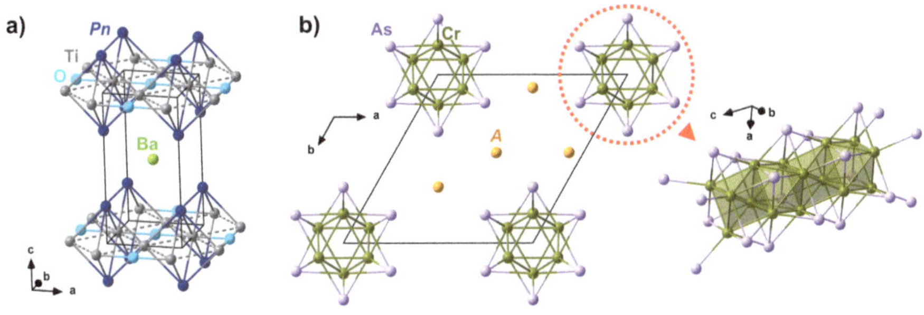

Figure 2.21 (a) Crystal structure of $BaTi_2Pn_2O$ (space group $P4/mmm$, Pearson code $tP6$, Pn = Sb, Bi). The square net of Ti atoms is highlighted by dashed lines. (b) Crystal structure of $A_2Cr_3As_3$ (space group $P\bar{6}m2$, Pearson code $hP16$, A = K, Rb, Cs, left) and a side view of the quasi-one-dimensional $[Cr_3As_3]$ column based on empty face-sharing Cr_6 octahedra (right).

pressure. Thus, superconductivity was discovered in the binary pnictides CrAs ($T_c \approx 1$ K at $P = 0.8$ GPa)[194] and MnP ($T_c \approx 1$ K under external pressure of $P = 8$ GPa),[195] both of which adopt helical magnetic ground states at ambient pressure.

Another notable family of intrinsic superconductors containing a magnetic element is $A_2Cr_3As_3$ (A = K, Rb, Cs), which forms with a non-centrosymmetric quasi-one-dimensional crystal structure (Figure 2.21(b)).[196–198] In contrast to the binary chromium arsenide CrAs, superconductivity in these materials is observed at ambient pressure. The superconducting critical temperature was found to decrease in the order K ($T_c = 6.1$ K) → Rb ($T_c = 4.8$ K) → Cs ($T_c = 2.2$ K). While these materials lack charge or magnetic ordering, short-range magnetic fluctuations have been detected by inelastic neutron scattering, suggesting that the $A_2Cr_3As_3$ phases are close to magnetic instability.[199] Subsequent studies uncovered a family of isostructural superconductors with the Cr heavier congener, Mo, $A_2Mo_3As_3$ (A = K, Rb, Cs), where the superconducting critical temperatures vary little upon change of the A metal and reach values of up to 11.5 K.[200–202]

While the interplay of magnetism and superconductivity remains an actively studied topic in condensed matter science, in recent years, the emergence of materials with non-trivial electronic band topologies has attracted significant interest since such materials can also exhibit unconventional superconductivity. We will discuss topological materials in more detail in Section 2.7.

2.6 Magnetism in Pnictides

The crystal structure diversity of pnictides provides a wide range of magnetic properties, observed in the representatives with d- or f-elements. A rich plethora of collinear and non-collinear magnetic structures, geometric frustration, as well as various types of magnetic excitations has been reported in these compounds. An important topic of the modern condensed matter science has been the elucidation of the interplay between superconductivity and magnetism in pnictide superconductors, predominantly iron arsenides. The topic of superconductivity in pnictides was covered in Section 2.5. Here, selected examples of magnetic materials belonging to the pnictide family will be discussed, focusing on the most well-studied structural classes and compounds with potential applications.

One of the largest structural families of ternary intermetallic compounds is constituted by the $ThCr_2Si_2$-type phases.[203] This structure class encompasses various chemical compositions, including numerous pnictides, and is hallmarked by a spectrum of physical properties exhibited by its representatives. While $ThCr_2Si_2$-type pnictides are mostly known in the context of superconductivity, many of these compounds have been extensively studied due to their interesting magnetic behaviour.

The ternary compound $BaMn_2As_2$ has been examined in detail because of its structural and compositional similarity to $BaFe_2As_2$, a parent phase

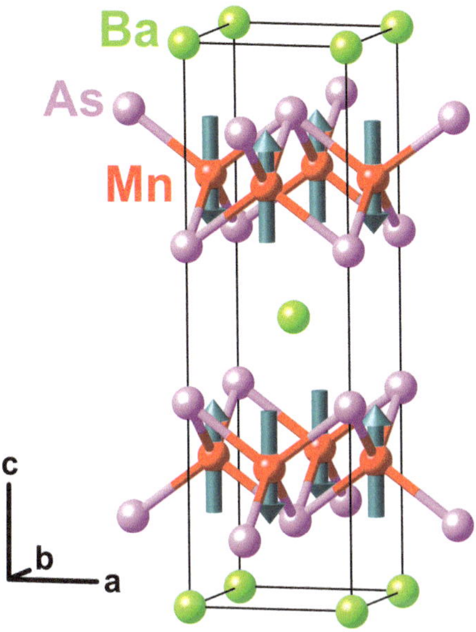

Figure 2.22 Magnetic structure of $BaMn_2As_2$. The unit cell is outlined in black.

of the 122 iron-based superconducting family. The structural and magnetic behaviour of $BaMn_2As_2$ is distinct from that of iron pnictides: $BaMn_2As_2$ enters a collinear, G-type antiferromagnetic state at $T_N = 625$ K (Figure 2.22) and does not undergo any orthorhombic distortion. The absence of the spin density wave order (SDW), typical for iron arsenides, is believed to be associated with strong nearest-neighbour magnetic interactions, whereas sizeable next-nearest neighbour exchange would be necessary to induce magnetic nematicity.[204,205]

The $ThCr_2Si_2$-type cobalt phosphides $AECo_2P_2$ (AE = alkaline-earth or rare-earth metal) have also received considerable attention due to itinerant magnetism of the cobalt sublattice that can be tuned by application of pressure, chemical doping, or the nature of the metal AE in the composition.

Thus, $LaCo_2P_2$ is a ferromagnet with the Curie temperature $T_C = 103$ K, whereas $CaCo_2P_2$ demonstrates antiferromagnetic ordering below $T_N = 113$ K. The crystal structure of the former compound shows a long interlayer P–P distance of about 3.2 Å. In contrast, $CaCo_2P_2$ adopts a 'collapsed' $ThCr_2Si_2$-type structure, similar to the high-pressure variant of $CaFe_2As_2$ (Section 2.5), with short P–P bonds of 2.6 Å between adjacent [CoP] layers.[206,207]

The Eu member, $EuCo_2P_2$, crystallising in an 'uncollapsed' structure, exhibits antiferromagnetism below 67 K. However, only the Eu^{2+} species carry localised magnetic moments, with the Co sublattice displaying Pauli paramagnetism. Upon application of pressure, $EuCo_2P_2$ undergoes a 'collapse' at $P = 3.1$ GPa, evidenced by the rapid change in the unit cell volume. This structural transformation is accompanied by the disappearance of the Eu magnetic moment due to the valence transition $Eu^{2+} \rightarrow Eu^{3+}$ and simultaneous emergence of itinerant antiferromagnetism in the Co substructure with $T_N = 260$ K.[208]

Another diverse structural family, which was described in Sections 2.1 and 2.4.1, is the Heusler compounds. Similar to the $ThCr_2Si_2$ type, Heusler phases encompass different chemical compositions, including pnictides.[5]

Arguably, one of the most exciting magnetic properties observed in a pnictide representative of this family is half-metallic ferromagnetism in the half-Heusler MnNiSb. Although this kind of behaviour has been found in several other phases, not limited to the Heusler compounds, MnNiSb was the first material theoretically shown to be a half-metallic ferromagnet, which was later confirmed experimentally.[209,210] The main property of such materials is the opening of an electronic band gap at the Fermi level for one of the spin channels in the electronic spectrum (Figure 2.23). This results in electrical conductance of electrons with a single spin direction, at least at very low temperatures, when thermally activated conductivity in the second channel is quenched. Half-metallic ferromagnets can be used in spintronic applications to produce spin-polarised electronic currents.[211]

Ferromagnetism, in general, is a valuable material property with many applications in fields such as permanent magnets and magnetocalorics. A number of ferromagnetic pnictides, mostly containing the abundant magnetic transition metals Mn and Fe, have been studied as promising materials capable of replacing ferromagnets based on rare-earth metals. One such pnictide is the binary manganese bismuthide MnBi. The low-temperature form, α-MnBi, stable below 613 K, crystallises in the NiAs structure type and displays high magnetocrystalline anisotropy.[213,214] Similar to hexaferrites, often employed in permanent magnets, the coercivity of MnBi increases with temperature.[214] Along with the high theoretical maximum energy product,

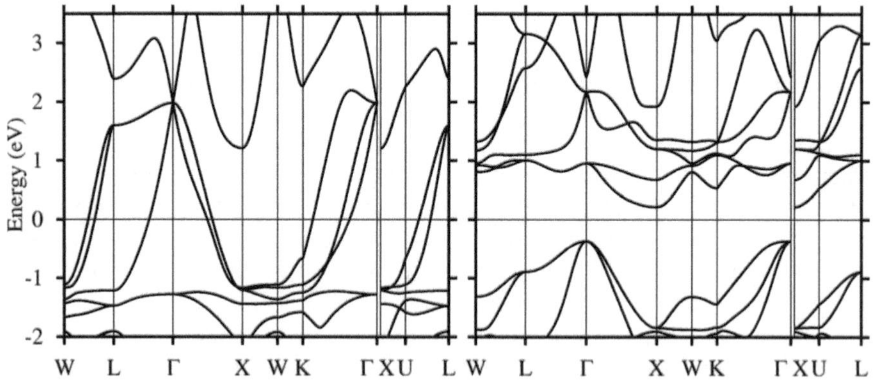

Figure 2.23 Electronic band structure of MnNiSb for the majority (left) and minority (right) spin channels. The Fermi level (E_F) is set at 0 eV. A pronounced band gap at E_F is seen for the minority spins. Reproduced from ref. 212 with permission from American Physical Society, Copright 2006.

these characteristics make MnBi a suitable candidate for a rare-earth-metal-free permanent magnet. However, commercial use of MnBi is hindered by the difficulty of producing a dense pure material due to the intrinsically low thermal stability and slow formation kinetics at low temperatures.[215]

The search for inexpensive ferromagnets has revealed that the iron phosphide Fe_2P (Figure 2.24) and its derivatives can be economically feasible alternatives to rare-earth-metal-based materials. The binary Fe_2P has a high magnetocrystalline anisotropy, but its low Curie temperature of only 214 K hampers its application.[216,217] However, the ordering temperature can be increased by doping on the cationic or anionic sites, although such substitutions often result in deterioration of other magnetic properties. Thus, the Co-doped compositions $Fe_{2-x}Co_xP$ demonstrate T_C values of up to *ca.* 440 K but exhibit rather low remanent magnetisation and magnetocrystalline anisotropy. In contrast, co-doping with Co and Si leads to high T_C and retains the important magnetic parameters, as in, *e.g.*, $Fe_{0.91}Co_{0.09}P_{0.78}Si_{0.22}$ with T_C of 506 K.[218]

The main proposed area of application for the Fe_2P-based ferromagnets is magnetic refrigeration since these materials demonstrate a strong magnetocaloric effect (MCE), *i.e.*, they are capable of lowering their temperature considerably upon adiabatic demagnetisation.[219] In addition, the Curie temperatures of these compounds can be tuned to be close to room temperature, which is required for operation at ambient conditions.

One of the most promising compositions for these applications is $MnFeP_{0.45}As_{0.55}$ with T_C of around 300 K.[220] Magnetocaloric materials are usually ranked according to the maximum entropy change that can be achieved upon going across the magnetic transition. For $MnFeP_{0.45}As_{0.55}$, this value reaches 18 J K^{-1} kg^{-1} under an applied field of 5 T.[220]

Some other Mn-containing pnictides have also been found to demonstrate good magnetocaloric properties. For example, in MnAs (T_C = 318 K), the

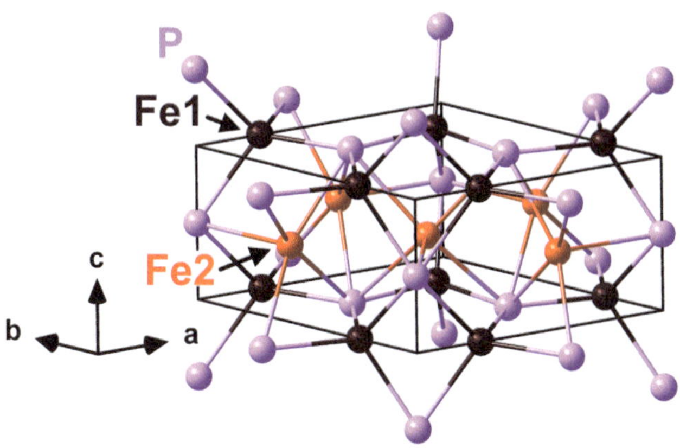

Figure 2.24 Crystal structure of Fe_2P (space group $P\bar{6}2m$, Pearson code *hP*9). Two types of iron sites adopt tetrahedral and square-pyramidal coordination of P atoms. The unit cell is outlined in black.

entropy change caused by a field of 5 T reaches 30 J K^{-1} kg^{-1}. Such a high value is related to the concomitant first-order crystallographic phase transition from the low-temperature NiAs-type modification to the high-temperature MnP-type form.[221] Notably, transition metal doping on the Mn site or application of pressure leads to a so-called colossal magnetocaloric effect (CMCE) in MnAs, with entropy changes exceeding 150 J kg^{-1} K^{-1}.[222,223]

The examples of materials given above cover only a small part of magnetic pnictides that have been investigated. Undoubtedly, this class of compounds can be interesting both from the fundamental science perspective and for applications in modern or future technologies.

2.7 Pnictides as Topological Materials

Physical phenomena related to non-trivial electronic band topology have been a major topic in condensed matter science since the first experimental realisation of a material with topology-related properties in 2007.[224,225] In that study, quantum wells of mercury telluride, HgTe, exhibited a quantum spin Hall effect (QSH), a special kind of topology-protected quantisation of Hall conductance.[225] The QSH phases, now called topological insulators (TIs), are a fundamentally new quantum state of matter. One of the most commonly observed properties of a TI is a gapped, *i.e.*, insulating, electronic structure in bulk and the presence of conducting edge or surface currents. These currents, being topologically protected, are less susceptible to local decoherence processes, which makes topological insulators promising materials for quantum computing.[224] Besides topological insulators, several other classes of materials exhibiting topologically non-trivial electronic bands are known, such as Dirac and Weyl semimetals. Unlike TIs, semimetals demonstrate a gapless bulk electronic state.[226]

In general, nontrivial electronic topologies result from an interplay of crystal symmetry and chemical bonding interactions and can arise in various groups of crystalline substances. Although topological materials do not constitute a particular chemical family, many pnictides have been identified as belonging to this large class.[227] Thus, the sodium bismuthide Na_3Bi was one of the first compounds confirmed as a 3D topological Dirac semimetal. In Dirac semimetals, the valence and conduction bands touch in a special point of the momentum space, called the Dirac point, and disperse linearly around this node (Figure 2.25(a)). Such an electronic pattern is well known for graphene, which hosts two-dimensional Dirac fermions. Theoretical predictions and angle-resolved photoelectron studies show that Na_3Bi, crystallising in a relatively simple hexagonal structure (Figure 2.25(b)), represents a 3D counterpart of graphene.[228,229]

Another binary pnictide, the cadmium arsenide Cd_3As_2, has also been found to be a topological semimetal. Cd_3As_2 hosts a pair of 3D Dirac fermions and demonstrates unusually high carrier mobility.[230] The ambient-temperature crystal structure of Cd_3As_2 was a matter of debate. The noncentrosymmetric space group $I4_1cd$ was assigned in the original crystallographic report.[231] The proposed lack of inversion symmetry could lift the spin-degeneracy of

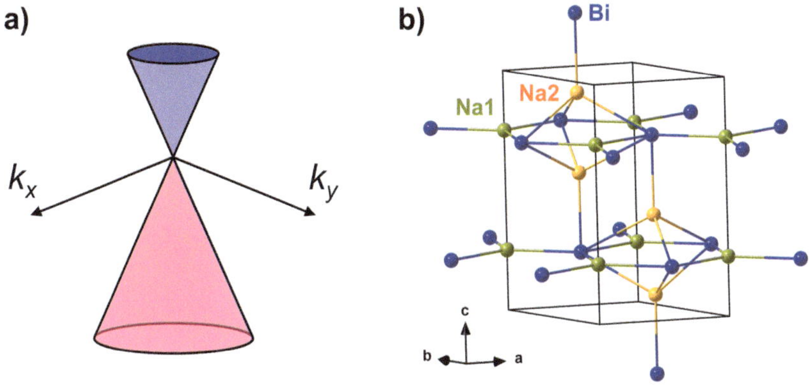

Figure 2.25 (a) Projection of a 3D Dirac fermion onto (k_x, k_y, E). The cone-like dispersion is emphasised. (b) Crystal structure of Na_3Bi (space group $P6_3/mmc$, Pearson code *hP*8). Two symmetry independent Na atoms are trigonally (Na1) and tetrahedrally (Na2) coordinated by Bi.

some electronic bands at the Fermi level and make Cd_3As_2 a different topological class of matter—a Weyl semimetal. However, a more recent study has shown that Cd_3As_2 crystallises in the centrosymmetric space group $I4_1/acd$ (Figure 2.26) and is therefore constrained by symmetry to be a Dirac semimetal.[232]

In contrast to Na_3Bi and Cd_3As_2, crystallising in centrosymmetric structures, the crystal structure of the semimetallic tantalum arsenide TaAs (space group $I4_1md$, Figure 2.27(a)) lacks inversion symmetry. For this reason, the Dirac node in TaAs splits into two nodes with opposite chirality, called Weyl nodes. The presence of the Weyl nodes reveals itself in the existence of Fermi arcs, which connect the Weyl nodes projections on the surface of the material. Experimental observation of this peculiar electronic structure in TaAs (Figure 2.27(b)) made this material the first confirmed Weyl semimetal.[233,234] Subsequent studies proposed realisation of the Weyl state in the whole family of isostructural pnictides MPn (M = Nb, Ta; Pn = P, As).[235–237] Interestingly, the conducting surface states in the MPn Weyl semimetals lend these materials high catalytic activity in the hydrogen evolution reaction (HER).[238]

While Dirac and Weyl semimetals are characterised by the presence of special points in their electronic structures (Dirac and Weyl points, respectively), another group of topological semimetals, nodal-line semimetals (NLSM), host a linearly dispersing band crossing along a continuous line in the momentum space.[239] A number of pnictides have been proposed as NLSMs based on theoretical calculations and experimental observations. Among them, the most studied representatives are CaAgAs and Ca_3P_2.[240–244]

In the recent years, materials combining non-trivial electronic band topologies and superconductivity or magnetism have become an actively studied topic due to the potential of accessing new quantum phenomena in such systems. For example, realisation of the elusive Majorana

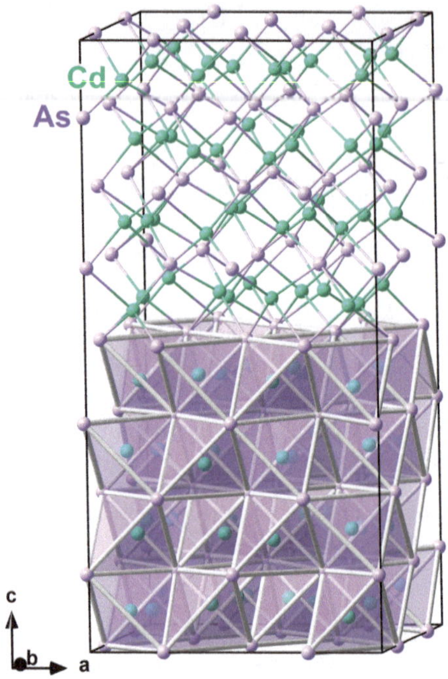

Figure 2.26 Crystal structure of Cd_3As_2 (space group $I4_1/acd$, Pearson code $tI176$). The network of corner- and edge-sharing $[CdAs_4]$ tetrahedra is emphasised. The unit cell is outlined in black.

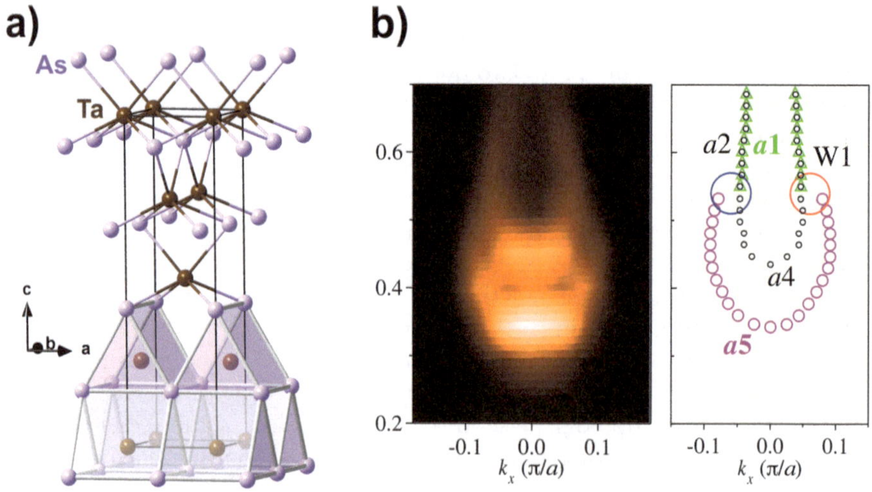

Figure 2.27 (a) Crystal structure of TaAs (space group $I4_1md$, Pearson code $tI8$) viewed as a 3D framework of corner- and edge-sharing $[TaAs_6]$ trigonal prisms. (b) Photoemission intensity plot at the Fermi level around Weyl point W1 (left) and the extracted Fermi arcs a_i (right). Reproduced from ref. 234, https://doi.org10.1103/PhysRevX.5.031013, under the terms of the CC-BY 3.0 License, https://creativecommons.org/licenses/by/3.0/.

fermions—quasiparticles being their own antiparticles—has been predicted to be possible on the surface of a topological superconductor.[224,245]

Although the number of known topological superconductors is so far limited, some superconducting pnictides, including the Fe_2P-type ZrRuAs (T_c = 8 K) and other MRuPn (M = Zr, Hf; Pn = P, As) have been proposed as realisations of such materials.[246,247] Due to the non-centrosymmetric crystal structure (space group $P\bar{6}2m$), some of the MRuPn phases host Weyl fermions.[246]

Recently, a new family of superconducting ternary antimonides with non-trivial band topology has been discovered. The compounds AV_3Sb_5 (A = K, Rb, Cs) crystallise in their own structure type, with the structures hosting kagomé layers of V atoms and honeycomb nets of Sb atoms (Figure 2.28).[248] Dirac-cone-like bands are present in the electronic spectrum, and superconductivity is observed below T_c = 0.93 K (A = K), 0.92 K (A = Rb), and 2.5 K (A = Cs).[249–251] High-resolution scanning tunneling microscopy studies indicate an unconventional CDW order in AV_3Sb_5.[252]

Materials combining topologically non-trivial electronic bands and magnetism are more common than topological superconductors. A large number of pnictides containing magnetic rare-earth metals demonstrate topological properties, including some Zintl phases.[106] Thus, several rare-earth metal (RE) monopnictides REPn (Pn = P, As, Sb, Bi) have been predicted and have been experimentally confirmed to display topological properties.[253,254] For example, NdSb has been proposed as an antiferromagnetic Dirac semimetal.[255]

Materials belonging to the large family of half-Heusler phases can potentially exhibit Weyl fermions due to the noncentrosymmetric crystal structure (space group $F\bar{4}3m$).[256] An interesting example in this family is GdPtBi, which shows emergence of Weyl nodes upon application of magnetic field.[257,258]

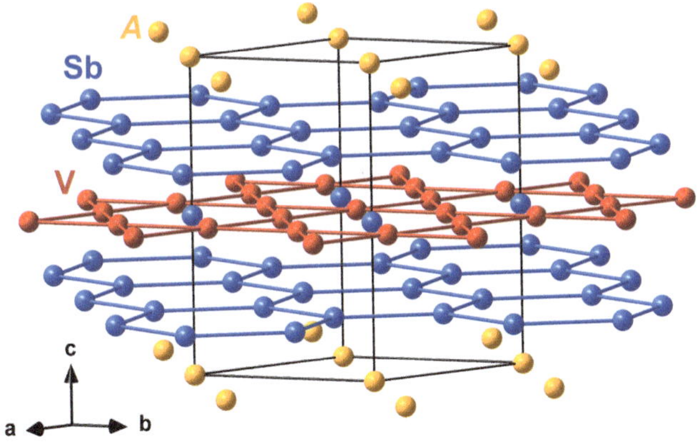

Figure 2.28 Crystal structure of AV_3Sb_5 (space group *P*6/*mmm*, Pearson code *hP*9, A = K, Rb, Cs). Kagomé layers of V atoms and honeycomb nets of Sb atoms are emphasised.

Several magnetic pnictides featuring square nets of pnictogen atoms have been reported to be topological semimetals. Two main families of such compounds are the layered $AMPn_2$ (A = alkaline-earth or rare-earth metal; M = transition metal; Pn = Sb, Bi) and mixed-anion pnictide-chalcogenides REChPn (RE = rare-earth metal; Ch = S, Se, Te; Pn = P, As, Sb, Bi).[259] The research on these materials was motivated by the discovery of highly anisotropic Dirac fermions in $SrMnBi_2$, a representative of the former family, crystallising in the $SmCuP_2$ crystal structure (space group *I*4/*mmm*, Figure 2.29(a)).[260] These two-dimensional Dirac fermions are believed to originate from the square Bi layers in the structure. Later, similar Dirac states were identified in the structurally related $CaMnBi_2$, adopting the ZrCuSiAs structure (space group *P*4/*nmm*, Figure 2.29(b)).[261,262] The interplay of magnetism and non-trivial band topology results in exotic properties in these materials, such as giant magnetoresistance.

With the advancement of the mathematical machinery and computational methods for analysis of topological materials, identification of new compounds with non-trivial band topology has become relatively simple, at least for nonmagnetic materials.[263] Experimental confirmation of these properties, on the contrary, remains challenging. Many new pnictides will likely be identified as candidates for realisation of topological phenomena. Availability of high-quality single crystals will be crucial for detailed experimental studies. With the considerable variety of crystal growth methods available for pnictides, as discussed in Section 2.2, this class of compounds may be a suitable playground for condensed matter research on topological phenomena.

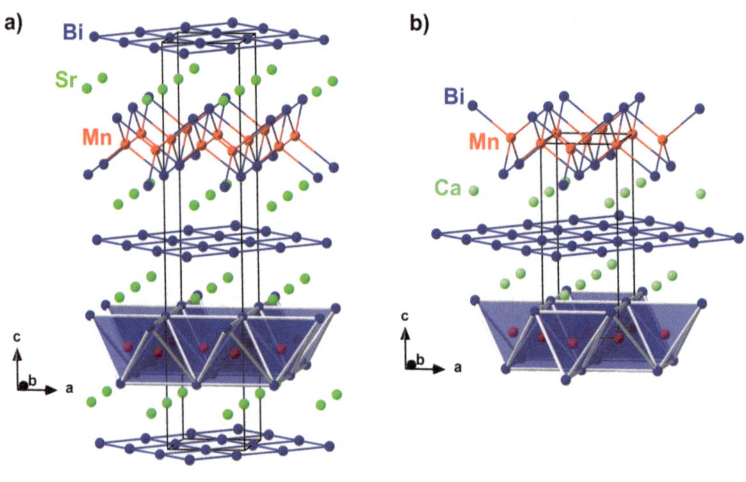

Figure 2.29 Crystal structures of $SrMnBi_2$ (a) (space group *I*4/*mmm*, Pearson code *tI*16) and $CaMnBi_2$ (b) (space group *P*4/*nmm*, Pearson code *tP*8). Layers of fused [$MnBi_4$] tetrahedra and square Bi nets are emphasised. The unit cells are outlined in black.

2.8 Catalytic Materials

Metal pnictides have interesting catalytic properties, particularly in the field of electrocatalysis and photocatalysis. The vast majority of them are binary metal-rich phosphides and their derivatives. Some of the common structure types adopted by these phases are shown in Figure 2.30. In practice, most of these catalytically active materials are used in the form of nanostructured composites with properties drastically different from those of the bulk phases.[264] Their syntheses often involve diverse innovative and unique techniques, as was briefly discussed in Section 2.2. Many controllable synthetic and engineering strategies are widely applied for the production of metal pnictide nanoparticles, and we refer the reader to the respective reviews on the structural design of metal phosphides for electrocatalysis.[264,265]

Most research efforts on metal-rich and P-rich pnictides were performed for the cathodic hydrogen evolution reaction (HER).[267,268] While Pt remains the best cathodic HER electrocatalyst, its cost and scarcity promote research

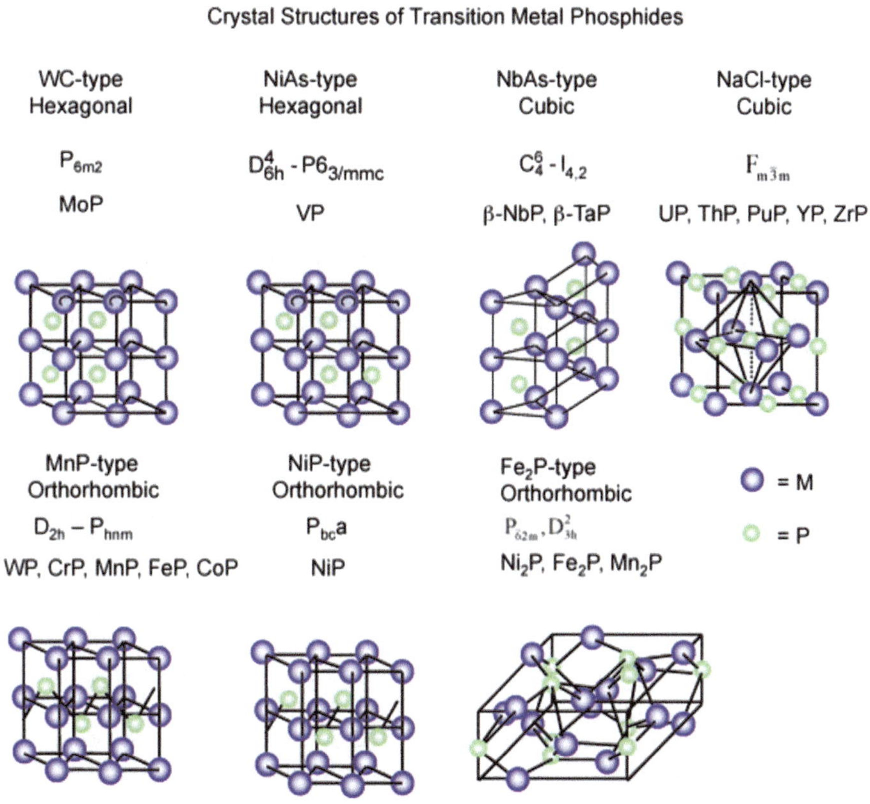

Figure 2.30 Crystal structures of some common binary transition metal phosphides. Reproduced from ref. 266 with permission from Elsevier, Copyright 2009.

efforts towards developing less expensive yet active and durable catalysts to satisfy an increasing demand in H_2 production on a massive scale.[269] Binary transition metal phosphides appear to be a class of promising catalysts for HER, due to their unique electronic structures.[267,268] One of the first reported transition metal phosphide nanomaterial for HER applications was nanosheets of FeP.[270] This material exhibits high electrocatalytic activity with a low overpotential of *ca.* 0.1 V. Other transition metal phosphide nanomaterials, such as CoP,[271] Ni_2P,[19] Cu_3P,[272] MoP[273] and WP[274] were also tested as Pt-free HER catalysts. In particular, nickel phosphide exhibits excellent performance, providing a nearly quantitative faradaic yield of H_2.[19]

For hydrogen and oxygen production, efficient water splitting involves efficacious and robust electrocatalysts with significantly reduced overpotentials for both the cathodic HER and anodic oxygen evolution reaction (OER).[275] In fact, many metal phosphides efficient as HER catalysts might also appear to be active for OER.[276,277] For instance, Pt_2P nanoparticles show improved catalytic performance in the oxygen reduction reaction (ORR).[278] However, for economical reasons, functional 3d-metal phosphide nanomaterials and composites appear to be preferable for such purposes.

Several phosphide-based composite systems are efficient catalysts for the HER and OER. Thus, P-doped MoS_2@CoP on carbon cloth exhibits bifunctional catalytic properties with reported overpotentials of 64 and 282 mV in alkaline solution for the HER and OER, respectively.[279] W and Mo co-doped NiCoP nanorods on Ni foam show a low overpotential of 249 mV for the HER at a substantial current density of 1000 mA cm^{-2}. At the same time, a significant enhancement for the OER was also observed. As a result, W,Mo–NiCoP/NF as both electrodes exhibited a low cell voltage of 1.85 V and a large current density of 500 mA cm^{-2}. With excellent long-term stability of 50 h, this material is currently one of the best reported bifunctional electrocatalysts.[280] Another bifunctional electrocatalyst for overall water splitting can be described as a nitrogen-doped carbon-decorated CoP/FeCoP composite.[281] These and several other studies indicate the great potential of bimetallic phosphides for electrochemical water splitting.[282–286]

Binary and ternary metal phosphides exhibit excellent catalytic activity in several hydrotreating reactions, such as hydrodeoxygenation, hydrodesulfurisation and hydrodenitrogenation.[266,287–292] Noble metal phosphide electrocatalysts are also shown to be a reasonable alternative for Pt in the process of the small organic molecule electrooxidation in direct liquid fuel cells.[293,294] Composite materials such as Pt_5P_2/C and Pt-Ni_2P/C enhance electrocatalytic properties of a Pt catalyst in the methanol oxidation reaction (MOR) at the anode of direct methanol fuel cells.[295,296] Among non-Pt catalysts, nickel phosphide nanospheres were also tested and showed excellent performance for water oxidation reaction (WOR) and MOR.[297] Other types of processes were also studied; among them, ethanol oxidation reaction (EOR), formic acid oxidation reaction (FAOR), ethylene glycol oxidation reaction (EGOR), hydrazine oxidation reaction (HOR), and ORR, to name a few. Phosphide electrocatalysts are very promising for the listed applications, and we refer the reader to the recent comprehensive review on this topic.[31]

Although transition metal phosphides are well-studied as electrocatalysts and hydrotreating catalysts, their application is less common in organic synthesis. Nevertheless, this area is rapidly developing, and multiple metal phosphide nanoparticle catalysts are being studied in several different processes. For instance, nickel phosphide nanoparticles, Ni_2P and Ni_5P_4, exhibit excellent activity in the hydrotreating reactions, particularly in the synthesis of hydrogen by C_2H_2 decomposition and the dry reforming of methane.[298] One of the most studied applications is hydrogenation catalysis in water.[299] Ni_2P and CoP nanoparticles combine their hydrogen-activating ability and surface acidity, which can be beneficial for the selective transformation of biofuranic aldehydes into diketones in water,[300] furfural into alcohols, such as cyclopentanol or 1,2,5-pentanetriol,[301] maltose to maltitol,[302] D-xylose hydrogenation to D-xylitol,[303] or D-glucose to D-sorbitol.[304] Cobalt phosphide nanorods, Co_2P, exhibit high-performance catalytic properties for the reductive amination of carbonyl compounds.[305] A cerium dioxide-supported nickel phosphide nanoalloy efficiently promotes the C-3 alkylation on oxidoles with alcohols.[306] Electrocatalytic reduction of NO_3^- to NH_3 with Ni_2P as a noble metal-free cathode results in high yields with a superior faradaic efficiency of 99.23%.[307]

Another emerging and exciting technology is biocatalysis, or microbial electrosynthesis, where autotrophic microorganisms uptake electrons on the cathode to reduce CO_2 to valuable chemicals. Modification of electrodes by metal pnictides enhances yields and production rates of butyrate and acetate, as was shown for the case of nickel phosphide catalyst deposited on the cathode.[308] Several metal phosphide catalysts, such as Ni_2P, CoP, Fe_xP, and Cu_3P were also tested as peroxymonosulfate activators to degrade organic dyes in water.[309–312]

Pnictide materials can serve as efficient catalysts for many processes. While covering the full range of studied processes and composite systems is beyond this chapter's scope, we are pointing out the rapidly developing field of nanostructured metal pnictide catalysts, which are efficient for different types of reactions. At the same time, the most recent and comprehensive review will be useful to immerse into the topic of metal phosphide heterogeneous catalysts.[313]

2.9 Pnictides for Electrochemical Applications

Modern society relies on electrochemical energy storage, focusing on lithium-ion batteries, which have become successful commercially for powering portable electronics and electric vehicles. The high stored volume and gravimetric energy density of Li batteries made these applications possible. However, further development is required to address several issues, such as cost, safety, stored energy density, service life, and charge–discharge rates.[314] Classic electrodes for rechargeable Li-ion batteries operate through lithium (de)insertion process involving cationic redox reactions of transition-metal ions.[314] Layered transition-metal oxides, such as $LiCoO_2$ or $LiMn_2O_4$, are used as efficient intercalation-type cathodes. They typically have specific capacities of 100–200 mA h g^{-1} and 3–5 V average voltage *vs.* Li/Li^+.[315] The most common commercially used anode material is lithiated graphite, Li_xC. The latter has a layered structure, where each carbon atom has three in-plane bonds and half-filled p_z orbitals perpendicular to the layers, which provides interaction with Li 2s orbitals. Although such a layered structure grants a reasonable capacity, the transfer of Li^+-ions through the solid–electrolyte interface layer limits the safe recharging rate of a carbon anode.[314] Therefore, replacing Li metal or lithiated graphite as the anode material in the electrochemical cell is an important task in contemporary research on energy storage.[316]

Binary and ternary pnictides are intensively studied to create novel anode materials. Although a lot of research has focused on nitrides, heavier pnictides were studied as well. From the application perspective, As-, Sb-, and Bi-based materials have several disadvantages, such as toxicity, flammability, limited abundance, and large atomic weight, which reduces specific capacity.[317]

Studies on electrochemical behaviour of transition metal pnictides MPn_y (Pn = P, As) upon lithiation, suggest that the following reaction takes place: $MPn_y + 3yLi = yLi_3Pn + M$. In the course of this process, MPn_y converts into a composite electrode, with metallic nanosized M particles embedded in a Li_3Pn matrix.[318] For a more detailed description, we can refer readers to a critical review on nanostructured binary pnictide materials for battery application, structural considerations for intermetallic electrode materials,[318,319] and a comprehensive review on the crystal structures and applications of lithium transition metal pnictides, which can also be insightful.[320]

Past research efforts were focused on 3d transition metal phosphides and polyphosphides, which were investigated for their electrochemical performance in Li-ion batteries. Binary composites and nanoparticles were among the most widely studied materials. They often are intermediate products of lithiation, although some multinary phases also exhibit excellent performance. For instance, several binary phosphides, such as MnP,[321] FeP_x (x = 1, 2, 4),[322] Cu_3P,[323] and Ni_5P_4[324] have been suggested to produce ternary phases in both the lithiation and delithiation processes. TiP_2/P composites showed an initial starting capacity of 1422 mA h g^{-1}, which decreased to 625 mA h g^{-1} after 80 cycles,[325] whereas black phosphorus@TiP_2–C nanocomposites exhibit much better capacity retention of *ca.* 925 mA h g^{-1} after 500 cycles.[326] NiP_2 goes through the formation of an intermediate $Li_{2.5}NiP_2$, which decomposes to a Li_3P/Ni nanocomposite.[327,328] Amorphous FeP_2, produced by low-temperature synthesis, shows remarkable performance toward electrochemical lithiation/delithiation, with reported capacities of 1258 and 766 mA h g^{-1} and discharge capacity of 906 mA h g^{-1} after 10 cycles.[329]

Cobalt phosphides were also studied for their electrochemical performance as anode materials. Co_2P was found to be a good candidate for alkaline rechargeable batteries, with a stable discharge capacity of 223.5 mA h g^{-1} after the 300th cycle.[330] Nanosized and carbon-coated Co_2P and CoP materials exhibit even better performance with reported capacities of 527 mA h g^{-1}

and 630 mA h g^{-1}, respectively, after 100 cycles.[331] An interesting case is observed for the skutterudite-type CoP_3, which intakes nine Li ions per formula unit, leading to an initial capacity of more than 1000 mA h g^{-1}. However, the first cycle results in the formation of Co and Li_3P, which reduces the capacity down to 487 mA h g^{-1}.[332,333] Isostructural antimonides sometimes exhibit comparable specific capacity, such as 770 mA h g^{-1} as reported for a $CoSb_3$/graphite composite anode.[333,334]

Phosphorus-rich CuP_2 undergoes an irreversible reaction with lithium, giving metal-rich Cu_3P and amorphous phosphorus. As a result, a high capacity of more than 1000 mA h g^{-1} was reported, yet the stability of the pure CuP_2 electrode is rather poor. However, the stability can be noticeably improved for a CuP_2/C composite anode, which exhibits an outstanding reversible capacity of more than 720 mA h g^{-1}.[335]

Another interesting study uncovered reversible low-potential intercalation of lithium in the manganese phosphide MnP_4. Upon lithiation, the phosphide converts into Li_7MnP_4 with an antifluorite-type structure, which is accompanied by breaking of the P–P bonds (Figure 2.31).[336] However, the initial high capacity of such a system (700 mA h g^{-1}) noticeably decreases in subsequent cycles because of the slow but gradual formation of Li_3P and Mn.

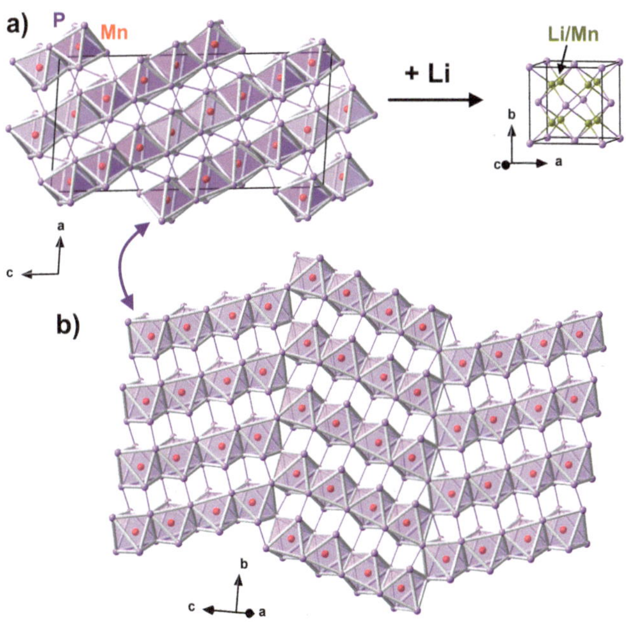

Figure 2.31 Crystal structure of MnP_4 (space group *C*2/*c*, Pearson code *mS*80) represented as layers of interconnected [MnP_6] octahedra and a network of P–P bonds ((a) left). The product of MnP_4 lithiation, Li_7MnP_4 (space group $Fm\bar{3}m$, Pearson code *cF*12) with disordered distribution of Li and Mn ((a), right). A single layer in the structure of MnP_4 (b).

Several other transition metal pnictides were studied in lithiation reactions as well. For instance, electrochemical Li^+ exchange in the transition metal pnictides Li_xMPn_4 (M = Ti, V; Pn = P, As; x = 3–11) involves a structural transformation mechanism that reconciles large electrode capacity and small electrode volume change.[337,338] These materials exhibit high specific capacities (C_{max} = 830 mA h g^{-1} for Li_9TiP_4) and a potential close to 1 V. Additional examinations on the lithium insertion mechanism were performed by electron energy-loss spectroscopy.[338] Recent studies revealed the applicability of phosphorus-rich phases, such as MnP_4, for sodium-ion batteries, although in this case, the electrochemical intercalation proceeds predominantly through complete reduction of the transition metal phosphide to elemental metal.[339]

Attempts to develop antimony-based negative electrodes are challenging. Li_3Sb is the common final product in the lithiation process.[340] For instance, lithiation of Cu_2Sb results in the simultaneous incorporation of Li and extrusion of Cu with the formation of LiCuSb, having a half-Heusler-type structure. Further lithiation increases Cu extrusion with the formation of the full Heusler-type $Li_{2-x}CuSb$, and, finally, Li_3Sb. Large volume changes for Sb-bearing intermetallics are often observed during the lithiation/delithiation process.[341] It results in cracks and degradation, which leads to contact problems and poor intercalation behaviour. We refer the reader to the comprehensive review of polyantimonides in battery science, which summarises many of the research efforts performed in the field of Sb-bearing materials.[342]

Several intermetallic antimony alloys were reported to be promising anode materials. Good performance has been reported for the TiSnSb phase, with a specific capacity of *ca.* 550 mA h g^{-1} after 240 cycles and 450 mA h g^{-1} after 400 cycles at 60 °C.[343,344] Studies on the Nb-substituted compositions $Ti_{1-x}Nb_xSnSb$ under lithiation conditions revealed the following reaction scheme MSnSb + 6.5Li = Li_3Sb + $0.5Li_7Sn_2$ + M (M = Ti + Nb).[345] Another antimonide, $ZnSnSb_2$, displays a high capacity of 615 mA h g^{-1} for 200 cycles and exhibits outstanding rate capability, long cycling life, and great resistance to the volume expansion.[346]

Due to the high price and scarcity of Li, a considerable amount of research has been devoted to developing Na-ion batteries (SIBs). Several pnictide anode materials have been proposed for these applications as well.[347–352] For instance, phosphorus and phosphides are auspicious anode materials for SIBs due to the natural abundance of P, low/moderate sodiation potential, and high theoretical capacity for Na storage.[347] Well-studied systems are metal phosphides, non-metal phosphides, and mixed-element phosphides in form of hierarchical structures, core–shell structures, and nanoarchitectures.[350,352] While several cited comprehensive reviews describe recent advances in the field, we would like to mention some selected systems. Among them, the binary phase SiP_2, with the pyrite structure, provides outstanding capacity in both lithium and sodium batteries with values up to 1000 mA h g^{-1} and 572 mA h g^{-1} sustained after 30

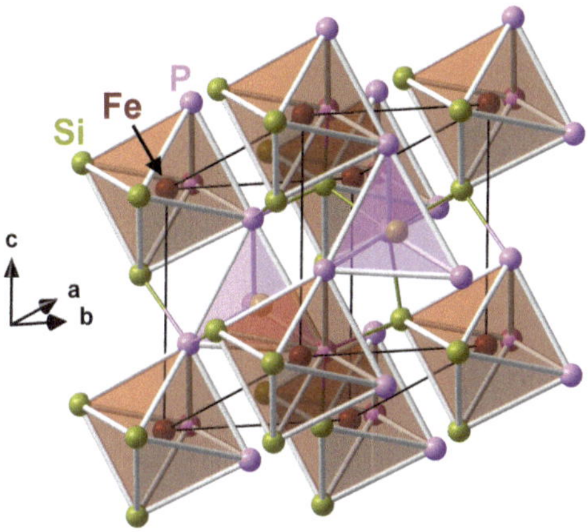

Figure 2.32 Crystal structure of $FeSi_4P_4$ (space group *P*1, Pearson code *aP*9) viewed as an interconnection of [$FeSi_3P_3$] octahedra (brown) and [SiP_4] tetrahedra (pink). The unit cell is outlined in black.

and 15 cycles, respectively.[353] A ZnP_2-C composite shows high electrochemical performance with a highly reversible capacity of 883 mA h g^{-1} after 130 cycles with no capacity deterioration.[354] Electrochemical performance of $FeSi_4P_4$ (Figure 2.32) shows great results for Li, with the retained capacity of 1120 mA h g^{-1} after 40 cycles, yet only moderate capacity of 350 mA h g^{-1} was achieved in sodiation/desodiation cycles.[355] In contrast, the difference is not that significant for the NiP_3 electrode, which possesses a promising capacity of *ca.* 1000 mA h g^{-1} after 50 cycles for Li and 900 mA h g^{-1} after 15 cycles for Na.[356]

One of the most promising materials for sodium batteries is the binary rhombohedral Sn_4P_3 phase (Figure 2.33(a)),[357] which shows a high reversible capacity of *ca.* 700 mA h g^{-1} after 100 cycles, further improved for Sn_4P_3/C composites (Figure 2.33(b)).[358–360] Negligible capacity fading is attributed to the confinement effect of Sn nanocrystallites in the amorphous phosphorus matrix forming in the sodiation process. Other reported products of sodiation are $Na_xSn_4P_3$, Na_3P, Na_5Sn_2, and $Na_{15}Sn_4$.[361,362]

A recent report indicates that a binder-free Si/NiSb alloy shows excellent cycle stability of 521 mA h g^{-1} upon 100 cycles suggesting the suitability of this antimonide-based material for SIBs.[363,364]

In conclusion, pnictide compounds clearly show potential for becoming reasonably good anode materials due to their encouraging properties. Although performance of pure pnictides is far from being optimal for Li-ion batteries, they can be used as components of nanostructured composite systems, which may enhance the properties of known systems. Importantly, pnictide-based systems show great promise for the rapidly growing field of Na-ion batteries, which may be important for future electrochemical energy storage technologies.[364–366]

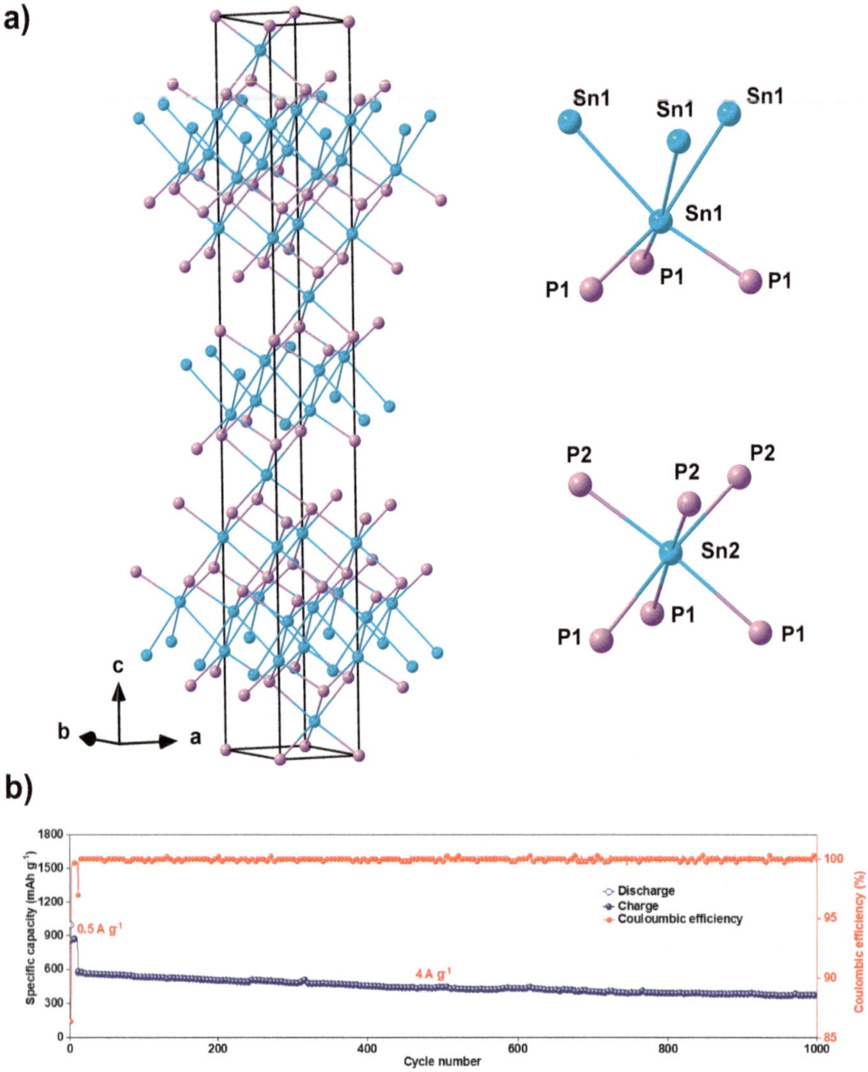

Figure 2.33 Crystal structure of Sn_4P_3 (a) (space group $R\bar{3}m$, Pearson code *hR*7, left) and local coordination of the two symmetry independent Sn positions ((a) right). The unit cell is outlined in black. Electrochemical data for a Sn_4P_3/C composite (b): long term cyclic performance and Coulombic efficiency for 1000 cycles at 4 A g^{-1} in 1 M $NaPF_6$ solution in 1,2-dimethoxyethane. Reproduced from ref. 359 with permission from Elsevier, Copyright 2022.

2.10 Other Applications

Some of the most common applications of pnictide materials were discussed in previous sections. However, there are many rising and exciting uses, which deserve separate brief attention. Most of the topics discussed below are also related to energy and sustainability.

A number of pnictides with noncentrosymmetric crystal structures have received attention due to potential nonlinear optical (NLO) applications in the infrared region.[367] In general, the absence of inversion symmetry is a prerequisite for emergence of nonlinear optical properties, such as second harmonic generation (SHG). NLO materials can upconvert incident light frequencies, but their development is challenging since many criteria must be met for a good NLO candidate. For instance, IR-NLO crystals should possess strong second harmonic generation intensity, significant laser damage threshold, wide IR transparency range, and thermal and chemical stability.

Nowadays, mostly chalcogenides are in focus for IR-NLO applications. However, some pnictides, in particular, the chalcopyrite-type $ZnGeP_2$, are commercially available for these purposes.[368] In addition to the compositionally and structurally related $ATtPn_2$ tetrel pnictides (A = Mg, Mn, Zn, Cd, Ba; Tt = Si, Ge; Pn = P, As),[368–372] several other binary and ternary noncentrosymmetric semiconductors, such as CdP_2, $Ba_2Si_3P_6$, $RuSi_4P_4$, $IrSi_3P_3$, Ca_2CdP_2, $BaGe_2Pn_2$ (Pn = P, As), $LaSiP_3$, and $LaSi_2P_6$, as well as the quaternary compounds $Ba_4AgGa_5P_8$ and $MgIn_3Si_2P_7$ were found to be promising IR-NLO materials.[373–379]

The antimonide Cs_3Sb and its derivatives are used as photocathodes for several applications, such as radio-frequency photoinjectors and photomultipliers.[380] Common photoemitters, Cs_3Sb and CsK_2Sb, are intrinsic p-type semiconductors with $E_g + E_a$ of around 2 eV, where E_g is the band gap and E_a is the electron affinity. Cs_3Sb was reported to crystallise in the partially disordered cubic NaTl structure, whereas CsK_2Sb adopts a full-Heusler structure.[381,382] The CsK_2Sb photocathode can generate a high-intensity and low-emittance electron beam with visible laser light. Since surface chemistry can profoundly affect photocathodes, many studies related to the growth and characterisation of oriented, epitaxial alkali antimonide films have been conducted.[383,384]

The cubic boron arsenide phase BAs, with the sphalerite type structure, has recently been shown to demonstrate ultra-high thermal conductivity (κ), which offers a great promise for power electronics. Reported values of *ca.* 1300 W m^{-1} K^{-1} were achieved at room temperatures for ultra-pure samples,[385] while predicted thermal conductivity exceeds 2000 W m^{-1} K^{-1}. The latter value is close to the room temperature κ of diamond—the material with the highest measured thermal conductivity.[386] The main challenge for producing BAs samples with really high κ is the formation of defects during synthesis, which hinder heat flow. However, controlled doping by tetrel elements, such as Si and Ge on As site is predicted to improve heat conductance in BAs.[387]

The discovery of novel materials with good microwave absorption can solve several security problems, such as radar shielding. Traditional microwave-absorbing materials include carbon-based nanomaterials and ferrite materials, with the primary mechanism of dipole rotations and magnetic domain resonance. Among pnictides, FeP nanoparticles were shown to be promising for microwave absorption applications.[388] Ni_2P nanocrystals coated on carbon nanotubes also exhibit excellent microwave-absorbing properties.[389] The reflection loss value as high as −35.43 dB was achieved for the modified complex composite materials made of Mo-doped Ni_2P on reduced graphene oxide.[390]

While most of the applications described above make use of the physical properties exhibited by pnictides, the chemical behaviour of these compounds also finds applications, sometimes in rather unexpected areas. Thus, several binary metal pnictides are utilised as toxic substances for pest control. In particular, Zn_3P_2, Ca_3P_2, Mg_3P_2, and AlP are used as rodenticides. Toxicity of these phosphides is related to the reaction with water or acid with the formation of the toxic phosphine gas, PH_3.[391,392]

2.11 Conclusions

In this chapter, we have highlighted some of the most common applications and actively studied properties of pnictide materials. Compounds of this class have already found usage in a variety of technologies and are enjoying considerable attention due to their potential applicability in emerging fields. The diversity of crystal structures and synthetic approaches provides high flexibility for design of new materials containing elements of Group 15. Recent developments in computational materials discovery will likely speed up the development of new pnictides with interesting properties. However, the crystal structural variability and complexity, which lend these materials their wide spectrum of qualities, makes purely theoretical predictions challenging, if not impossible in certain cases. For this reason, exploratory synthesis, aided by *ab initio* approaches and such simple, yet well-established, chemical guidelines as the Zintl approach will likely remain the main tool for discovery of new pnictides.

References

1. L. Öhrström and J. Reedijk, *Pure Appl. Chem.*, 2016, **88**, 1225.
2. A. Ovchinnikov, V. Smetana and A.-V. Mudring, *J. Phys.: Condens. Matter*, 2020, **32**, 243002.
3. R. Nesper, *Z. Anorg. Allg. Chem.*, 2014, **640**, 2639.
4. A. Ovchinnikov and S. Bobev, *J. Solid State Chem.*, 2019, **270**, 346.
5. T. Graf, C. Felser and S. S. P. Parkin, *Prog. Solid State Chem.*, 2011, **39**, 1.
6. G. A. Papoian and R. Hoffmann, *Angew. Chem., Int. Ed.*, 2000, **39**, 2408.
7. F. Gascoin and S. C. Sevov, *Angew. Chem., Int. Ed.*, 2002, **41**, 1232.
8. S. Baranets, H. He and S. Bobev, *Acta Crystallogr., Sect. C: Struct. Chem.*, 2018, **74**, 623.
9. R. Frankovsky, A. Marchuk, R. Pobel and D. Johrendt, *Solid State Commun.*, 2012, **152**, 632.
10. G. Akopov, J. Mark, G. Viswanathan, S. J. Lee, B. C. McBride, J. Won, F. A. Perras, A. L. Paterson, B. Yuan, S. Sen, A. N. Adeyemi, F. Zhang,

C.-Z. Wang, K.-M. Ho, G. J. Miller and K. Kovnir, *Dalton Trans.*, 2021, **50**, 6463.

11. S. M. Clarke, J. P. S. Walsh, M. Amsler, C. D. Malliakas, T. Yu, S. Goedecker, Y. Wang, C. Wolverton and D. E. Freedman, *Angew. Chem., Int. Ed.*, 2016, **55**, 13446.
12. K. Guo, L. Akselrud, M. Bobnar, U. Burkhardt, M. Schmidt, J.-T. Zhao, U. Schwarz and Y. Grin, *Angew. Chem., Int. Ed.*, 2017, **56**, 5620.
13. J. P. S. Walsh, S. M. Clarke, Y. Meng, S. D. Jacobsen and D. E. Freedman, *ACS Cent. Sci.*, 2016, **2**, 867.
14. P. C. Canfield and Z. Fisk, *Philos. Mag. B*, 1992, **65**, 1117.
15. M. G. Kanatzidis, R. Pöttgen and W. Jeitschko, *Angew. Chem., Int. Ed.*, 2005, **44**, 6996.
16. D. C. Johnston, *Adv. Phys.*, 2010, **59**, 803.
17. F. Steckel, M. Roslova, R. Beck, I. Morozov, S. Aswartham, D. Evtushinsky, C. G. F. Blum, M. Abdel-Hafiez, D. Bombor, J. Maletz, S. Borisenko, A. V. Shevelkov, A. U. B. Wolter, C. Hess, S. Wurmehl and B. Büchner, *Phys. Rev. B:Condens. Matter Mater. Phys.*, 2015, **91**, 184516.
18. M. Binnewies, R. Glaum, M. Schmidt and P. Schmidt, *Chemical Vapor Transport Reactions*, De Gruyter, Berlin Boston, 2012.
19. E. J. Popczun, J. R. McKone, C. G. Read, A. J. Biacchi, A. M. Wiltrout, N. S. Lewis and R. E. Schaak, *J. Am. Chem. Soc.*, 2013, **135**, 9267.
20. A. E. Henkes, Y. Vasquez and R. E. Schaak, *J. Am. Chem. Soc.*, 2007, **129**, 1896.
21. A. Datta and G. S. Nolas, *Eur. J. Inorg. Chem.*, 2012, **2012**, 55.
22. M. A. White, M. J. Thompson, G. J. Miller and J. Vela, *Chem. Commun.*, 2016, **52**, 3497.
23. R. J. Wilson, B. Weinert and S. Dehnen, *Dalton Trans.*, 2018, **47**, 14861.
24. R. J. Wilson, N. Lichtenberger, B. Weinert and S. Dehnen, *Chem. Rev.*, 2019, **119**, 8506.
25. M. J. Moses, J. C. Fettinger and B. W. Eichhorn, *Science*, 2003, **300**, 778.
26. H. Ohno, A. Shen, F. Matsukura, A. Oiwa, A. Endo, S. Katsumoto and Y. Iye, *Appl. Phys. Lett.*, 1996, **69**, 363.
27. P. Y. Yu and M. Cardona, *Fundamentals of Semiconductors*, Springer, Berlin Heidelberg, Germany, 4th edn, 2010.
28. E. K. Arushanov, *Prog. Cryst. Growth Charact. Mater.*, 1992, **25**, 131.
29. V. M. Trukhan, A. D. Izotov and T. V. Shoukavaya, *Inorg. Mater.*, 2014, **50**, 868.
30. E. J. Luber, M. H. Mobarok and J. M. Buriak, *ACS Nano*, 2013, **7**, 8136.
31. A. M. Hermann, A. Madan, M. W. Wanlass, V. Badri, R. Ahrenkiel, S. Morrison and C. Gonzalez, *Sol. Energy Mater. Sol. Cells*, 2004, **82**, 241.
32. J. S. Blakemore, *J. Appl. Phys.*, 1982, **53**, R123.
33. M. A. Green and A. W. Y. Ho-Baillie, *ACS Energy Lett.*, 2019, **4**, 1639.
34. M. A. Green, E. D. Dunlop, J. Hohl-Ebinger, M. Yoshita, N. Kopidakis and X. Hao, *Prog. Photovoltaics*, 2021, **29**, 657.

35. M. R. Krames, O. B. Shchekin, R. Mueller-Mach, G. O. Mueller, L. Zhou, G. Harbers and M. G. Craford, *J. Disp. Technol.*, 2007, **3**, 160.
36. E. G. Gillan, in *Comprehensive Inorganic Chemistry II*, ed. J. Reedijk and K. Poeppelmeier, Elsevier, Amsterdam, the Netherlands, 2nd edn, 2013, pp. 969–1000.
37. H. Anabestani, R. Shazzad, M. F. A. Fattah, J. Therrien and D. Ban, *Mater. Today Commun.*, 2021, **28**, 102542.
38. I. Hayashi and M. B. Panish, *J. Appl. Phys.*, 1970, **41**, 150.
39. Z. Chung-En, R. Bhat, B. N. Pathak, F. Favire, L. Wei, M. C. Wang, N. C. Andreadakis, D. M. Hwang, M. A. Koza, L. Tein-Pei, W. Zheng, D. Darby, D. Flanders and J. J. Heieh, *IEEE J. Quantum Electron.*, 1994, **30**, 511.
40. E. Hulicius and V. Kubeček, in *Lasers for Medical Applications*, ed. H. Jelínková, Woodhead Publishing Series in Electronic and Optical Materials, Elsevier, Amsterdam, the Netherlands, 2013, pp. 222–250.
41. *Gallium Arsenide IC Applications Handbook*, ed. D. Fisher and I. Bahl, Elsevier, Amsterdam, the Netherlands, 1995.
42. R. N. Nottenburg, J. Bischoff, M. B. Panish and H. Temkin, *IEEE Electron Device Lett.*, 1987, **8**, 282.
43. R. L. Bates, M. Campbell, S. D'Auria, C. Da Vià, V. O'Shea, C. Raine and K. M. Smith, *Nucl. Instrum. Methods Phys. Res., Sect. A*, 1997, **392**, 269.
44. D. G. Avery, D. W. Goodwin and M. A. E. Rennie, *J. Sci. Instrum.*, 1957, **34**, 394.
45. S. C. Abrahams and J. L. Bernstein, *J. Chem. Phys.*, 1971, **55**, 796.
46. S. C. Abrahams and J. L. Bernstein, *J. Chem. Phys.*, 1970, **52**, 5607.
47. A. D. Martinez, A. N. Fioretti, E. S. Toberer and A. C. Tamboli, *J. Mater. Chem. A*, 2017, **5**, 11418.
48. G. He, I. Rozahun, Z. Li, J. Zhang and M.-H. Lee, *Chem. Phys.*, 2019, **518**, 101.
49. J. He and T. M. Tritt, *Science*, 2017, **357**, eaak9997.
50. G. Tan, L.-D. Zhao and M. G. Kanatzidis, *Chem. Rev.*, 2016, **116**, 12123.
51. T. Zhu, Y. Liu, C. Fu, J. P. Heremans, J. G. Snyder and X. Zhao, *Adv. Mater.*, 2017, **29**, 1605884.
52. J. Yang, L. Xi, W. Qiu, L. Wu, X. Shi, L. Chen, J. Yang, W. Zhang, C. Uher and D. J. Singh, *npj Comput. Mater.*, 2016, **2**, 15015.
53. C. Xiao, Z. Li, K. Li, P. Huang and Y. Xie, *Acc. Chem. Res.*, 2014, **47**, 1287.
54. G. A. Slack, in *Handbook of Thermoelectrics*, ed. D. M. Rowe, CRC Press, Boca Raton, FL, 1995, pp. 407–440.
55. H. C. Kandpal, C. Felser and R. Seshadri, *J. Phys. D: Appl. Phys.*, 2006, **39**, 776.
56. K. Xia, C. Hu, C. Fu, X. Zhao and T. Zhu, *Appl. Phys. Lett.*, 2021, **118**, 140503.
57. S. Chen and Z. Ren, *Mater. Today*, 2013, **16**, 387.
58. J. Yu, C. Fu, Y. Liu, K. Xia, U. Aydemir, T. C. Chasapis, G. J. Snyder, X. Zhao and T. Zhu, *Adv. Energy Mater.*, 2018, **8**, 1701313.

59. Y. Liu, C. Fu, K. Xia, J. Yu, X. Zhao, H. Pan, C. Felser and T. Zhu, *Adv. Mater.*, 2018, **30**, 1800881.
60. Z.-Y. Liu, J.-L. Zhu, X. Tong, S. Niu and W.-Y. Zhao, *J. Adv. Ceram.*, 2020, **9**, 647.
61. G. Rogl and P. Rogl, *Curr. Opin. Green Sustainable Chem.*, 2017, **4**, 50.
62. X. Tong, Z. Liu, J. Zhu, T. Yang, Y. Wang and A. Xia, *Front. Mater. Sci.*, 2021, **15**, 317.
63. P.-X. Lu, Z.-G. Shen and X. Hu, *Phys. B*, 2010, **405**, 1740.
64. B. C. Sales, D. Mandrus, B. C. Chakoumakos, V. Keppens and J. R. Thompson, *Phys. Rev. B:Condens. Matter Mater. Phys.*, 1997, **56**, 15081.
65. B. C. Sales, D. Mandrus and R. K. Williams, *Science*, 1996, **272**, 1325.
66. S. Peng, J. Sun, B. Cui, X. Meng, D. Qin, Z. Liu and W. Cai, *Inorg. Chem. Front.*, 2017, **4**, 1697.
67. S. Ballikaya, N. Uzar, S. Yildirim, H. Chi, X. Su, G. Tan, X. Tang and C. Uher, *J. Electron. Mater.*, 2013, **42**, 1622.
68. T. Dahal, S. Gahlawat, Q. Jie, K. Dahal, Y. Lan, K. White and Z. Ren, *J. Appl. Phys.*, 2015, **117**, 055101.
69. G. Rogl, A. Grytsiv, P. Rogl, N. Peranio, E. Bauer, M. Zehetbauer and O. Eibl, *Acta Mater.*, 2014, **63**, 30.
70. G. Rogl, A. Grytsiv, K. Yubuta, S. Puchegger, E. Bauer, C. Raju, R. C. Mallik and P. Rogl, *Acta Mater.*, 2015, **95**, 201.
71. S. Wang, J. R. Salvador, J. Yang, P. Wei, B. Duan and J. Yang, *NPG Asia Mater.*, 2016, **8**, e285.
72. X. Shi, J. Yang, J. R. Salvador, M. Chi, J. Y. Cho, H. Wang, S. Bai, J. Yang, W. Zhang and L. Chen, *J. Am. Chem. Soc.*, 2011, **133**, 7837.
73. G. Rogl and P. Rogl, *Mater. Today Phys.*, 2017, **3**, 48.
74. T. Takabatake, K. Suekuni, T. Nakayama and E. Kaneshita, *Rev. Mod. Phys.*, 2014, **86**, 669.
75. H. Kleinke, *Chem. Mater.*, 2010, **22**, 604.
76. *The Physics and Chemistry of Inorganic Clathrates*, ed. G. S. Nolas, Springer, Dordrecht, the Netherlands, 2014.
77. J.-A. Dolyniuk, B. Owens-Baird, J. Wang, J. V. Zaikina and K. Kovnir, *Mater. Sci. Eng., R*, 2016, **108**, 1.
78. H. He, A. Zevalkink, Z. M. Gibbs, G. J. Snyder and S. Bobev, *Chem. Mater.*, 2012, **24**, 3596.
79. T. Cox, V. Gvozdetskyi, M. Bertolami, S. Lee, K. Shipley, O. I. Lebedev and J. V. Zaikina, *Angew. Chem., Int. Ed.*, 2021, **60**, 415.
80. B. Owens-Baird, P. Yox, S. Lee, X. B. Carroll, S. G. Wang, Y.-S. Chen, O. I. Lebedev and K. Kovnir, *Chem. Sci.*, 2020, **11**, 10255.
81. J. Fulmer, O. I. Lebedev, V. V. Roddatis, D. C. Kaseman, S. Sen, J.-A. Dolyniuk, K. Lee, A. V. Olenev and K. Kovnir, *J. Am. Chem. Soc.*, 2013, **135**, 12313.
82. J. Dolyniuk, P. S. Whitfield, K. Lee, O. I. Lebedev and K. Kovnir, *Chem. Sci.*, 2017, **8**, 3650.
83. J. Wang, O. I. Lebedev, K. Lee, J.-A. Dolyniuk, P. Klavins, S. Bux and K. Kovnir, *Chem. Sci.*, 2017, **8**, 8030.

84. B. Owens-Baird, J. Wang, S. G. Wang, Y.-S. Chen, S. Lee, D. Donadio and K. Kovnir, *J. Am. Chem. Soc.*, 2020, **142**, 2031.
85. J. Wang, J. Voyles, S. Grzybowski and K. Kovnir, *J. Appl. Phys.*, 2020, **127**, 055104.
86. J. Wang, Y. He, N. E. Mordvinova, O. I. Lebedev and K. Kovnir, *Chem*, 2018, **4**, 1465.
87. S.-J. Kim, S. Hu, C. Uher and M. G. Kanatzidis, *Chem. Mater.*, 1999, **11**, 3154.
88. S.-J. Kim, J. R. Ireland, C. R. Kannewurf and M. G. Kanatzidis, *J. Solid State Chem.*, 2000, **155**, 55.
89. N. Kazem and S. M. Kauzlarich, in *Handbook on the Physics and Chemistry of Rare Earths*, ed. J.-C. G. Bünzli and V. K. Pecharsky, Elsevier, 2016, vol. 50, pp. 177–208.
90. E. S. Toberer, A. F. May and G. J. Snyder, *Chem. Mater.*, 2010, **22**, 624.
91. S. M. Kauzlarich, A. Zevalkink, E. Toberer and G. J. Snyder, in *Thermoelectric Materials and Devices*, ed. I. Nandhakumar, N. M. White and S. Beeby, The Royal Society of Chemistry, Cambridge, UK, 2017, pp. 1–26.
92. K.-F. Liu and S.-Q. Xia, *J. Solid State Chem.*, 2019, **270**, 252.
93. J. Shuai, J. Mao, S. Song, Q. Zhang, G. Chen and Z. Ren, *Mater. Today Phys.*, 2017, **1**, 74.
94. S. R. Brown, S. M. Kauzlarich, F. Gascoin and G. J. Snyder, *Chem. Mater.*, 2006, **18**, 1873.
95. J. H. Grebenkemper, Y. Hu, D. Barrett, P. Gogna, C.-K. Huang, S. K. Bux and S. M. Kauzlarich, *Chem. Mater.*, 2015, **27**, 5791.
96. E. S. Toberer, C. A. Cox, S. R. Brown, T. Ikeda, A. F. May, S. M. Kauzlarich and G. J. Snyder, *Adv. Funct. Mater.*, 2008, **18**, 2795.
97. S. R. Brown, E. S. Toberer, T. Ikeda, C. A. Cox, F. Gascoin, S. M. Kauzlarich and G. J. Snyder, *Chem. Mater.*, 2008, **20**, 3412.
98. E. S. Toberer, S. R. Brown, T. Ikeda, S. M. Kauzlarich and G. J. Snyder, *Appl. Phys. Lett.*, 2008, **93**, 062110.
99. A. P. Justl, G. Cerretti, S. K. Bux and S. M. Kauzlarich, *J. Appl. Phys.*, 2019, **126**, 165106.
100. W. Peng, S. Chanakian and A. Zevalkink, *Inorg. Chem. Front.*, 2018, **5**, 1744.
101. X.-J. Wang, M.-B. Tang, H.-H. Chen, X.-X. Yang, J.-T. Zhao, U. Burkhardt and Y. Grin, *Appl. Phys. Lett.*, 2009, **94**, 092106.
102. J. Shuai, Z. Liu, H. S. Kim, Y. Wang, J. Mao, R. He, J. Sui and Z. Ren, *J. Mater. Chem. A*, 2016, **4**, 4312.
103. C. L. Condron, S. M. Kauzlarich, F. Gascoin and G. J. Snyder, *J. Solid State Chem.*, 2006, **179**, 2252.
104. H. Tamaki, H. K. Sato and T. Kanno, *Adv. Mater.*, 2016, **28**, 10182.
105. J. Zhang, L. Song, S. H. Pedersen, H. Yin, L. T. Hung and B. B. Iversen, *Nat. Commun.*, 2017, **8**, 13901.
106. S. Baranets, A. Ovchinnikov and S. Bobev, in *Handbook on the Physics and Chemistry of Rare Earths*, ed. J.-C. G. Bünzli and V. K. Pecharsky, Elsevier, Amsterdam, the Netherlands, 2021, vol. 60, pp. 227–324.
107. S. Ohno, U. Aydemir, M. Amsler, J.-H. Pöhls, S. Chanakian, A. Zevalkink, M. A. White, S. K. Bux, C. Wolverton and G. J. Snyder, *Adv. Funct. Mater.*, 2017, **27**, 1606361.

108. E. S. Toberer, A. Zevalkink, N. Crisosto and G. J. Snyder, *Adv. Funct. Mater.*, 2010, **20**, 4375.
109. A. Zevalkink, G. S. Pomrehn, S. Johnson, J. Swallow, Z. M. Gibbs and G. J. Snyder, *Chem. Mater.*, 2012, **24**, 2091.
110. A. Zevalkink, E. S. Toberer, W. G. Zeier, E. Flage-Larsen and G. J. Snyder, *Energy Environ. Sci.*, 2011, **4**, 510.
111. A. Zevalkink, W. G. Zeier, G. Pomrehn, E. Schechtel, W. Tremel and G. J. Snyder, *Energy Environ. Sci.*, 2012, **5**, 9121.
112. C. Chen, W. Xue, S. Li, Z. Zhang, X. Li, X. Wang, Y. Liu, J. Sui, X. Liu, F. Cao, Z. Ren, C.-W. Chu, Y. Wang and Q. Zhang, *Proc. Natl. Acad. Sci. U. S. A.*, 2019, **116**, 2831.
113. C. Chen, X. Li, W. Xue, F. Bai, Y. Huang, H. Yao, S. Li, Z. Zhang, X. Wang, J. Sui, X. Liu, F. Cao, Y. Wang and Q. Zhang, *Nano Energy*, 2020, **73**, 104771.
114. Q. Liu, X.-C. Liu, J. Zhang, K.-F. Liu and S.-Q. Xia, *ACS Appl. Mater. Interfaces*, 2021, **13**, 17809.
115. R. Pothin, R. M. Ayral, A. Berche, D. Granier, F. Rouessac and P. Jund, *Chem. Eng. J.*, 2016, **299**, 126.
116. M. Chitroub, F. Besse and H. Scherrer, *J. Alloys Compd.*, 2008, **460**, 90.
117. G. J. Snyder, M. Christensen, E. Nishibori, T. Caillat and B. B. Iversen, *Nat. Mater.*, 2004, **3**, 458.
118. T. Caillat, J. P. Fleurial and A. Borshchevsky, *J. Phys. Chem. Solids*, 1997, **58**, 1119.
119. Y. Mozharivskyj, A. O. Pecharsky, S. Bud'ko and G. J. Miller, *Chem. Mater.*, 2004, **16**, 1580.
120. C.-W. T. Lo, B. R. Ortiz, E. S. Toberer, A. He, V. Svitlyk, D. Chernyshov, T. Kolodiazhnyi, S. Lidin and Y. Mozharivskyj, *Chem. Mater.*, 2017, **29**, 5249.
121. J. Wang and K. Kovnir, *J. Am. Chem. Soc.*, 2015, **137**, 12474.
122. Y. Mozharivskyj, Y. Janssen, J. L. Harringa, A. Kracher, A. O. Tsokol and G. J. Miller, *Chem. Mater.*, 2006, **18**, 822.
123. C. Okamura, T. Ueda and K. Hasezaki, *Mater. Trans.*, 2010, **51**, 860.
124. D. Tang, W. Zhao, J. Yu, P. Wei, H. Zhou, W. Zhu and Q. Zhang, *J. Alloys Compd.*, 2014, **601**, 50.
125. S. Yoshioka, K. Hayashi, A. Yokoyama, W. Saito, H. Li, T. Takamatsu and Y. Miyazaki, *J. Mater. Chem. C*, 2020, **8**, 9205.
126. P. Sun, N. Oeschler, S. Johnsen, B. B. Iversen and F. Steglich, *Dalton Trans.*, 2010, **39**, 1012.
127. Q. Du and C. Petrovic, *Appl. Phys. Lett.*, 2021, **118**, 233901.
128. A. Bentien, S. Johnsen, G. K. H. Madsen, B. B. Iversen and F. Steglich, *EPL*, 2007, **80**, 17008.
129. W. J. de Haas, E. van Aubel and J. Voogd, *Proc. – Acad. Sci. Amsterdam*, 1929, **32**, 730.
130. W. J. de Haas and F. Jurriaanse, *Naturwissenschaften*, 1931, **19**, 706.
131. I. Shirotani, T. Uchiumi, K. Ohno, C. Sekine, Y. Nakazawa, K. Kanoda, S. Todo and T. Yagi, *Phys. Rev. B:Condens. Matter Mater. Phys.*, 1997, **56**, 7866.

132. I. Shirotani, K. Ohno, C. Sekine, T. Yagi, T. Kawakami, T. Nakanishi, H. Takahashi, J. Tang, A. Matsushita and T. Matsumoto, *Phys. B*, 2000, **281–282**, 1021.
133. H. Sugawara, S. Osaki, S. R. Saha, Y. Aoki, H. Sato, Y. Inada, H. Shishido, R. Settai, Y. Ōnuki, H. Harima and K. Oikawa, *Phys. Rev. B:Condens. Matter Mater. Phys.*, 2002, **66**, 220504.
134. B. Goshchitskii, S. Naumov, N. Kostromitina and A. Karkin, *Phys. C*, 2007, **460–462**, 691.
135. Y. Iwahashi, H. Sugawara, K.-i. Magishi, T. Saito, K. Koyama, R. Settai, Y. Ōnuki, G. Giester and P. Rogl, *J. Phys. Soc. Jpn.*, 2008, **77**, 219.
136. H. M. Tütüncü, E. Karaca and G. P. Srivastava, *Phys. Rev. B:Condens. Matter Mater. Phys.*, 2017, **95**, 214514.
137. M. B. Maple, P. C. Ho, V. S. Zapf, N. A. Frederick, E. D. Bauer, W. M. Yuhasz, F. M. Woodward and J. W. Lynn, *J. Phys. Soc. Jpn.*, 2002, **71**, 23.
138. Y. Kamihara, T. Watanabe, M. Hirano and H. Hosono, *J. Am. Chem. Soc.*, 2008, **130**, 3296.
139. G. R. Stewart, *Rev. Mod. Phys.*, 2011, **83**, 1589.
140. A. A. Kordyuk, *Low Temp. Phys.*, 2012, **38**, 888.
141. C. Wang, L. Li, S. Chi, Z. Zhu, Z. Ren, Y. Li, Y. Wang, X. Lin, Y. Luo, S. Jiang, X. Xu, G. Cao and Z. a. Xu, *EPL*, 2008, **83**, 67006.
142. R. Zhi-An, L. Wei, Y. Jie, Y. Wei, S. Xiao-Li, C. Zheng, C. Guang-Can, D. Xiao-Li, S. Li-Ling, Z. Fang and Z. Zhong-Xian, *Chin. Phys. Lett.*, 2008, **25**, 2215.
143. G. Wu, Y. L. Xie, H. Chen, M. Zhong, R. H. Liu, B. C. Shi, Q. J. Li, X. F. Wang, T. Wu, Y. J. Yan, J. J. Ying and X. H. Chen, *J. Phys.: Condens. Matter*, 2009, **21**, 142203.
144. P. Cheng, B. Shen, G. Mu, X. Zhu, F. Han, B. Zeng and H.-H. Wen, *EPL*, 2009, **85**, 67003.
145. S. Matsuishi, Y. Inoue, T. Nomura, H. Yanagi, M. Hirano and H. Hosono, *J. Am. Chem. Soc.*, 2008, **130**, 14428.
146. C. Wang, Z.-C. Wang, Y.-X. Mei, Y.-K. Li, L. Li, Z.-T. Tang, Y. Liu, P. Zhang, H.-F. Zhai, Z.-A. Xu and G.-H. Cao, *J. Am. Chem. Soc.*, 2016, **138**, 2170.
147. J. Zhao, Q. Huang, C. de la Cruz, S. Li, J. W. Lynn, Y. Chen, M. A. Green, G. F. Chen, G. Li, Z. Li, J. L. Luo, N. L. Wang and P. Dai, *Nat. Mater.*, 2008, **7**, 953.
148. M. Rotter, M. Tegel and D. Johrendt, *Phys. Rev. Lett.*, 2008, **101**, 107006.
149. K. Kihou, T. Saito, S. Ishida, M. Nakajima, Y. Tomioka, H. Fukazawa, Y. Kohori, T. Ito, S.-i. Uchida, A. Iyo, C.-H. Lee and H. Eisaki, *J. Phys. Soc. Jpn.*, 2010, **79**, 124713.
150. S. Nandi, M. G. Kim, A. Kreyssig, R. M. Fernandes, D. K. Pratt, A. Thaler, N. Ni, S. L. Bud'ko, P. C. Canfield, J. Schmalian, R. J. McQueeney and A. I. Goldman, *Phys. Rev. Lett.*, 2010, **104**, 057006.
151. M. Rotter, M. Pangerl, M. Tegel and D. Johrendt, *Angew. Chem., Int. Ed.*, 2008, **47**, 7949.
152. H. Chen, Y. Ren, Y. Qiu, W. Bao, R. H. Liu, G. Wu, T. Wu, Y. L. Xie, X. F. Wang, Q. Huang and X. H. Chen, *EPL*, 2009, **85**, 17006.

153. P. L. Alireza, Y. T. C. Ko, J. Gillett, C. M. Petrone, J. M. Cole, G. G. Lonzarich and S. E. Sebastian, *J. Phys.: Condens. Matter*, 2008, **21**, 012208.
154. A. Kreyssig, M. A. Green, Y. Lee, G. D. Samolyuk, P. Zajdel, J. W. Lynn, S. L. Bud'ko, M. S. Torikachvili, N. Ni, S. Nandi, J. B. Leão, S. J. Poulton, D. N. Argyriou, B. N. Harmon, R. J. McQueeney, P. C. Canfield and A. I. Goldman, *Phys. Rev. B:Condens. Matter Mater. Phys.*, 2008, **78**, 184517.
155. W. Yu, A. A. Aczel, T. J. Williams, S. L. Bud'ko, N. Ni, P. C. Canfield and G. M. Luke, *Phys. Rev. B:Condens. Matter Mater. Phys.*, 2009, **79**, 020511.
156. A. Iyo, K. Kawashima, T. Kinjo, T. Nishio, S. Ishida, H. Fujihisa, Y. Gotoh, K. Kihou, H. Eisaki and Y. Yoshida, *J. Am. Chem. Soc.*, 2016, **138**, 3410.
157. K. Kawashima, S. Ishida, H. Fujihisa, Y. Gotoh, K. Kihou, Y. Yoshida, H. Eisaki, H. Ogino and A. Iyo, *J. Phys. Chem. Lett.*, 2018, **9**, 868.
158. Y. Liu, Y.-B. Liu, Z.-T. Tang, H. Jiang, Z.-C. Wang, A. Ablimit, W.-H. Jiao, Q. Tao, C.-M. Feng, Z.-A. Xu and G.-H. Cao, *Phys. Rev. B:Condens. Matter Mater. Phys.*, 2016, **93**, 214503.
159. K. Iida, Y. Nagai, S. Ishida, M. Ishikado, N. Murai, A. D. Christianson, H. Yoshida, Y. Inamura, H. Nakamura, A. Nakao, K. Munakata, D. Kagerbauer, M. Eisterer, K. Kawashima, Y. Yoshida, H. Eisaki and A. Iyo, *Phys. Rev. B:Condens. Matter Mater. Phys.*, 2019, **100**, 014506.
160. J.-K. Bao, K. Willa, M. P. Smylie, H. Chen, U. Welp, D. Y. Chung and M. G. Kanatzidis, *Cryst. Growth Des.*, 2018, **18**, 3517.
161. Z.-C. Wang, C.-Y. He, S.-Q. Wu, Z.-T. Tang, Y. Liu and G.-H. Cao, *Chem. Mater.*, 2017, **29**, 1805.
162. Z. Wang, C. He, Z. Tang, S. Wu and G. Cao, *Sci. China Mater.*, 2017, **60**, 83.
163. X. C. Wang, Q. Q. Liu, Y. X. Lv, W. B. Gao, L. X. Yang, R. C. Yu, F. Y. Li and C. Q. Jin, *Solid State Commun.*, 2008, **148**, 538.
164. S. V. Borisenko, V. B. Zabolotnyy, D. V. Evtushinsky, T. K. Kim, I. V. Morozov, A. N. Yaresko, A. A. Kordyuk, G. Behr, A. Vasiliev, R. Follath and B. Büchner, *Phys. Rev. Lett.*, 2010, **105**, 067002.
165. M. J. Pitcher, T. Lancaster, J. D. Wright, I. Franke, A. J. Steele, P. J. Baker, F. L. Pratt, W. T. Thomas, D. R. Parker, S. J. Blundell and S. J. Clarke, *J. Am. Chem. Soc.*, 2010, **132**, 10467.
166. Z. Xu, G. Dai, Y. Li, Z. Yin, Y. Rong, L. Tian, P. Liu, H. Wang, L. Xing, Y. Wei, R. Kajimoto, K. Ikeuchi, D. L. Abernathy, X. Wang, C. Jin, X. Lu, G. Tan and P. Dai, *npj Quantum Mater.*, 2020, **5**, 11.
167. D. R. Parker, M. J. Pitcher, P. J. Baker, I. Franke, T. Lancaster, S. J. Blundell and S. J. Clarke, *Chem. Commun.*, 2009, 2189.
168. D. R. Parker, M. J. P. Smith, T. Lancaster, A. J. Steele, I. Franke, P. J. Baker, F. L. Pratt, M. J. Pitcher, S. J. Blundell and S. J. Clarke, *Phys. Rev. Lett.*, 2010, **104**, 057007.
169. S. Li, C. de la Cruz, Q. Huang, G. F. Chen, T. L. Xia, J. L. Luo, N. L. Wang and P. Dai, *Phys. Rev. B:Condens. Matter Mater. Phys.*, 2009, **80**, 020504.

170. I. Todorov, D. Y. Chung, H. Claus, C. D. Malliakas, A. P. Douvalis, T. Bakas, J. He, V. P. Dravid and M. G. Kanatzidis, *Chem. Mater.*, 2010, **22**, 3916.
171. N. Katayama, K. Kudo, S. Onari, T. Mizukami, K. Sugawara, Y. Sugiyama, Y. Kitahama, K. Iba, K. Fujimura, N. Nishimoto, M. Nohara and H. Sawa, *J. Phys. Soc. Jpn.*, 2013, **82**, 123702.
172. H. Yakita, H. Ogino, T. Okada, A. Yamamoto, K. Kishio, T. Tohei, Y. Ikuhara, Y. Gotoh, H. Fujihisa, K. Kataoka, H. Eisaki and J.-i. Shimoyama, *J. Am. Chem. Soc.*, 2014, **136**, 846.
173. K. Kudo, Y. Kitahama, K. Fujimura, T. Mizukami, H. Ota and M. Nohara, *J. Phys. Soc. Jpn.*, 2014, **83**, 093705.
174. N. Ni, J. M. Allred, B. C. Chan and R. J. Cava, *Proc. Natl. Acad. Sci. U. S. A.*, 2011, **108**, E1019.
175. S. Kakiya, K. Kudo, Y. Nishikubo, K. Oku, E. Nishibori, H. Sawa, T. Yamamoto, T. Nozaka and M. Nohara, *J. Phys. Soc. Jpn.*, 2011, **80**, 093704.
176. T. Stürzer, C. Stürzer and D. Johrendt, *Phys. Status Solidi B*, 2017, **254**, 1600417.
177. E. Paris, T. Wakita, O. Proux, T. Yokoya, K. Kudo, D. Mitsuoka, T. Kimura, K. Fujimura, N. Nishimoto, S. Ioka, M. Nohara, T. Mizokawa and N. L. Saini, *Phys. Rev. B:Condens. Matter Mater. Phys.*, 2017, **96**, 224507.
178. T. Stürzer, G. Derondeau, E.-M. Bertschler and D. Johrendt, *Solid State Commun.*, 2015, **201**, 36.
179. X. Zhu, F. Han, G. Mu, P. Cheng, B. Shen, B. Zeng and H.-H. Wen, *Phys. Rev. B:Condens. Matter Mater. Phys.*, 2009, **79**, 220512.
180. S. Sato, H. Ogino, N. Kawaguchi, Y. Katsura, K. Kishio, J.-i. Shimoyama, H. Kotegawa and H. Tou, *Supercond. Sci. Technol.*, 2010, **23**, 045001.
181. M. Tegel, I. Schellenberg, F. Hummel, R. Pöttgen and D. Jorendt, *Z. Naturforsch., B:J. Chem. Sci.*, 2009, **64**, 815.
182. H. Ogino, S. Sato, Y. Matsumura, N. Kawaguchi, K. Ushiyama, Y. Katsura, S. Horii, K. Kishio and J. Shimoyama, *Phys. C*, 2010, **470**, S280.
183. H. Ogino, S. Sato, K. Kishio, J.-i. Shimoyama, T. Tohei and Y. Ikuhara, *Appl. Phys. Lett.*, 2010, **97**, 072506.
184. Y.-L. Sun, H. Jiang, H.-F. Zhai, J.-K. Bao, W.-H. Jiao, Q. Tao, C.-Y. Shen, Y.-W. Zeng, Z.-A. Xu and G.-H. Cao, *J. Am. Chem. Soc.*, 2012, **134**, 12893.
185. C.-H. Lee, A. Iyo, H. Eisaki, H. Kito, M. Teresa Fernandez-Diaz, T. Ito, K. Kihou, H. Matsuhata, M. Braden and K. Yamada, *J. Phys. Soc. Jpn.*, 2008, **77**, 083704.
186. Y. Mizuguchi, Y. Hara, K. Deguchi, S. Tsuda, T. Yamaguchi, K. Takeda, H. Kotegawa, H. Tou and Y. Takano, *Supercond. Sci. Technol.*, 2010, **23**, 054013.
187. Y. Kamihara, H. Hiramatsu, M. Hirano, R. Kawamura, H. Yanagi, T. Kamiya and H. Hosono, *J. Am. Chem. Soc.*, 2006, **128**, 10012.
188. T. Mine, H. Yanagi, T. Kamiya, Y. Kamihara, M. Hirano and H. Hosono, *Solid State Commun.*, 2008, **147**, 111.

189. T. Yajima, K. Nakano, F. Takeiri, T. Ono, Y. Hosokoshi, Y. Matsushita, J. Hester, Y. Kobayashi and H. Kageyama, *J. Phys. Soc. Jpn.*, 2012, **81**, 103706.
190. P. Doan, M. Gooch, Z. Tang, B. Lorenz, A. Möller, J. Tapp, P. C. W. Chu and A. M. Guloy, *J. Am. Chem. Soc.*, 2012, **134**, 16520.
191. T. Yajima, K. Nakano, Y. Nozaki and H. Kageyama, *Phys. C*, 2014, **504**, 36.
192. Q. Song, Y. J. Yan, Z. R. Ye, M. Q. Ren, D. F. Xu, S. Y. Tan, X. H. Niu, B. P. Xie, T. Zhang, R. Peng, H. C. Xu, J. Jiang and D. L. Feng, *Phys. Rev. B:Condens. Matter Mater. Phys.*, 2016, **93**, 024508.
193. T. Taguchi, Y. Wang, X. Yang, H. Li, Y. Takabayashi, K. Hayashi, T. Miyazaki, Y.-F. Liao, H. Ishii, H. Goto, R. Eguchi and Y. Kubozono, *Inorg. Chem.*, 2021, **60**, 3585.
194. W. Wu, J. Cheng, K. Matsubayashi, P. Kong, F. Lin, C. Jin, N. Wang, Y. Uwatoko and J. Luo, *Nat. Commun.*, 2014, **5**, 5508.
195. J. G. Cheng, K. Matsubayashi, W. Wu, J. P. Sun, F. K. Lin, J. L. Luo and Y. Uwatoko, *Phys. Rev. Lett.*, 2015, **114**, 117001.
196. J.-K. Bao, J.-Y. Liu, C.-W. Ma, Z.-H. Meng, Z.-T. Tang, Y.-L. Sun, H.-F. Zhai, H. Jiang, H. Bai, C.-M. Feng, Z.-A. Xu and G.-H. Cao, *Phys. Rev. X*, 2015, **5**, 011013.
197. Z.-T. Tang, J.-K. Bao, Y. Liu, Y.-L. Sun, A. Ablimit, H.-F. Zhai, H. Jiang, C.-M. Feng, Z.-A. Xu and G.-H. Cao, *Phys. Rev. B:Condens. Matter Mater. Phys.*, 2015, **91**, 020506.
198. Z.-T. Tang, J.-K. Bao, Z. Wang, H. Bai, H. Jiang, Y. Liu, H.-F. Zhai, C.-M. Feng, Z.-A. Xu and G.-H. Cao, *Sci. China Mater.*, 2015, **58**, 16.
199. K. M. Taddei, Q. Zheng, A. S. Sefat and C. de la Cruz, *Phys. Rev. B:Condens. Matter Mater. Phys.*, 2017, **96**, 180506.
200. Q.-G. Mu, B.-B. Ruan, K. Zhao, B.-J. Pan, T. Liu, L. Shan, G.-F. Chen and Z.-A. Ren, *Sci. Bull.*, 2018, **63**, 952.
201. K. Zhao, Q.-G. Mu, B.-B. Ruan, M.-H. Zhou, Q.-S. Yang, T. Liu, B.-J. Pan, S. Zhang, G.-F. Chen and Z.-A. Ren, *Chin. Phys. Lett.*, 2020, **37**, 097401.
202. K. Zhao, Q.-G. Mu, B.-B. Ruan, T. Liu, B.-J. Pan, M.-H. Zhou, S. Zhang, G.-F. Chen and Z.-A. Ren, *APL Mater.*, 2020, **8**, 031103.
203. M. Shatruk, *J. Solid State Chem.*, 2019, **272**, 198.
204. Y. Singh, A. Ellern and D. C. Johnston, *Phys. Rev. B:Condens. Matter Mater. Phys.*, 2009, **79**, 094519.
205. Y. Singh, M. A. Green, Q. Huang, A. Kreyssig, R. J. McQueeney, D. C. Johnston and A. I. Goldman, *Phys. Rev. B:Condens. Matter Mater. Phys.*, 2009, **80**, 100403.
206. M. Reehuis, C. Ritter, R. Ballou and W. Jeitschko, *J. Magn. Magn. Mater.*, 1994, **138**, 85.
207. M. Reehuis, W. Jeitschko, G. Kotzyba, B. Zimmer and X. Hu, *J. Alloys Compd.*, 1998, **266**, 54.
208. M. Chefki, M. M. Abd-Elmeguid, H. Micklitz, C. Huhnt, W. Schlabitz, M. Reehuis and W. Jeitschko, *Phys. Rev. Lett.*, 1998, **80**, 802.
209. R. A. de Groot, F. M. Mueller, P. G. v. Engen and K. H. J. Buschow, *Phys. Rev. Lett.*, 1983, **50**, 2024.

210. K. E. H. M. Hanssen, P. E. Mijnarends, L. P. L. M. Rabou and K. H. J. Buschow, *Phys. Rev. B:Condens. Matter Mater. Phys.*, 1990, **42**, 1533.
211. S. A. Wolf, D. D. Awschalom, R. A. Buhrman, J. M. Daughton, S. von Molnár, M. L. Roukes, A. Y. Chtchelkanova and D. M. Treger, *Science*, 2001, **294**, 1488.
212. A. Yamasaki, L. Chioncel, A. I. Lichtenstein and O. K. Andersen, *Phys. Rev. B:Condens. Matter Mater. Phys.*, 2006, **74**, 024419.
213. C. Tu and W. Stutius, *IEEE Trans. Magn.*, 1974, **10**, 581.
214. X. Guo, X. Chen, Z. Altounian and J. O. Ström-Olsen, *Phys. Rev. B:Condens. Matter Mater. Phys.*, 1992, **46**, 14578.
215. J. Cui, M. Kramer, L. Zhou, F. Liu, A. Gabay, G. Hadjipanayis, B. Balasubramanian and D. Sellmyer, *Acta Mater.*, 2018, **158**, 118.
216. H. Fujii, T. Hokabe, T. Kamigaichi and T. Okamoto, *J. Phys. Soc. Jpn.*, 1977, **43**, 41.
217. L. Caron, M. Hudl, V. Höglin, N. H. Dung, C. P. Gomez, M. Sahlberg, E. Brück, Y. Andersson and P. Nordblad, *Phys. Rev. B:Condens. Matter Mater. Phys.*, 2013, **88**, 094440.
218. Y. He, P. Adler, S. Schneider, I. Soldatov, Q. Mu, H. Borrmann, W. Schnelle, R. Schaefer, B. Rellinghaus, G. H. Fecher and C. Felser, *Adv. Funct. Mater.*, 2022, **32**, 2107513.
219. O. Gutfleisch, M. A. Willard, E. Brück, C. H. Chen, S. G. Sankar and J. P. Liu, *Adv. Mater.*, 2011, **23**, 821.
220. O. Tegus, E. Brück, K. H. J. Buschow and F. R. de Boer, *Nature*, 2002, **415**, 150.
221. H. Wada and Y. Tanabe, *Appl. Phys. Lett.*, 2001, **79**, 3302.
222. A. de Campos, D. L. Rocco, A. M. G. Carvalho, L. Caron, A. A. Coelho, S. Gama, L. M. da Silva, F. C. G. Gandra, A. O. dos Santos, L. P. Cardoso, P. J. von Ranke and N. A. de Oliveira, *Nat. Mater.*, 2006, **5**, 802.
223. S. Gama, A. A. Coelho, A. de Campos, A. M. G. Carvalho, F. C. G. Gandra, P. J. von Ranke and N. A. de Oliveira, *Phys. Rev. Lett.*, 2004, **93**, 237202.
224. X.-L. Qi and S.-C. Zhang, *Rev. Mod. Phys.*, 2011, **83**, 1057.
225. M. König, S. Wiedmann, C. Brüne, A. Roth, H. Buhmann, L. W. Molenkamp, X.-L. Qi and S.-C. Zhang, *Science*, 2007, **318**, 766.
226. N. P. Armitage, E. J. Mele and A. Vishwanath, *Rev. Mod. Phys.*, 2018, **90**, 015001.
227. H. Lin, L. A. Wray, Y. Xia, S. Xu, S. Jia, R. J. Cava, A. Bansil and M. Z. Hasan, *Nat. Mater.*, 2010, **9**, 546.
228. Z. Wang, Y. Sun, X.-Q. Chen, C. Franchini, G. Xu, H. Weng, X. Dai and Z. Fang, *Phys. Rev. B:Condens. Matter Mater. Phys.*, 2012, **85**, 195320.
229. Z. K. Liu, B. Zhou, Y. Zhang, Z. J. Wang, H. M. Weng, D. Prabhakaran, S. K. Mo, Z. X. Shen, Z. Fang, X. Dai, Z. Hussain and Y. L. Chen, *Science*, 2014, **343**, 864.
230. Z. K. Liu, J. Jiang, B. Zhou, Z. J. Wang, Y. Zhang, H. M. Weng, D. Prabhakaran, S. K. Mo, H. Peng, P. Dudin, T. Kim, M. Hoesch, Z. Fang, X. Dai, Z. X. Shen, D. L. Feng, Z. Hussain and Y. L. Chen, *Nat. Mater.*, 2014, **13**, 677.
231. G. A. Steigmann and J. Goodyear, *Acta Crystallogr., Sect. B*, 1968, **24**, 1062.

232. M. N. Ali, Q. Gibson, S. Jeon, B. B. Zhou, A. Yazdani and R. J. Cava, *Inorg. Chem.*, 2014, **53**, 4062.
233. S.-Y. Xu, I. Belopolski, N. Alidoust, M. Neupane, G. Bian, C. Zhang, R. Sankar, G. Chang, Z. Yuan, C.-C. Lee, S.-M. Huang, H. Zheng, J. Ma, D. S. Sanchez, B. Wang, A. Bansil, F. Chou, P. P. Shibayev, H. Lin, S. Jia and M. Z. Hasan, *Science*, 2015, **349**, 613.
234. B. Q. Lv, H. M. Weng, B. B. Fu, X. P. Wang, H. Miao, J. Ma, P. Richard, X. C. Huang, L. X. Zhao, G. F. Chen, Z. Fang, X. Dai, T. Qian and H. Ding, *Phys. Rev. X*, 2015, **5**, 031013.
235. Z. K. Liu, L. X. Yang, Y. Sun, T. Zhang, H. Peng, H. F. Yang, C. Chen, Y. Zhang, Y. F. Guo, D. Prabhakaran, M. Schmidt, Z. Hussain, S. K. Mo, C. Felser, B. Yan and Y. L. Chen, *Nat. Mater.*, 2016, **15**, 27.
236. S.-Y. Xu, I. Belopolski, D. S. Sanchez, C. Zhang, G. Chang, C. Guo, G. Bian, Z. Yuan, H. Lu, T.-R. Chang, P. P. Shibayev, M. L. Prokopovych, N. Alidoust, H. Zheng, C.-C. Lee, S.-M. Huang, R. Sankar, F. Chou, C.-H. Hsu, H.-T. Jeng, A. Bansil, T. Neupert, V. N. Strocov, H. Lin, S. Jia and M. Z. Hasan, *Sci. Adv.*, 2015, **1**, e1501092.
237. I. Belopolski, S.-Y. Xu, D. S. Sanchez, G. Chang, C. Guo, M. Neupane, H. Zheng, C.-C. Lee, S.-M. Huang, G. Bian, N. Alidoust, T.-R. Chang, B. Wang, X. Zhang, A. Bansil, H.-T. Jeng, H. Lin, S. Jia and M. Z. Hasan, *Phys. Rev. Lett.*, 2016, **116**, 066802.
238. C. R. Rajamathi, U. Gupta, N. Kumar, H. Yang, Y. Sun, V. Süß, C. Shekhar, M. Schmidt, H. Blumtritt, P. Werner, B. Yan, S. Parkin, C. Felser and C. N. R. Rao, *Adv. Mater.*, 2017, **29**, 1606202.
239. A. A. Burkov, M. D. Hook and L. Balents, *Phys. Rev. B:Condens. Matter Mater. Phys.*, 2011, **84**, 235126.
240. A. Yamakage, Y. Yamakawa, Y. Tanaka and Y. Okamoto, *J. Phys. Soc. Jpn.*, 2015, **85**, 013708.
241. X.-B. Wang, X.-M. Ma, E. Emmanouilidou, B. Shen, C.-H. Hsu, C.-S. Zhou, Y. Zuo, R.-R. Song, S.-Y. Xu, G. Wang, L. Huang, N. Ni and C. Liu, *Phys. Rev. B:Condens. Matter Mater. Phys.*, 2017, **96**, 161112.
242. D. Takane, K. Nakayama, S. Souma, T. Wada, Y. Okamoto, K. Takenaka, Y. Yamakawa, A. Yamakage, T. Mitsuhashi, K. Horiba, H. Kumigashira, T. Takahashi and T. Sato, *npj Quantum Mater.*, 2018, **3**, 1.
243. L. S. Xie, L. M. Schoop, E. M. Seibel, Q. D. Gibson, W. Xie and R. J. Cava, *APL Mater.*, 2015, **3**, 083602.
244. Y. H. Chan, C.-K. Chiu, M. Y. Chou and A. P. Schnyder, *Phys. Rev. B:Condens. Matter Mater. Phys.*, 2016, **93**, 205132.
245. M. Sato and Y. Ando, *Rep. Prog. Phys.*, 2017, **80**, 076501.
246. Y. Qian, S. Nie, C. Yi, L. Kong, C. Fang, T. Qian, H. Ding, Y. Shi, Z. Wang, H. Weng and Z. Fang, *npj Comput. Mater.*, 2019, **5**, 121.
247. D. Das, D. T. Adroja, M. R. Lees, R. W. Taylor, Z. S. Bishnoi, V. K. Anand, A. Bhattacharyya, Z. Guguchia, C. Baines, H. Luetkens, G. B. G. Stenning, L. Duan, X. Wang and C. Jin, *Phys. Rev. B:Condens. Matter Mater. Phys.*, 2021, **103**, 144516.
248. B. R. Ortiz, L. C. Gomes, J. R. Morey, M. Winiarski, M. Bordelon, J. S. Mangum, I. W. H. Oswald, J. A. Rodriguez-Rivera, J. R. Neilson,

S. D. Wilson, E. Ertekin, T. M. McQueen and E. S. Toberer, *Phys. Rev. Mater.*, 2019, **3**, 094407.
249. B. R. Ortiz, S. M. L. Teicher, Y. Hu, J. L. Zuo, P. M. Sarte, E. C. Schueller, A. M. M. Abeykoon, M. J. Krogstad, S. Rosenkranz, R. Osborn, R. Seshadri, L. Balents, J. He and S. D. Wilson, *Phys. Rev. Lett.*, 2020, **125**, 247002.
250. Q. Yin, Z. Tu, C. Gong, Y. Fu, S. Yan and H. Lei, *Chin. Phys. Lett.*, 2021, **38**, 037403.
251. B. R. Ortiz, P. M. Sarte, E. M. Kenney, M. J. Graf, S. M. L. Teicher, R. Seshadri and S. D. Wilson, *Phys. Rev. Mater.*, 2021, **5**, 034801.
252. Y.-X. Jiang, J.-X. Yin, M. M. Denner, N. Shumiya, B. R. Ortiz, G. Xu, Z. Guguchia, J. He, M. S. Hossain, X. Liu, J. Ruff, L. Kautzsch, S. S. Zhang, G. Chang, I. Belopolski, Q. Zhang, T. A. Cochran, D. Multer, M. Litskevich, Z.-J. Cheng, X. P. Yang, Z. Wang, R. Thomale, T. Neupert, S. D. Wilson and M. Z. Hasan, *Nat. Mater.*, 2021, **20**, 1353.
253. J. Nayak, S.-C. Wu, N. Kumar, C. Shekhar, S. Singh, J. Fink, E. E. D. Rienks, G. H. Fecher, S. S. P. Parkin, B. Yan and C. Felser, *Nat. Commun.*, 2017, **8**, 13942.
254. X. Duan, F. Wu, J. Chen, P. Zhang, Y. Liu, H. Yuan and C. Cao, *Commun. Phys.*, 2018, **1**, 71.
255. M. Neupane, M. Mofazzel Hosen, I. Belopolski, N. Wakeham, K. Dimitri, N. Dhakal, J.-X. Zhu, M. Z. Hasan, E. D. Bauer and F. Ronning, *J. Phys.: Condens. Matter*, 2016, **28**, 23LT02.
256. B. Yan and C. Felser, *Annu. Rev. Condens. Matter Phys.*, 2017, **8**, 337.
257. C. Shekhar, N. Kumar, V. Grinenko, S. Singh, R. Sarkar, H. Luetkens, S.-C. Wu, Y. Zhang, A. C. Komarek, E. Kampert, Y. Skourski, J. Wosnitza, W. Schnelle, A. McCollam, U. Zeitler, J. Kübler, B. Yan, H. H. Klauss, S. S. P. Parkin and C. Felser, *Proc. Natl. Acad. Sci. U. S. A.*, 2018, **115**, 9140.
258. M. Hirschberger, S. Kushwaha, Z. Wang, Q. Gibson, S. Liang, C. A. Belvin, B. A. Bernevig, R. J. Cava and N. P. Ong, *Nat. Mater.*, 2016, **15**, 1161.
259. S. Klemenz, S. Lei and L. M. Schoop, *Annu. Rev. Mater. Res.*, 2019, **49**, 185.
260. J. Park, G. Lee, F. Wolff-Fabris, Y. Y. Koh, M. J. Eom, Y. K. Kim, M. A. Farhan, Y. J. Jo, C. Kim, J. H. Shim and J. S. Kim, *Phys. Rev. Lett.*, 2011, **107**, 126402.
261. K. Wang, D. Graf, L. Wang, H. Lei, S. W. Tozer and C. Petrovic, *Phys. Rev. B:Condens. Matter Mater. Phys.*, 2012, **85**, 041101.
262. Y. Feng, Z. Wang, C. Chen, Y. Shi, Z. Xie, H. Yi, A. Liang, S. He, J. He, Y. Peng, X. Liu, Y. Liu, L. Zhao, G. Liu, X. Dong, J. Zhang, C. Chen, Z. Xu, X. Dai, Z. Fang and X. J. Zhou, *Sci. Rep.*, 2014, **4**, 5385.
263. B. Bradlyn, L. Elcoro, J. Cano, M. G. Vergniory, Z. Wang, C. Felser, M. I. Aroyo and B. A. Bernevig, *Nature*, 2017, **547**, 298.
264. S.-H. Li, M.-Y. Qi, Z.-R. Tang and Y.-J. Xu, *Chem. Soc. Rev.*, 2021, **50**, 7539.
265. Y. Zhang and L. Song, *ChemCatChem*, 2020, **12**, 3621.
266. S. T. Oyama, T. Gott, H. Zhao and Y.-K. Lee, *Catal. Today*, 2009, **143**, 94.

267. Y. Shi and B. Zhang, *Chem. Soc. Rev.*, 2016, **45**, 1529.
268. J. Cai, Y. Song, Y. Zang, S. Niu, Y. Wu, Y. Xie, X. Zheng, Y. Liu, Y. Lin, X. Liu, G. Wang and Y. Qian, *Sci. Adv.*, 2020, **6**, eaaw8113.
269. X. Zou and Y. Zhang, *Chem. Soc. Rev.*, 2015, **44**, 5148.
270. Y. Xu, R. Wu, J. Zhang, Y. Shi and B. Zhang, *Chem. Commun.*, 2013, **49**, 6656.
271. E. J. Popczun, C. G. Read, C. W. Roske, N. S. Lewis and R. E. Schaak, *Angew. Chem., Int. Ed.*, 2014, **53**, 5427.
272. J. Tian, Q. Liu, N. Cheng, A. M. Asiri and X. Sun, *Angew. Chem., Int. Ed.*, 2014, **53**, 9577.
273. P. Xiao, M. A. Sk, L. Thia, X. Ge, R. J. Lim, J.-Y. Wang, K. H. Lim and X. Wang, *Energy Environ. Sci.*, 2014, **7**, 2624.
274. F. Li, C. Wang, X. Han, X. Feng, Y. Qu, J. Liu, W. Chen, L. Zhao, X. Song, H. Zhu, H. Chen, M. Zhao, Z. Deng, J. Wu, P. Zhang and L. Gao, *ACS Appl. Mater. Interfaces*, 2020, **12**, 22741.
275. Z. W. Seh, J. Kibsgaard, C. F. Dickens, I. Chorkendorff, J. K. Nørskov and T. F. Jaramillo, *Science*, 2017, **355**, eaad4998.
276. S. Jin, *ACS Energy Lett.*, 2017, **2**, 1937.
277. M. Rana, S. Mondal, L. Sahoo, K. Chatterjee, P. E. Karthik and U. K. Gautam, *ACS Appl. Mater. Interfaces*, 2018, **10**, 33737.
278. R. Guo, W. Bi, K. Zhang, Y. Liu, C. Wang, Y. Zheng and M. Jin, *Chem. Mater.*, 2019, **31**, 8205.
279. Y. Hu, H. Yu, L. Qi, J. Dong, P. Yan, T. Taylor Isimjan and X. Yang, *ChemSusChem*, 2021, **14**, 1565.
280. X. Guo, M. Li, L. He, S. Geng, F. Tian, Y. Song, W. Yang and Y. Yu, *Nanoscale*, 2021, **13**, 14179.
281. J. Shi, F. Qiu, W. Yuan, M. Guo and Z.-H. Lu, *Chem. Eng. J.*, 2021, **403**, 126312.
282. J. Li, H. Zheng, C. Xu, Z. Su, X. Li and J. Sun, *Inorg. Chem.*, 2021, **60**, 1624.
283. L. Chai, S. Liu, S. Pei and C. Wang, *Chem. Eng. J.*, 2021, **420**, 129686.
284. H. Liang, A. N. Gandi, D. H. Anjum, X. Wang, U. Schwingenschlögl and H. N. Alshareef, *Nano Lett.*, 2016, **16**, 7718.
285. B. You, N. Jiang, M. Sheng, M. W. Bhushan and Y. Sun, *ACS Catal.*, 2016, **6**, 714.
286. Y. Tan, H. Wang, P. Liu, Y. Shen, C. Cheng, A. Hirata, T. Fujita, Z. Tang and M. Chen, *Energy Environ. Sci.*, 2016, **9**, 2257.
287. S. T. Oyama, *J. Catal.*, 2003, **216**, 343.
288. S. T. Oyama, X. Wang, Y. K. Lee, K. Bando and F. G. Requejo, *J. Catal.*, 2002, **210**, 207.
289. C. Stinner, R. Prins and T. Weber, *J. Catal.*, 2001, **202**, 187.
290. P. M. de Souza, C. V. M. Inocêncio, V. I. Perez, R. C. Rabelo-Neto, V. O. O. Gonçalves, G. Jacobs, F. Richard, V. T. da Silva and F. B. Noronha, *Catal. Today*, 2020, **356**, 366.
291. C. V. M. Inocêncio, P. M. de Souza, R. C. Rabelo-Neto, V. T. da Silva and F. B. Noronha, *Catal. Today*, 2021, **381**, 133.

292. M. A. Golubeva, E. M. Zakharyan and A. L. Maximov, *Pet. Chem.*, 2020, **60**, 1109.
293. J. Zhang, L. Wu, L. Xu, D. Sun, H. Sun and Y. Tang, *Nanoscale*, 2021, **13**, 16052.
294. Z. Yu, X.-K. Wei, J. Xu, Y. Li, A. Araujo, J. L. Faria, R. E. Dunin-Borkowski and L. Liu, *ACS Appl. Energy Mater.*, 2021, **4**, 1593.
295. J. Chang, L. Feng, C. Liu, W. Xing and X. Hu, *Energy Environ. Sci.*, 2014, **7**, 1628.
296. M. Li, Y. Fang, G. Zhang, P. Cui, Z. Yang and J. He, *J. Mater. Chem. A*, 2020, **8**, 10433.
297. C. Wang, L. Chai, X. Cui, Z. Zhou and S. Liu, *J. Phys. Chem. C*, 2021, **125**, 21443.
298. H. Kim, A. W. Robertson, G.-h. Kwon, J.-w. O, J. H. Warner and J. M. Kim, *ACS Appl. Energy Mater.*, 2019, **2**, 8649.
299. S. Fujita, S. Yamaguchi, J. Yamasaki, K. Nakajima, S. Yamazoe, T. Mizugaki and T. Mitsudome, *Chem. – Eur. J.*, 2021, **27**, 4439.
300. S. Fujita, K. Nakajima, J. Yamasaki, T. Mizugaki, K. Jitsukawa and T. Mitsudome, *ACS Catal.*, 2020, **10**, 4261.
301. Z. Tong, X. Li, J. Dong, R. Gao, Q. Deng, J. Wang, Z. Zeng, J.-J. Zou and S. Deng, *ACS Catal.*, 2021, **11**, 6406.
302. S. Yamaguchi, S. Fujita, K. Nakajima, S. Yamazoe, J. Yamasaki, T. Mizugaki and T. Mitsudome, *ACS Sustainable Chem. Eng.*, 2021, **9**, 6347.
303. S. Yamaguchi, T. Mizugaki and T. Mitsudome, *Eur. J. Inorg. Chem.*, 2021, **2021**, 3327.
304. S. Yamaguchi, S. Fujita, K. Nakajima, S. Yamazoe, J. Yamasaki, T. Mizugaki and T. Mitsudome, *Green Chem.*, 2021, **23**, 2010.
305. M. Sheng, S. Fujita, S. Yamaguchi, J. Yamasaki, K. Nakajima, S. Yamazoe, T. Mizugaki and T. Mitsudome, *JACS Au*, 2021, **1**, 501.
306. S. Fujita, K. Imagawa, S. Yamaguchi, J. Yamasaki, S. Yamazoe, T. Mizugaki and T. Mitsudome, *Sci. Rep.*, 2021, **11**, 10673.
307. Q. Yao, J. Chen, S. Xiao, Y. Zhang and X. Zhou, *ACS Appl. Mater. Interfaces*, 2021, **13**, 30458.
308. G. Wang, Q. Huang, T.-s. Song and J. Xie, *Energy Fuels*, 2020, **34**, 8666.
309. X. Wan, D. Qian, L. Ai and J. Jiang, *Ind. Eng. Chem. Res.*, 2020, **59**, 22040.
310. R. Luo, C. Liu, J. Li, J. Wang, X. Hu, X. Sun, J. Shen, W. Han and L. Wang, *J. Hazard. Mater.*, 2017, **329**, 92.
311. Y. Xie, X. Wang, W. Tong, W. Hu, P. Li, L. Dai, Y. Wang and Y. Zhang, *Chem. Eng. J.*, 2020, **396**, 125321.
312. S. Feng, B. Xiao, M. Wu, Y. Wang, R. Chen and H. Liu, *Sep. Purif. Technol.*, 2020, **248**, 117004.
313. Y. Liu, A. J. McCue and D. Li, *ACS Catal.*, 2021, **11**, 9102.
314. J. B. Goodenough and Y. Kim, *Chem. Mater.*, 2010, **22**, 587.
315. N. Nitta, F. Wu, J. T. Lee and G. Yushin, *Mater. Today*, 2015, **18**, 252.
316. G. Assat and J.-M. Tarascon, *Nat. Energy*, 2018, **3**, 373.
317. N. Nitta and G. Yushin, *Part. Part. Syst. Charact.*, 2014, **31**, 317.

318. L. Monconduit, *Nanosci. Nanotechnol. Lett.*, 2012, **4**, 118.
319. M. M. Thackeray, J. T. Vaughey, C. S. Johnson, A. J. Kropf, R. Benedek, L. M. L. Fransson and K. Edstrom, *J. Power Sources*, 2003, **113**, 124.
320. F. Winter, R. Pöttgen, M. Greiwe and T. Nilges, *Rev. Inorg. Chem.*, 2015, **35**, 1.
321. S. Sim and J. Cho, *J. Electrochem. Soc.*, 2012, **159**, A669.
322. S. Boyanov, D. Zitoun, M. Ménétrier, J. C. Jumas, M. Womes and L. Monconduit, *J. Phys. Chem. C*, 2009, **113**, 21441.
323. B. Mauvernay, M. L. Doublet and L. Monconduit, *J. Phys. Chem. Solids*, 2006, **67**, 1252.
324. Y. Lu, J.-P. Tu, Q.-Q. Xiong, J.-Y. Xiang, Y.-J. Mai, J. Zhang, Y.-Q. Qiao, X.-L. Wang, C.-D. Gu and S. X. Mao, *Adv. Funct. Mater.*, 2012, **22**, 3927.
325. S.-g. Woo, J.-H. Jung, H. Kim, M. G. Kim, C. K. Lee, H.-J. Sohn and B. W. Cho, *J. Electrochem. Soc.*, 2006, **153**, A1979.
326. F. Zhou, X.-S. Yang, J. Liu, J. Liu, R. Hu, L. Ouyang and M. Zhu, *J. Power Sources*, 2021, **499**, 229979.
327. F. Gillot, S. Boyanov, L. Dupont, M. L. Doublet, M. Morcrette, L. Monconduit and J. M. Tarascon, *Chem. Mater.*, 2005, **17**, 6327.
328. S. Boyanov, J. Bernardi, E. Bekaert, M. Ménétrier, M. L. Doublet and L. Monconduit, *Chem. Mater.*, 2009, **21**, 298.
329. J. W. Hall, N. Membreno, J. Wu, H. Celio, R. A. Jones and K. J. Stevenson, *J. Am. Chem. Soc.*, 2012, **134**, 5532.
330. W. Peng, L. Jiao, Q. Huan, L. Li, J. Yang, Q. Zhao, Q. Wang, H. Du, G. Liu, Y. Si, Y. Wang and H. Yuan, *J. Alloys Compd.*, 2012, **511**, 198.
331. D. Yang, J. Zhu, X. Rui, H. Tan, R. Cai, H. E. Hoster, D. Y. W. Yu, H. H. Hng and Q. Yan, *ACS Appl. Mater. Interfaces*, 2013, **5**, 1093.
332. V. Pralong, D. C. S. Souza, K. T. Leung and L. F. Nazar, *Electrochem. Commun.*, 2002, **4**, 516.
333. R. Alcántara, J. L. Tirado, J. C. Jumas, L. Monconduit and J. Olivier-Fourcade, *J. Power Sources*, 2002, **109**, 308.
334. C. H. Mi, Y. X. Cao, X. G. Zhang and H. L. Li, *Solid State Commun.*, 2009, **149**, 781.
335. S.-O. Kim and A. Manthiram, *ACS Appl. Mater. Interfaces*, 2017, **9**, 16221.
336. D. C. S. Souza, V. Pralong, A. J. Jacobson and L. F. Nazar, *Science*, 2002, **296**, 2012.
337. M. P. Bichat, F. Gillot, L. Monconduit, F. Favier, M. Morcrette, F. Lemoigno and M. L. Doublet, *Chem. Mater.*, 2004, **16**, 1002.
338. V. Mauchamp, P. Moreau, L. Monconduit, M.-L. Doublet, F. Boucher and G. Ouvrard, *J. Phys. Chem. C*, 2007, **111**, 3996.
339. K.-H. Kim and S.-H. Hong, *Adv. Energy Mater.*, 2021, **11**, 2003609.
340. S. Matsuno, M. Noji, T. Kashiwagi, M. Nakayama and M. Wakihara, *J. Phys. Chem. C*, 2007, **111**, 7548.
341. R. Benedek and M. M. Thackeray, *J. Power Sources*, 2002, **110**, 406.
342. M. Greiwe and T. Nilges, *Prog. Solid State Chem.*, 2014, **42**, 191.
343. C. Marino, M. T. Sougrati, B. Gerke, R. Pöttgen, H. Huo, M. Ménétrier, C. P. Grey and L. Monconduit, *Chem. Mater.*, 2012, **24**, 4735.

344. G. Coquil, M. T. Sougrati, S. Biscaglia, D. Aymé-Perrot, P.-F. Girard and L. Monconduit, *Electrochim. Acta*, 2018, **281**, 619.
345. C. Marino, M. T. Sougrati, A. Darwiche, J. Fullenwarth, B. Fraisse, J. C. Jumas and L. Monconduit, *J. Power Sources*, 2013, **244**, 736.
346. G. Coquil, B. Fraisse, S. Biscaglia, D. Aymé-Perrot, M. T. Sougrati and L. Monconduit, *J. Power Sources*, 2019, **441**, 227165.
347. Z. Li and H. Zhao, *J. Mater. Chem. A*, 2018, **6**, 24013.
348. Q. Li, D. Yang, H. Chen, X. Lv, Y. Jiang, Y. Feng, X. Rui and Y. Yu, *SusMat*, 2021, **1**, 359.
349. W. Liu, H. Zhi and X. Yu, *Energy Storage Mater.*, 2019, **16**, 290.
350. W. Zhang, T. Liu, Y. Wang, Y. Liu, J. Nai, L. Zhang, O. Sheng and X. Tao, *Nano Energy*, 2021, **90**, 106475.
351. Q. Xia, W. Li, Z. Miao, S. Chou and H. Liu, *Nano Res.*, 2017, **10**, 4055.
352. G. Chang, Y. Zhao, L. Dong, D. P. Wilkinson, L. Zhang, Q. Shao, W. Yan, X. Sun and J. Zhang, *J. Mater. Chem. A*, 2020, **8**, 4996.
353. D. Duveau, S. S. Israel, J. Fullenwarth, F. Cunin and L. Monconduit, *J. Mater. Chem. A*, 2016, **4**, 3228.
354. K.-H. Nam, Y. Hwa and C.-M. Park, *ACS Appl. Mater. Interfaces*, 2020, **12**, 15053.
355. G. Coquil, J. Fullenwarth, G. Grinbom, M. T. Sougrati, L. Stievano, D. Zitoun and L. Monconduit, *J. Power Sources*, 2017, **372**, 196.
356. J. Fullenwarth, A. Darwiche, A. Soares, B. Donnadieu and L. Monconduit, *J. Mater. Chem. A*, 2014, **2**, 2050.
357. Y. Kim, Y. Kim, A. Choi, S. Woo, D. Mok, N.-S. Choi, Y. S. Jung, J. H. Ryu, S. M. Oh and K. T. Lee, *Adv. Mater.*, 2014, **26**, 4139.
358. X. Fan, T. Gao, C. Luo, F. Wang, J. Hu and C. Wang, *Nano Energy*, 2017, **38**, 350.
359. Z. Hao, N. Dimov, J.-K. Chang and S. Okada, *J. Energy Chem.*, 2022, **64**, 463.
360. J. Choi, W.-S. Kim, K.-H. Kim and S.-H. Hong, *J. Mater. Chem. A*, 2018, **6**, 17437.
361. S. C. Jung, J.-H. Choi and Y.-K. Han, *J. Mater. Chem. A*, 2018, **6**, 1772.
362. L. Zheng, R. A. Dunlap and M. N. Obrovac, *J. Electrochem. Soc.*, 2016, **163**, A1188.
363. X.-M. Zheng, J.-H. You, J.-J. Fan, G.-P. Tu, W.-Q. Rong, W.-J. Li, Y.-X. Wang, S. Tao, P.-Y. Zhang, S.-Y. Zhang, S.-Y. Shen, J.-T. Li, L. Huang and S.-G. Sun, *Nano Energy*, 2020, **77**, 105123.
364. S. Sarkar and S. C. Peter, *J. Mater. Chem. A*, 2021, **9**, 5164.
365. M. D. Slater, D. Kim, E. Lee and C. S. Johnson, *Adv. Funct. Mater.*, 2013, **23**, 947.
366. N. Yabuuchi, K. Kubota, M. Dahbi and S. Komaba, *Chem. Rev.*, 2014, **114**, 11636.
367. H.-D. Yang, M.-Y. Ran, W.-B. Wei, X.-T. Wu, H. Lin and Q.-L. Zhu, *Chem. – Asian J.*, 2021, **16**, 3299.
368. G. D. Boyd, E. Buehler and F. G. Storz, *Appl. Phys. Lett.*, 1971, **18**, 301.
369. G. Boyd, E. Buehler, F. Storz and J. Wernick, *IEEE J. Quantum Electron.*, 1972, **8**, 419.

370. C. Zhu, G. A. Verozubova, Y. P. Mironov, Z. Lei, L. Song, T. Ma, A. O. Okunev and C. Yang, *J. Cryst. Growth*, 2016, **455**, 129.
371. K. E. Woo, J. Wang, K. Wu, K. Lee, J.-A. Dolyniuk, S. Pan and K. Kovnir, *Adv. Funct. Mater.*, 2018, **28**, 1801589.
372. T. Yu, S. Wang, X. Zhang, C. Li, J. Qiao, N. Jia, B. Han, S.-Q. Xia and X. Tao, *Chem. Mater.*, 2019, **31**, 2010.
373. S. Lee, S. L. Carnahan, G. Akopov, P. Yox, L.-L. Wang, A. J. Rossini, K. Wu and K. Kovnir, *Adv. Funct. Mater.*, 2021, **31**, 2010293.
374. J. Mark, J. Wang, K. Wu, J. G. Lo, S. Lee and K. Kovnir, *J. Am. Chem. Soc.*, 2019, **141**, 11976.
375. J. Chen, C. Lin, F. Xu, S. Yang, Y. Sun, X. Zhao, X. Jiang, B. Li, T. Yan and N. Ye, *Chem. Mater.*, 2020, **32**, 10246.
376. Y. Sun, C. Lin, J. Chen, F. Xu, S. Yang, B. Li, G. Yang, M. Luo and N. Ye, *Inorg. Chem.*, 2021, **60**, 7553.
377. J. Chen, C. Lin, G. Peng, F. Xu, M. Luo, S. Yang, S. Shi, Y. Sun, T. Yan, B. Li and N. Ye, *Chem. Mater.*, 2019, **31**, 10170.
378. M. Pan, Z. Ma, X. Liu, S. Xia, X. Tao and K. Wu, *J. Mater. Chem. C*, 2015, **3**, 9695.
379. J. Chen, H. Chen, F. Xu, L. Cao, X. Jiang, S. Yang, Y. Sun, X. Zhao, C. Lin and N. Ye, *J. Am. Chem. Soc.*, 2021, **143**, 10309.
380. P. Michelato, *Nucl. Instrum. Methods Phys. Res., Sect. A*, 1997, **393**, 455.
381. K. H. Jack and M. M. Wachtel, *Nature*, 1956, **178**, 1408.
382. W. H. McCarroll, *J. Phys. Chem. Solids*, 1965, **26**, 191.
383. A. Galdi, W. J. I. DeBenedetti, J. Balajka, L. Cultrera, I. V. Bazarov, J. M. Maxson and M. A. Hines, *J. Chem. Phys.*, 2020, **153**, 144705.
384. H.-S. Jeong, K. Keller and B. Culkin, *J. Korean Phys. Soc.*, 2017, **70**, 610.
385. J. S. Kang, M. Li, H. Wu, H. Nguyen and Y. Hu, *Science*, 2018, **361**, 575.
386. L. Lindsay, D. A. Broido and T. L. Reinecke, *Phys. Rev. Lett.*, 2013, **111**, 025901.
387. M. Fava, N. H. Protik, C. Li, N. K. Ravichandran, J. Carrete, A. van Roekeghem, G. K. H. Madsen, N. Mingo and D. Broido, *npj Comput. Mater.*, 2021, **7**, 54.
388. M. Green, L. Tian, P. Xiang, J. Murowchick, X. Tan and X. Chen, *Mater. Chem. Front.*, 2018, **2**, 1119.
389. B. Bateer, X. Wang, C. Tian, Y. Xie, K. Pan, W. Ping and H. Fu, *Carbon*, 2020, **161**, 51.
390. Y. Ma, B. Quan, Z. Zeng, Y. Zhang, L. Zhang, Y. Wang and X. Huang, *Inorg. Chem. Front.*, 2021, **8**, 1810.
391. A. T. Proudfoot, *Clin. Toxicol.*, 2009, **47**, 89.
392. G. S. Bumbrah, K. Krishan, T. Kanchan, M. Sharma and G. S. Sodhi, *Forensic Sci. Int.*, 2012, **214**, 1.

CHAPTER 3

Two-dimensional Transition Metal Carbides (MXenes)

LINFAN CUI[a†], MIN GAO[b,c†], JING WANG[b,c*] AND CHUANFANG (JOHN) ZHANG[d*]

[a]Department of Electronics and Nanoengineering, Aalto University, Espoo 02150, Finland; [b]Institute of Environmental Engineering, ETH Zürich, Zürich 8093, Switzerland; [c]Laboratory for Advanced Analytical Technologies, Empa, Swiss Federal Laboratories for Materials Science and Technology, Dübendorf 8600, Switzerland; [d]College of Materials Science & Engineering, Sichuan University, Chengdu, Sichuan, 610065, China
*E-mail: jing.wang@ifu.baug.ethz.ch; chuanfang.zhang@scu.edu.cn

3.1 Introduction

Carbides are a class of chemical compounds containing carbon and metallic or semi-metallic elements. More specifically, ionic carbides are formed when carbon ions combine with highly electropositive elements such as alkali metals or alkaline earth metals (*e.g.*, calcium carbide).[1] Covalent or molecular carbides such as boron carbide and silicon carbide are formed if the elements have a similar electronegativity to carbon.[2,3] By contrast, when the transition metals are combined with carbon they tend to form interstitial carbides (group 4–6 transition metals)[4] or intermediate transition metal carbides (*e.g.*, iron carbides[5] with multiple stoichiometries). The transition metal carbides constitute a diverse class of materials, including simple

[†]These authors contributed equally.

Inorganic Materials Series
Non-oxide and Heteroanionic Materials
Edited by Duncan H. Gregory
A catalogue record for this book is available from the British Library

single-metal compounds and complex multi-metallic compositions.[6] Particularly, the early transition metal (group 4–6) carbides are the most important class of inorganic refractory compounds with high melting points, extreme hardness and excellent strength at elevated temperatures.[7] With a number of valuable properties, the traditional metal carbides are also promising as electronic, magnetic, optical, and catalytic materials to be applied in various fields of technology.[8–11] In the nanotechnology era, materials with atomically layered structures, *e.g.*, two-dimensional (2D) materials, can be created with unusual physicochemical properties to meet the demand of new technologies.[12–22] The concept of nanostructure design has also been proposed to further explore the potential physicochemical properties and applications of transition metal carbides.

MXenes, known as a new family of 2D nanometre-thin carbides, nitrides, and/or carbonitrides of the early transition metals, jumped into the limelight in 2011, expanding the family of few-atom-thick crystals.[23–25] MXenes have the general formula of $M_{n+1}X_nT_x$. In a 2D flake of MXene, $n + 1$ (n = 1–4) layers of early transition metals (M: Sc, Ti, V, Cr, Y, Zr, Nb, Mo, Hf, Ta, or W) are interleaved with n layers of carbon and/or nitrogen (X). Various surface terminations (T) are bonded on the surface metal layers during the synthesis process, such as –O, –OH, –F, –Cl, –I, –Br, –S, –Se, –Te, $-NH_2$, and x represents the number of surface functional groups.[26,27] MXenes are made from bulk ceramics, MAX (A: an element from Group 13 or 14 in the Periodic Table), which are layered ternary carbides or nitrides (discovered in the 1930s to 1960s).[28] Ever since the first MXene, titanium carbide ($Ti_3C_2T_x$) was discovered,[23] the family of 2D MXenes has grown with impressive speed.[29–32] To date, there are more than 30 different experimentally made MXenes (Figure 3.1),[33] while an infinite number of compositions are predicted to be theoretically possible.

The atomic thickness of MXene offers maximum mechanical flexibility and optical transparency. The electronic properties range from metallic

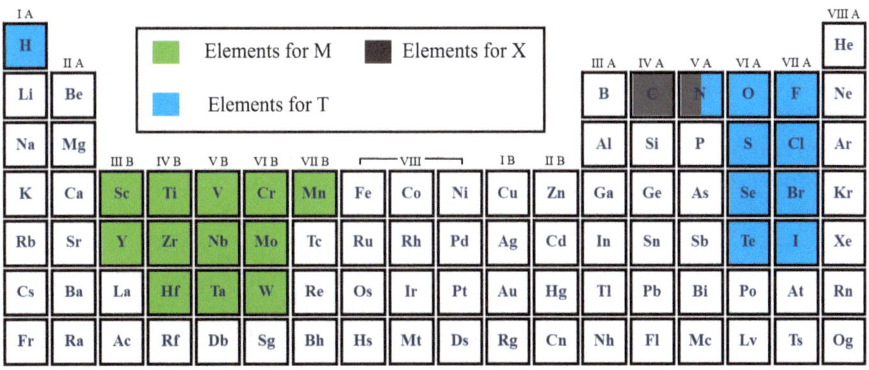

Figure 3.1 The Periodic Table showcasing the constituent elements of known MXene phases that have been experimentally reported so far.

to semi-metal, semiconducting to insulating, *etc.*[29–31] Among all the applications of MXenes to date, energy-related topics are the most extensively reported.[26] In addition, MXenes are widely used in electronic, optical, environmental, catalytic, electromagnetic, biomedical and sensor applications, among others (Figure 3.2).[29–31]

An overview of 2D MXenes is given in this chapter, covering synthesis, structure, composition, and physicochemical properties, which are related in detail to their current and potential applications. The challenges and prospects for the development of MXenes have also been discussed, while providing potential routes for MXene material design.

3.2 Synthesis

MXenes are usually obtained by etching the A elements from MAX phase precursors. To date, most of the precursors are Al-containing MAX phases,[50] while there are also several exceptions, with certain Si-containing precursors proven to be suitable for etching.[51,52] Some MXene structures, such as Mo_2CT_x and $Zr_3C_2T_x$, were synthesised from non-MAX precursors.[53,54] Concentrated HF was initially used to etch the precursors and proved to be an efficient etchant for the synthesis of various MXene components. Several other etching strategies were then developed towards safer and more facile MXene synthesis approaches. Recently, layered transition metal carbides have also been synthesised by following a bottom-up strategy.[55] The overall development of the MXene synthesis routes is summarised in Figure 3.3.

The etching of MAX phases: separating a single or few layers from the layered bulk components is the commonly used approach for acquiring 2D materials. Since the interlayer bonding is much weaker than the planar intralayer bonding,[66] some 2D materials, such as graphene,[67] can be mechanically exfoliated from their 3D counterparts. However, MAX phases cannot be efficiently exfoliated into single-layer MX flakes *via* such mechanical approaches, due to the strong metallic bonding forces between the M and A elements.[68] Since the M–X bonding is stronger than that of M–A, the A elements can be selectively etched away from the MAX precursor. The desired etchant should be able to break the M–A bonds in the layered MAX precursors selectively without destroying the ionic/covalent bonding between M and X elements.[69–71] To date, several etching methods have been developed, which can be mainly categorised into HF-based processes and non-HF involving processes (Figure 3.4 and Table 3.1), and the detail of the representative approaches will be discussed one by one over the course of this chapter.

Intercalation and delamination: generally, after the etching process, the A element of a given MAX phase is replaced by certain surface terminations and the multilayers remain bonded together by hydrogen and/or van der Waals bonds. The obtained MX multilayered structures, like other 2D-layered materials, demonstrate intercalation of various species, including organic molecules and cations. Flakes comprised of a single layer or a few layers are then exfoliated into the colloidal suspensions by intercalation,

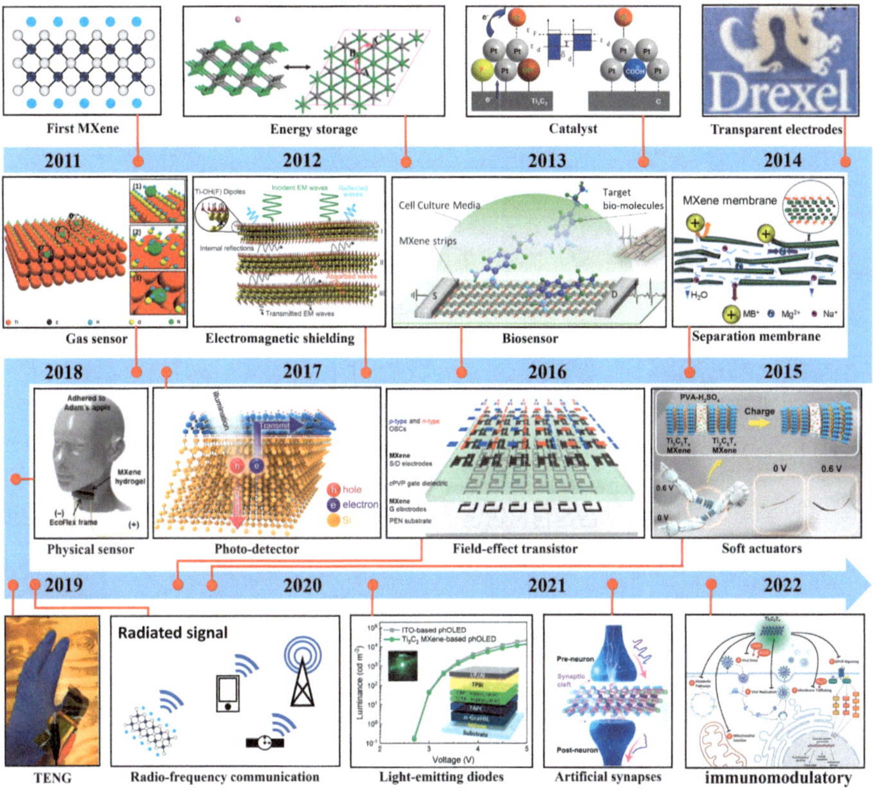

Figure 3.2 The development of MXenes in various applications over approximately one decade, including: energy storage (reproduced from ref. 34 with permission American Chemical Society, Copyright 2012), catalysis (reproduced from ref. 35 with permission from the Royal Society of Chemistry), transparent electrodes (reproduced from ref. 36 with permission from American Chemical Society, Copyright 2014), separation membranes (reproduced from ref. 37 with permission from American Chemical Society, Copyright 2015), biosensors (reproduced from ref. 38 with permission from John Wiley & Sons, Copyright 2016), electromagnetic shielding (reproduced from ref. 39 with permission from American Association for the Advancement of Science, Copyright 2016), photodetectors (reproduced from ref. 40 with permission from John Wiley & Sons, Copyright 2017), gas sensors (reproduced from ref. 41 with permission from American Chemical Society, Copyright 2017), physical sensors (reproduced from ref. 42 with permission from American Association for the Advancement of Science, Copyright 2018), field-effect transistors (reproduced from ref. 43 with permission from American Chemical Society, Copyright 2019), soft actuators (reproduced from ref. 44 with permission from American Chemical Society, Copyright 2019), triboelectric nanogenerators (TENGs) (reproduced from ref. 45 with permission from Elsevier, Copyright 2018), radio-frequency communication (first reported from ref. 46), light-emitting diodes (reproduced from ref. 47 with permission from Wiley & Sons, Copyright 2020), artificial synapses (reproduced from ref. 48 with permission from John Wiley & Sons, Copyright 2021) and immunomodulatory (reproduced from ref. 49 with permission from Elsevier, Copyright 2021).

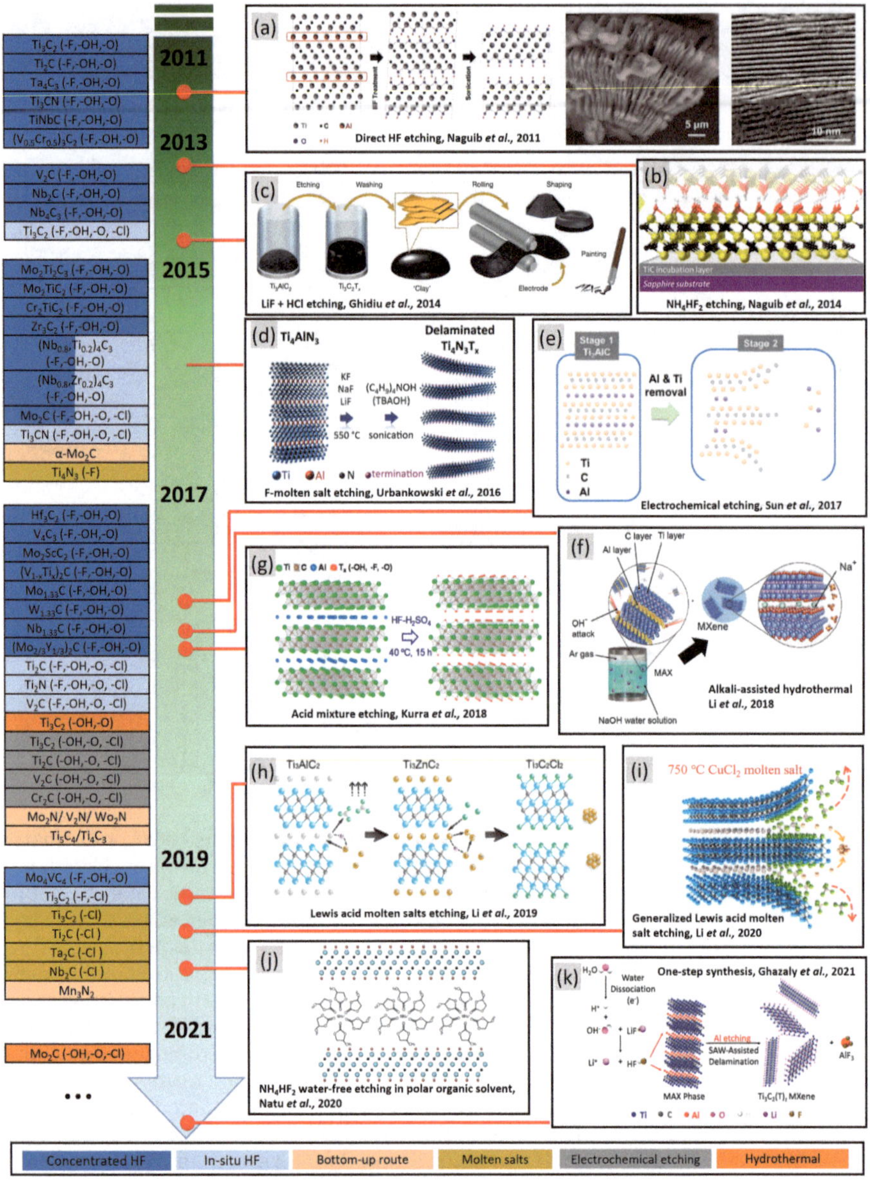

Figure 3.3 Timeline showing the development of MXene etching strategies and the respective MXenes synthesised over the same period (with the corresponding methods of their synthesis, including MAX phase etching and other routes). (a) Reproduced from ref. 56 with permission from John Wiley & Sons, Copyright 2011. (b) Reproduced from ref. 36, https://doi.org/10.1021/cm500641a, under the terms of the CC BY 4.0 Licence, https://creativecommons.org/licenses/by/4.0/deed.en. (c) Reproduced from ref. 57 with permission from Springer Nature, Copyright 2014. (d) Reproduced from ref. 58 with permission from the Royal Society of Chemistry. (e) Reproduced from ref. 59 with permission from the Royal

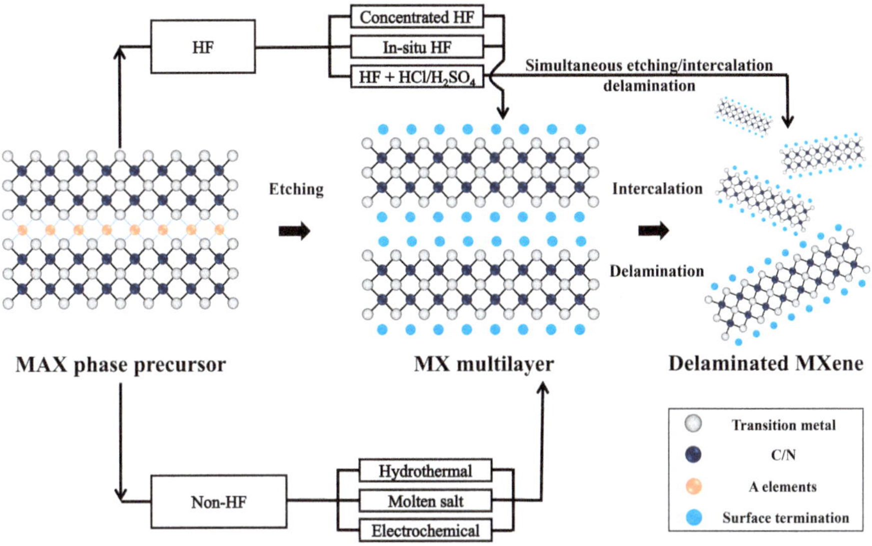

Figure 3.4 Scheme of MXene synthesis *via* the etching of A elements from the corresponding MAX phase.

shaking and sonication. The exfoliation process utilised depends on the etching method used and the composition of the MXenes.[72] Regarding the conventional HF etching methods, pre-intercalation with large organic molecules or ions, such as dimethyl sulfoxide (DMSO), tetrabutylammonium hydroxide (TBAOH), tetramethylammonium hydroxide (TMAOH), hydrazine monohydrate (HM), *etc.*, are required to diminish the interlayer interactions and release the single layer flakes from one another.[73,74] Experimental results show that certain intercalants only work for particular MXenes: for example, DMSO can be used to successfully exfoliate $Ti_3C_2T_x$[73] and $(Mo_{2/3}Ti_{1/3})_3C_2T_x$,[75] yet it does not work for other MXene compositions.[76] TBAOH was found to be capable of exfoliating many MXene composites, but cannot be successfully employed with $Ti_3C_2T_x$.[53,77,78] Interestingly, TMAOH works for $Ti_3C_2T_x$.[72] In addition, the size of the intercalant molecules or ions affects the MXene interlayer spacing, which in turn can dramatically change the charge carrier behaviour among the interlayers.[79] However, regarding the *in situ* HF

Society of Chemistry. (f) Reproduced from ref. 60 with permission from John Wiley & Sons, Copyright 2018. (g) Reproduced from ref. 61 with permission from American Chemical Society, Copyright 2018. (h) Reproduced from ref. 62 with permission from American Chemical Society, Copyright 2019. (i) Reproduced from ref. 63 with permission from Springer Nature, Copyright 2020. (j) Reproduced from ref. 64 with permission from Elsevier, Copyright 2020, (k) Reproduced from ref. 65, https://doi.org10.1021/acsnano.0c07242, under the terms of the CC BY 4.0 Licence, https://creativecommons.org/licenses/by/4.0/deed.en.

Table 3.1 Typical etching methods used to obtain MXenes from MAX phases.

Type of method	Etchant	MXenes	Temperature/°C	Breakthrough/milestone	References
HF-based	Concentrated HF	Ti_3C_2	RT	First MXene reported	105
	HCl + LiF	Ti_3C_2	40	*In situ* HF, single-step	80
	NH_4HF_2	Ti_3C_2	RT/35	*In situ* HF, single-step, water-free synthesis possible	83 and 88
	HF + HCl/H_2SO_4	Ti_3C_2	40	Lower concentration of HF (5%), various surface chemistries	89
Hydrothermal	NaOH	Ti_3C_2	270	Alkali-assisted etching, F-free	94
	HCl	Mo_2C	140	Acid-assisted etching, F-free, Ga/Al-containing MAX	95
Molten salts	KF + LiF + NaF	Ti_4N_3	550	First molten salt etching, first nitride MXene	77
	$ZnCl_2$	Ti_3C_2/Ti_2C	550	F-free, water-free, Cl-terminated MXene	102
	$CuCl_2$	$Ti_3C_2/Ti_2C/Ta_2C$	750	Generalised Lewis etching route to various MAX phases	96
	$CdCl_2/CdBr_2$	$Ti_3C_2/Ti_2C/Nb_2C$	300–600	Controllable and varied terminations	99
Electrochemical	HCl	Ti_2C	NA	First fluoride-free wet etching process	97
	NH_4Cl + TMAOH	Ti_3C_2	NA	Self-intercalation, high yield electrochemical etching	98
	HCl	$Ti_2C/Cr_2C/V_2C$	Up to 50	Universal thermal-assisted E-etching for various MXenes	106

methods, HF and intercalant cations are produced on site, and thus, etching and intercalation/delamination can be achieved simultaneously.[72]

3.2.1 HF-based Etching

As an effective synthesis strategy, concentrated HF has been widely used by researchers at the initial research stage of MXene synthesis (until ~2014).[80] The reactions between HF and MAX precursors (*e.g.*, Ti_3AlC_2) have been postulated to be:[68]

$$Ti_3AlC_2 + 3HF \rightarrow AlF_3 + 3/2H_2 + Ti_3C_2 \tag{3.1}$$

$$Ti_3C_2 + 2H_2O \rightarrow Ti_3C_2(OH)_2 + H_2 \tag{3.2}$$

$$Ti_3C_2 + 2HF \rightarrow Ti_3C_2F_2 + H_2 \tag{3.3}$$

Eqn (3.1) demonstrates how the Al elements of MAX phases are etched away and MX phase components are formed, while eqn (3.2) and (3.3) account for the subsequent formation of surface groups. The concentration of HF was found to be one of the crucial factors towards obtaining the desired final product, since it influences not only the MX formation process from eqn (3.1) and but also the ratio of surface groups in the final products as shown in eqn (3.2) and (3.3). The required exposure times in HF vary from a few hours to days, while etching temperatures can be employed from room temperature up to 60 °C.[68,81]

Although HF has been proven to be an effective etchant for selectively removing the A element from MAX phases, its corrosive and hazardous characteristics raise significant concerns towards the large-scale fabrication of MXene materials and further hinder extensive application of MXene-based devices, which eventually drives researchers to find alternative routes. Thus, following the etching method using highly concentrated HF directly, several other HF-based etching routes have been developed to mitigate the usage of concentrated HF directly, such as *in situ* HF production (a mixture of fluoride salts and HCl,[80,82] or bifluoride salts such as NH_4HF_2[83]) or mixed acid (HF with additional HCl or H_2SO_4) methods.[84,85]

Notably, *in situ* formation of HF is also effective for the selective removal of certain A elements, and can also be beneficial for intercalation and delamination (*e.g.*, Li^+ in case of etching with LiF + HCl). The *in situ* HF etching method was originally introduced to etch Ti_3AlC_2 with NH_4HF_2, where Al can be successfully etched away from the MAX precursors by the HF produced *in situ via* a much less hazardous route.[83] Later, the mixture of LiF and HCl was found to be able to etch the Ti_3AlC_2 MAX phase at 40 °C.[80] Similar strategies are then successfully applied to synthesise Mo_2C[53] and Ti_2C with various fluoride salts (LiF, NaF, KF and NH_4F).[82] These synthesis strategies are also named as MILD (minimally intensive layer delamination) methods.[86] In addition, the MILD methods also show better ability to control the flake size and quality.[87] To date, the MILD methods are widely utilised in those abovementioned MXene syntheses, especially for $Ti_3C_2T_x$, which dominates a great portion of MXene research so far.

NH_4HF_2 etching methods can be further modified and used in polar organic solvents, providing water-free synthesis routes, by which the structural instability caused by the intercalating water from the etching process can be diminished.[88] As another milder etching strategy, etching MAX phases with mixture of HF and additional HCl or H_2SO_4 offers an alternative route to reduce the concentration of HF used directly down to 5%, which can result in high quality MXenes with fewer defects, a wider range of surface chemistries and ultimately providing better performance for certain applications, such as energy storage.[89,90]

The abovementioned milder HF-based etching processes provide safer ways to synthesise various MXenes, yet the –F terminations on MXenes are inevitable, which are non-favourable surface functional groups for some MXene properties (*e.g.*, electronic conductivity,[91] surface tenability,[92] and electrochemical properties[93]). With this in mind, several emerging etching methods without any traces of HF have been developed more recently, offering the possibility of a F-free synthesis strategy, such as (1) hydrothermal etching (Figure 3.5);[94,95] (2) molten salt etching (Figure 3.6);[77,96] or (3) electrochemical etching (Figure 3.7).[97,98]

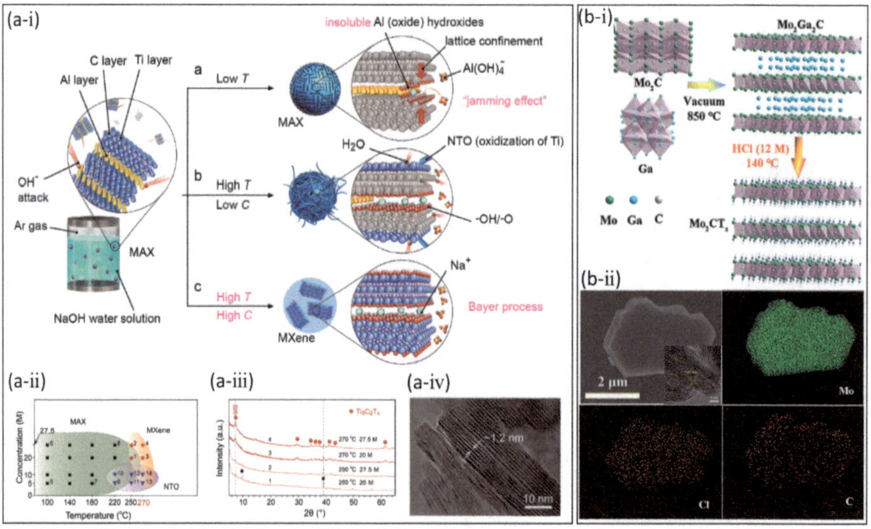

Figure 3.5 Hydrothermal etching approaches for MXene synthesis. (a-i) The reaction between Ti_3AlC_2 and aqueous NaOH as performed under different synthetic conditions, (a-ii) the dominant products obtained as a function of concentration and temperature (red circles: MXene; black squares: MAX; blue triangles: Na/K–Ti–O compounds.), (a-iii) XRD patterns and (a-iv) TEM images of the synthesised MXenes. Reproduced from ref. 94 with permission from John Wiley & Sons, Copyright 2018. (b-i) Schematics of the HCl-assisted approach for Mo_2CT_x synthesis and (b-ii) morphology and element maps of the synthesised Mo_2CT_x. Reproduced from ref. 95 with permission from John Wiley & Sons, Copyright 2021.

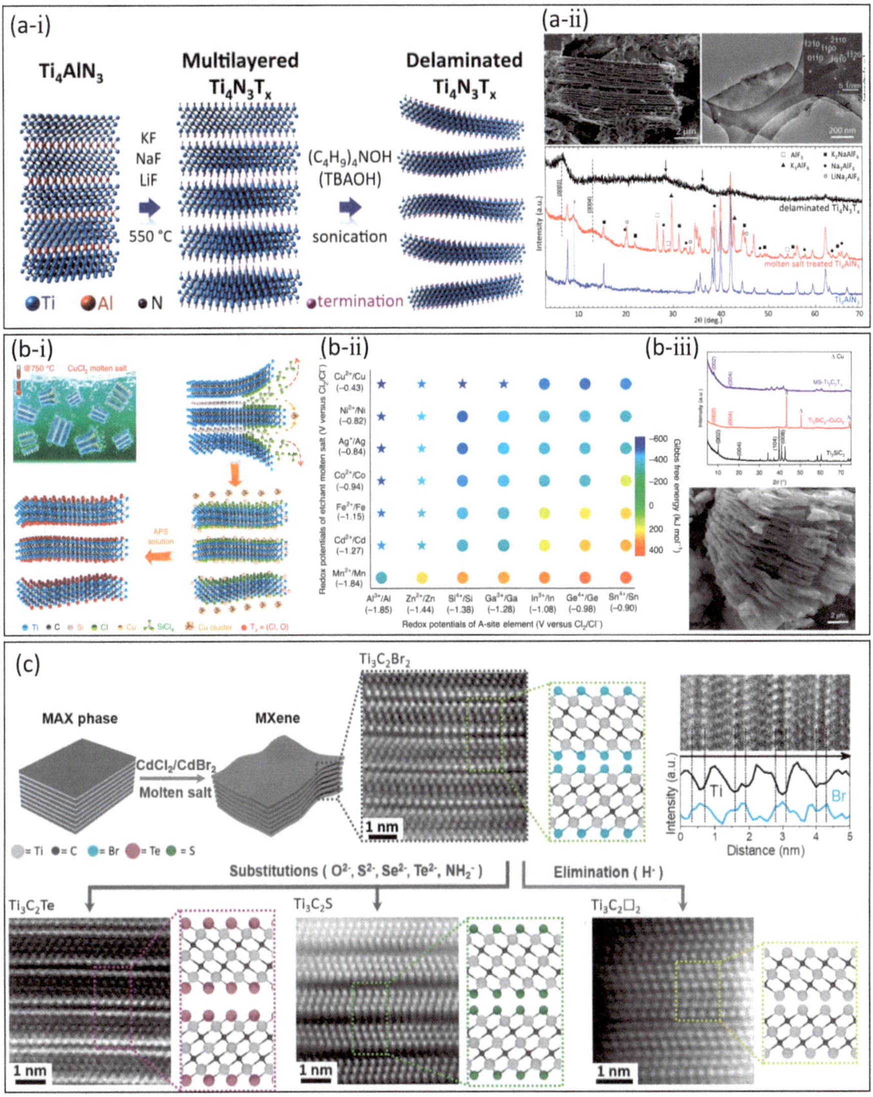

Figure 3.6 Molten salt etching approaches for MXene synthesis: (a-i) synthesis of $Ti_4N_3T_x$ by the F-containing molten salt (KF, NaF or LiF) treatment of Ti_4AlN_3 at 550 °C and (a-ii) the characterisation of the synthesised MXenes. Reproduced from ref. 77 with permission Royal Society of Chemistry. (b-i) General Lewis acid process using $CuCl_2$ at 750 °C, (b-ii) Gibbs free energy mapping (700 °C), which can guide the selection of Lewis acid Cl salts (stars were experimentally verified while the spots remain theoretical predictions) and (b-iii) characterisation of synthesised $Ti_3C_2T_x$. Reproduced from ref. 63 with permission from Springer Nature, Copyright 2020. (c) Surface reactions of MXenes in molten inorganic salts for tunable termination. Reproduced from ref. 103 with permission from American Association for the Advancement of Science, Copyright 2020.

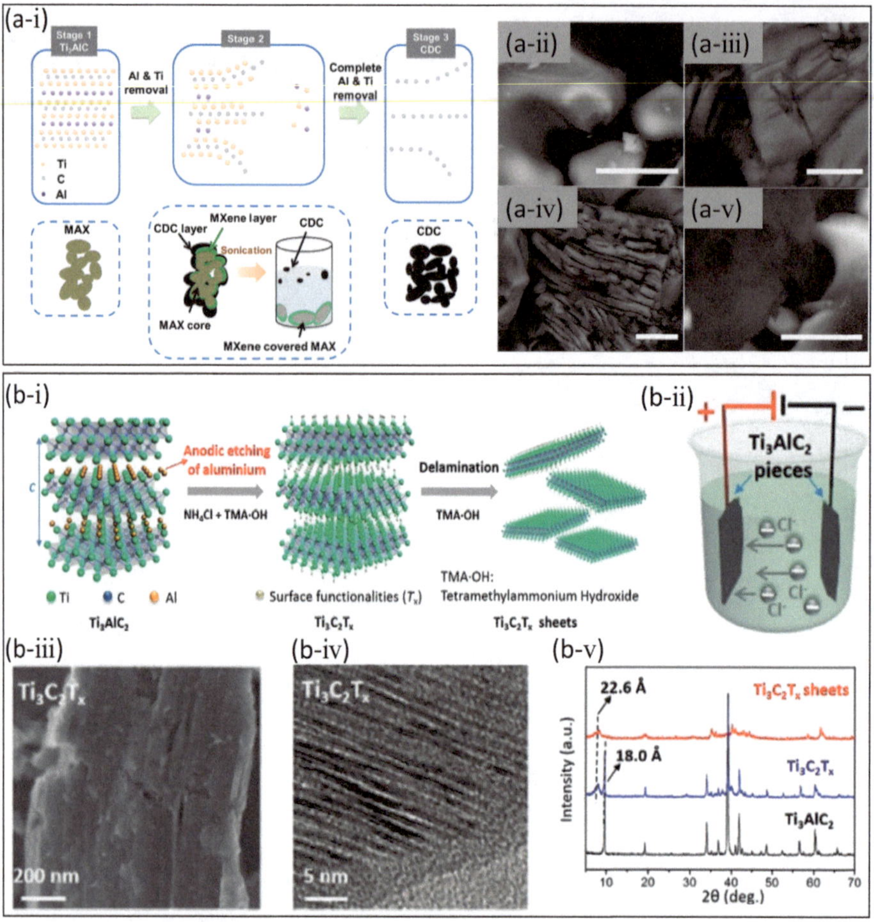

Figure 3.7 Electrochemical etching approaches for MXene synthesis: (a-i) mechanism for electrochemical etching of Ti_2AlC in HCl aqueous electrolyte and the SEM images of (a-ii) the respective MAX phase, and the result after etching with (a-iii) 1 M-0.6 V-1 day, (a-iv) 2 M-0.6 V-5 days and (a-v) 2 M-0.6 V-14 days. Reproduced from ref. 97 with permission from Royal Society of Chemistry. (b-i) Schematic of the Ti_3AlC_2 etching and delamination process with NH_4Cl and TMAOH, (b-ii) the configuration of electrochemical cell, (b-iii) SEM image, (b-iv) TEM image and (b-v) the XRD pattern of the synthesised $Ti_3C_2T_x$ MXenes. Reproduced from ref. 98 with permission John Wiley & Sons, Copyright 2018.

3.2.2 Hydrothermal Etching

Alkali-assisted hydrothermal etching originally utilised sodium hydroxide to extract the Al layers from Ti_3AlC_2, due to the strong binding propensity of OH^- and Al elements.[94] Notably, the final etching products are highly dependent on the processing temperature and the concentration of the alkaline reagent. High-yield MXene production can only be achieved with high alkaline concentration at high temperature. Performing the etching process at low temperature is inefficient in dissolve the Al component of the MAX phase. Conversely, etching with low alkaline concentration at high temperature can only produce MXenes on the surface layer of the parent MAX phase, because the formed Al hydroxides (oxides) hinder the etchant from penetrating deeper into the bulk during the subsequent etching process. When the concentration of alkali increases, the excess alkali dissolves the Al hydroxides (oxides), while the remaining alkali is able to permeate further into the bulk to etch the MAX phase layers below. Li *et al.* showed that etching Ti_3AlC_2 with 27.5 M NaOH at 270 °C can successfully produce multilayer F-free $Ti_3C_2T_x$ with 92 wt% purity (Figure 3.5(a)).[94] Lately, inspired by the mechanism of the alkali-assisted etching strategy and the strong binding ability of Cl^- with various A elements such as Al and Ga, an HCl-based hydrothermal etching strategy was reported to obtain Mo_2C and Cr_2C from Mo_2Ga_2C and Cr_2Al_2C, respectively (Figure 3.5(b)).[95] This thus extends the scope of hydrothermal etching methods for synthesising a variety of fluoride-free MXenes from a widening selection of MAX phase precursors.

3.2.3 Molten Salt Etching

Since, unlike for carbide MAX phases, aqueous HF etching methods fail to etch the A element from the nitride MAX phases,[77] an etching strategy with fluoride-containing molten salts (KF, LiF, and NaF) was initially used for synthesis of $Ti_4N_3T_x$, which is the first-reported nitride-based MXene (Figure 3.6(a)). Since the type of surface termination for the final MXene product is solely determined by the salts involved, it followed that developing an F-free etching process based on a molten salts strategy would be advantageous.[99] Although, as has been reported, carbide MXenes can be effectively synthesised by multiple wet etching methods, various carbide MXenes (Ti_2C, Ti_3C_2 and Nb_2C) have also been synthesised *via* a molten salts etching strategy more recently, with the intention of taking advantage of the non-aqueous process and controllable termination.[100,101] Lately, Ti_3C_2 and Ti_2C MXenes with pure chlorine terminations were reported by etching the corresponding Al-containing MAX counterpart in Lewis acidic melts composed of $ZnCl_2$.[102] Notably, a low ratio of the Lewis acid in the starting mixture with Al-MAX (Ti_3AlC_2–$ZnCl_2$ = 2 : 3) will result in the final product of Zn-MAX (Ti_3ZnC_2) *via* a substitution reaction of Al by Zn, while a higher ratio of $ZnCl_2$ (Ti_3AlC_2–$ZnCl_2$ = 1 : 6) in the starting mixture further transforms the Zn-MAX into the $Ti_3C_2Cl_2$ MXenes and excess $ZnCl_2$. In addition, this strategy can be further generalised to become applicable to several Ga-, and Si-containing MAX phases when used together with molten $CuCl_2$ at 750 °C (Figure 3.6(b)).[96] As a result, Cl-terminated MXenes are obtained exclusively *via* this method, which are expected to be more stable than the F-terminated MXenes.[59,62] Moreover, various functional groups, *e.g.*, –NH, –S, –Cl, –Se, –Br, –Te, can also be utilised as the surface terminations *via* substitution reactions using the appropriate reagents, offering a significant tunability to the surface chemistry of MXenes (Figure 3.6(c)).[99] Overall, the moderate and F-free processing conditions offer a green and viable method for the synthesis of MXenes. Nevertheless, molten salt methods require a high processing temperature (550 °C–750 °C) and the yields still need to be improved.

3.2.4 Electrochemical Etching

An electrochemical etching process, in which A elements are selectively removed in chloride-containing electrolytes by regulating time and reaction potential so as to break the M–A bond, has emerged as another practical approach for fluoride-free synthesis. Scaling up this method to mass production volumes, however, may still present a challenge, since the efficacy of this technique is restricted by the formation of carbide-derived carbon (CDC) on the MAX phase surface layer (Figure 3.7(a)).[97] Employing intercalants (*e.g.*, TMA^+) is an effective method of mitigating the restriction imposed by the surface-covered CDC layer to continue the etching process in the MAX phase beneath. Intercalation serves to expand the interlayer distance of MXenes, enabling the etchant to reach deeper layers of the precursors (Figure 3.7(b)) and eventually improve the yields for large-scale production.[98] However, the toxicity of the utilised intercalants still raises safety concerns. To solve this problem, Pang *et al.* proposed a more generalised three-dimensional (3D) electrode-assisted electrochemical etching method using dilute HCl at temperatures up to 50 °C and avoiding the use of the toxic intercalants. This strategy is able to fabricate not only Ti-based MXenes but also other metal-based MXenes such as Cr_2C and V_2C through a safer and rapid synthesis process.[104]

3.2.5 Bottom-up Synthesis Strategy

In contrast to the methods discussed earlier, bottom-up strategies, aiming to construct MXene-like structures from the precursor molecular or atomic arrangements, using methods such as chemical vapour deposition (CVD) (Figure 3.8(a)),[55,107,108] and templating (Figure 3.8(b) and (c)), have been reported.[109–111] α-Mo_2C was the first reported multilayer transition metal carbide structure to be synthesised *via* CVD at 1085 °C, where Mo atoms are able to diffuse into the surface layer of the molten Cu and react with the carbon resource (methane).[55] Lately, 2D Ti_4C_3 and Ti_5C_4 were synthesised by growing a hexagonal TiC layer on the defunctionalised surface of Ti_3C_2 at a temperature over 500 °C.[112] Additionally, various nitrides (V_2N, W_2N, and Mn_3N_2) have been prepared *via* a salt-templated method involving the reaction of ammonia with the templated salt precursors at temperatures of 650–750 °C.[111,113] Bottom-up routes enable precise control of the size distribution, morphology and surface termination of MXene-like structures, as well as enabling the synthesis of various defect-free structures that are not obtained *via* conventional etching strategies. Yet, notably, synthesis of single layer MXene-like

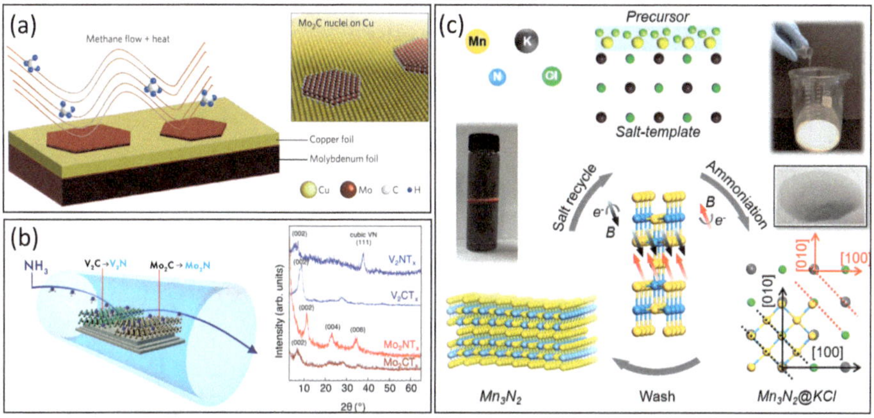

Figure 3.8 Bottom-up synthesis routes for MXene-like structures. (a) Schematics of 2D Mo_2C crystal produced by chemical vapor deposition (CVD). Reproduced from ref. 107 with permission from Springer Nature, Copyright 2015. (b) Ammonisation synthesis using Mo_2C and V_2C to form their nitride counterparts. Reproduced from ref. 109 with permission from Royal Society of Chemistry. (c) Salt templated method to transform $MnCl_2$@KCl into 2D Mn_3N_2@KCl. Reproduced from ref. 111 with permission from John Wiley & Sons, Copyright 2019.

structures rather than ultrathin films with few layers *via* these bottom-up approaches still faces challenges. To date, only a few 2D transition metal carbides have been reported *via* bottom-up approaches.

3.2.6 Scalable Synthesis

Scalable synthesis methods for MXenes that are both fast and safe are of critical significance to take MXenes from the laboratory to industrial settings (Figure 3.9(a)). Top-down approaches, by etching the MAX precursors, intrinsically show more feasibility for industrial level manufacturing. A recent study showed that MXene yield can be scaled up to 50 g from 1 g per cycle without fundamentally changing the physical and chemical properties or sacrificing performance (Figure 3.9(b)).[114] Moreover, a prototype of a kg-level production-line is under-construction, further paving the way for industrial mass-production (Figure 3.9(c)). Researchers are also trying to increase the MXene yields per cycle to reduce the time consumption requirement for MXene synthesis, which is another aspect necessary for the viable scale up of MXene production. To date, several fast synthesis routes are proposed,[115,116] where the major idea is to achieve the etching, intercalation and delamination processes simultaneously. For example, Ghazaly *et al.* recently, integrated a localised surface acoustic wave (SAW) method with *in situ* HF etching to achieve an ultrafast, one-step MXene synthesis, highlighting the possibility of increasing MXene production yields (Figure 3.9(d)).[115] With the production of MXenes at an industrial scale continuing to present

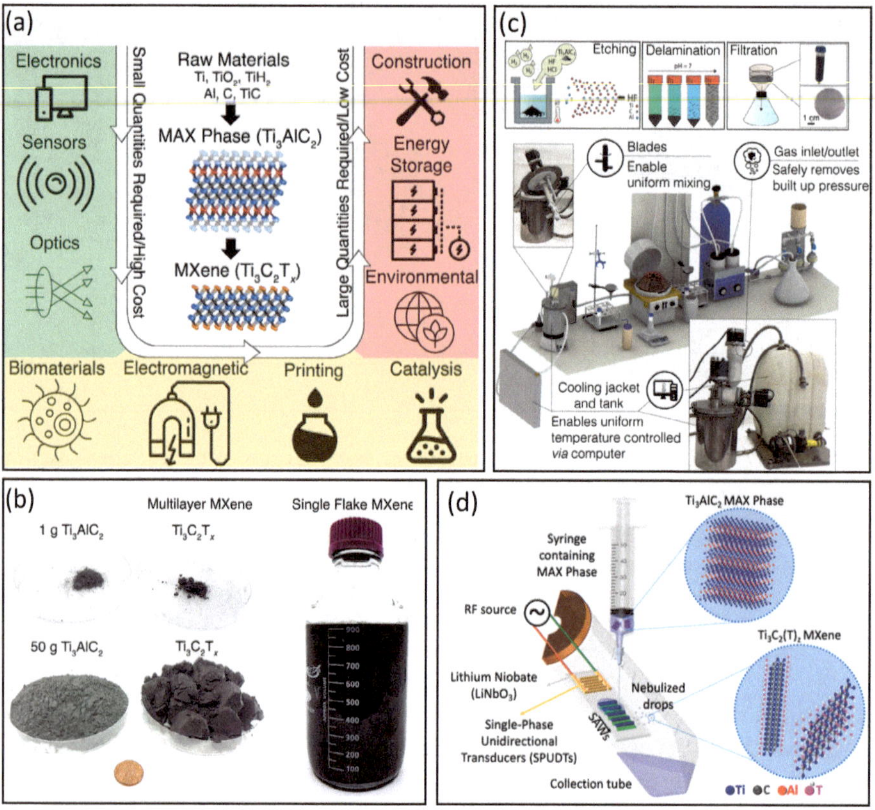

Figure 3.9 Bringing MXenes from the laboratory to industry. (a) Various applications of MXenes with the corresponding different quantity requirements. Reproduced from ref. 117 with permission from Elsevier, Copyright 2020. (b) Scalable synthesis of the $Ti_3C_2T_x$ MXene. Reproduced from ref. 114 with permission from John Wiley & Sons, Copyright 2020. (c) A prototype production line for kg-level MXene manufacture. Reproduced from ref. 114 with permission from John Wiley & Sons, Copyright 2020. (d) Ultra-fast, one-step MXene synthesis employing an integrated SAWs/*in situ* HF (HCl + LiF) etching strategy. Reproduced from ref. 115, https://doi.org/10.1021/acsnano.0c07242, under the terms of the CC BY 4.0 license, https://creativecommons.org/licenses/by/4.0/deed.en.

challenges, significant work is still needed to bring MXenes from the laboratory to industrial production.

3.3 Structure and Composition

Mxenes, typically written as $M_{n+1}X_nT_x$ ($n = 1–4$) (Figure 3.10(a)), consists of n + 1 layers of the early transition metals (M) interleaved with n layers of carbon or nitrogen (X) in a hexagonal symmetry inherited from the MAX precursor

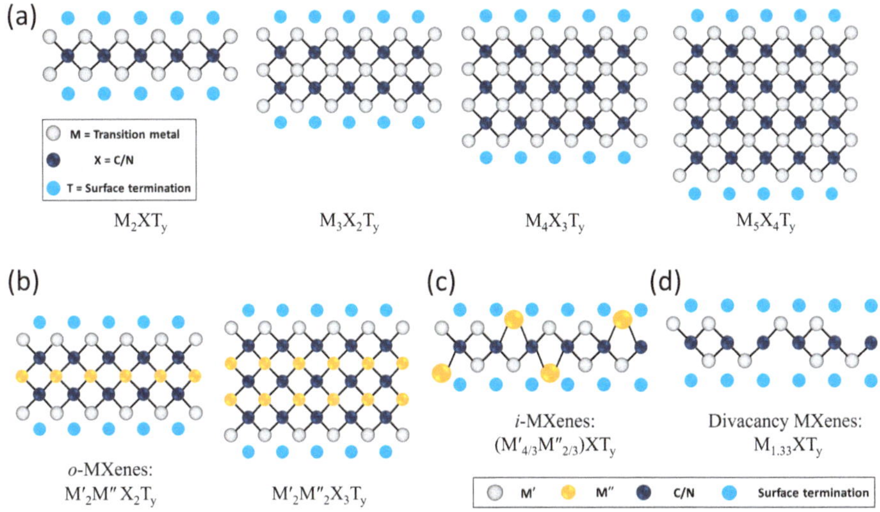

Figure 3.10 Atomic structures of (a) typical MXenes, (b) out-of-plane MXenes (o-MXenes), (c) in-plane MXenes (i-MXenes) and (d) divacancy MXenes.

phases. The surface terminators (T), such as –O, –OH, –F, and –Cl, –Br, –S, –Se and –Te (depending on the etching approaches involved and the MXene composition) are bonded on the surface of the metal layers during the synthesis process.[99] Following $Ti_3C_2T_x$, the first experimentally synthesised MXene (in 2011),[105] more than 30 MXenes have been reported experimentally (as indicated in Figure 3.3) and over 100 stoichiometric MXene compositions have been predicted.[118,119] Moreover, apart from those MXenes consisting of only one M metal and one X (either carbon or nitrogen), MXenes containing two or more M metals and more than one X element are also achievable. When combined with the various single or mixed surface terminations, the compositional diversity of the MXene based 2D materials is enhanced tremendously.[120] To date, MXenes with transition metal elements of Ti, V, Nb, Mo, Cr, Zr, Hf, Sc, Ta, W, Y and Mn have been reported, where Sc, W and Y have only been present as components together with other metals.

MXenes with more than one M element exist in two major categories: solid solutions (where a random distribution of the two different transition metals is observed in the M layers), and ordered phases (where the two transition metals have an ordered distribution). Ordered MXenes with certain combinations of transition metals are predicted to be more stable than their counterparts in a solid solution type.[75] In the ordered MXenes, two transition metals atoms (M′ and M″) either exist as a regularly-ordered mixture in the same M layer, or form structures in which layers of the M″ elements are sandwiched between layers of M′. The former structures are named the in-plane MXenes (i-MXenes) while the latter structures are identified as out-of-plane MXenes (o-MXenes). The structures of o-MXenes and i-MXenes are schematically illustrated in Figure 3.10(b) and (c). To date, all the i-MXenes reported follow a formula of $(M'_{4/3}M''_{2/3})XT_x$, where the two constituent transition metal elements (M′ and M″) are present in a ratio of 2 : 1. In i-MXenes, the size difference of M′ and M″ should be at least 0.2 Å, where the smaller M′ atoms form a hexagon lattice and the larger M″ atoms are located at the hexagon centre and extend out from the metal layers.[121] Additionally, due to the weaker bonding between M″ and the carbon sites as compared to M′–C bonds, the M″ elements such as Sc and Y, can be selectively etched away, leaving ordered divacancy MXenes as $M'_{1.33}XT_x$ (Figure 3.10(d)).[122,123] To date, M_2CT_x is the only reported template for divacancy MXenes, giving the final compositions as $M_{1.33}CT_x$ including $Mo_{1.33}CT_x$,[122] $W_{1.33}CT_x$,[123] and $Nb_{1.33}CT_x$.[124] The o-MXenes, however, exist with formulae of either $M'_2M''X_2T_x$ where the M′ is the outer layer, or $M'_2M''_2X_3T_x$.[120] Likewise, besides pure carbide[105] and nitride[77] MXenes, a carbonitride-based MXene (Ti_3CN) has also been synthesised, where the C and N atoms occupy the X sites randomly.[68] Thus far, Ti_3CN is the only reported carbonitride-based MXene.

As illustrated in the synthesis section, different surface terminations can be introduced with single or mixed functional groups on the MXenes outer layers, which ultimately have a profound influence on the MXene surface chemistry.[91] The type of surface terminations depend on the specific synthesis methods, and can be modified during the synthesis or even thoroughly removed by post-synthetic processing.[99] For instance, traditional HF etching process gives dominant –F terminations in combination with –OH and –O. Conversely *in situ* HF (HCl + LiF) etching method can reduce the percentage of –F terminations among the surface groups, and possibly introduce –Cl terminations. F-free MXenes are attainable *via* electrochemical etching in chloride containing solution,[97,98] *via* hydrothermal processes with NaOH or HCl[94,95] or by molten Lewis acid etching methods.[96] Notably, Lewis acidic etching using $ZnCl_2$ results in purely Cl-terminated MXenes, which are expected to be more stable than the F-terminated MXenes. More recently, NH_4HF_2 together with a polar organic solvent was used for a water-free etching process, resulting in highly F-terminated structures where both –O and –OH are absent.[88]

3.4 Solution Processing

The applications of MXenes are heavily dependent on their processing methods and manufacturing strategies. Since MXenes are mainly synthesised by solution-based methods, liquid-phase processing techniques such as vacuum filtration,[125] coating (*e.g.*, dip-,[126] spray-,[127] and spin-coating[128]) and printing (*e.g.*, inkjet printing,[129] extrusion printing,[129] screen printing[130]), are preferable in terms of device fabrication (Figure 3.11). Through these techniques, MXene dispersions are usually processed into thin films by interfacial assembly. Each processing technique requires an MXene dispersion with specific properties, such as the degree of stability and the basic rheological properties, which influence final MXene-based microstructures. The stability and rheology of a MXene dispersion heavily depend on the particle size, surface

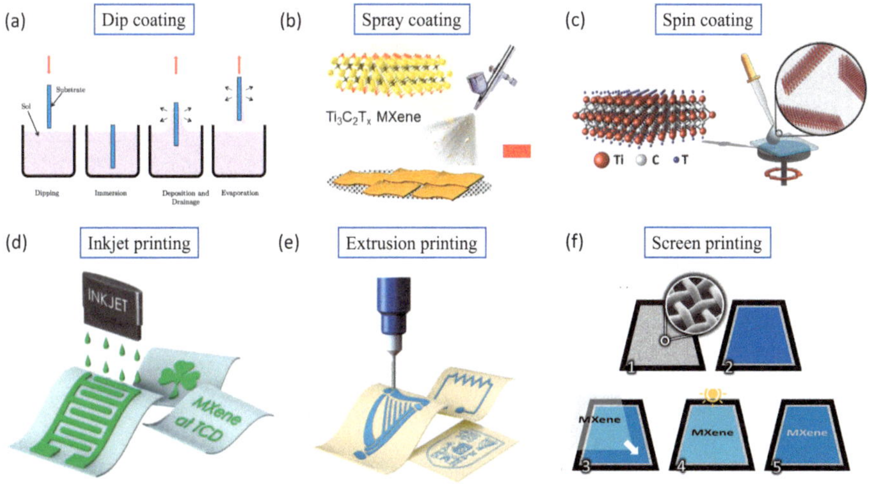

Figure 3.11 Processing methods for fabricating MXene films. (a) Dip coating. Reproduced from ref. 126, https://doi.org/10.3390/ma9090728, under the terms of the CC BY 4.0 license, https://creativecommons.org/licenses/by/4.0/deed.en. (b) Spray coating. Reproduced from ref. 127 with permission from John Wiley & Sons, Copyright 2019. (c) Spin coating. Reproduced from ref. 128 with permission from John Wiley & Sons, Copyright 2016. (d) Inkjet printing, and (e) extrusion printing, both reproduced from ref. 129 https://doi.org/10.1038/s41467-019-09398-1, under the terms of the CC BY 4.0 license, http://creativecommons.org/licenses/by/4.0/. (f) Screen printing. Reproduced from ref. 130 with permission from Elsevier, Copyright 2021.

charge and chemistry of the MXenes, which can be tuned through both the synthesis routes (the type and concentration of etchants, intercalation, and delamination methods, *etc.*) and post-processing (*e.g.*, high temperature vacuum annealing).[131]

MXenes possess rich surface terminations, superior hydrophilicity, high negative surface charge (ζ-potential of −30 to −80 mV) and strong electrostatic repulsive forces, making them readily dispersible and stable in aqueous media without the need for additives.[132] MXenes exhibit versatile dispersion behaviours in different organic solvents, from polar protic solvents and polar aprotic solvents to less polar or non-polar solvents (Figure 3.12(a)).[132] For instance, $Ti_3C_2T_x$ with polar terminations is more easily dispersed in polar protic and aprotic solvents; the less polar solvents give little or even no dispersions of $Ti_3C_2T_x$; $Ti_3C_2T_x$ dispersions in dimethyl sulfoxide (DMSO), *N*,*N*-dimethylformamide (DMF), *N*-methy-2-pyrrolidone (NMP), propylene carbonate (PC) and ethanol (EtOH) exhibit similar stability as a $Ti_3C_2T_x$ colloidal solution in water. The dispersion stability not only relies on the surface groups of the MXenes, but also depends on the properties of selected dispersion media, such as surface tension, boiling

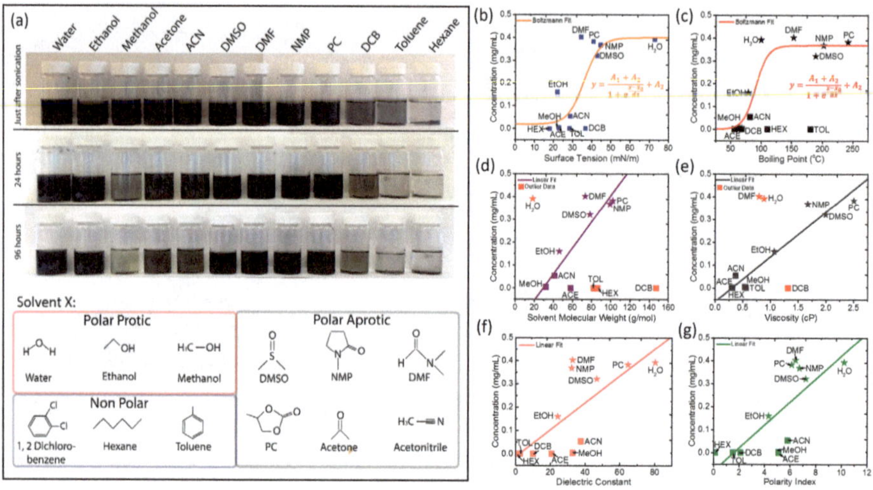

Figure 3.12 (a) Dispersions of $Ti_3C_2T_x$ in different types of solvents; (b)–(g) concentrations of MXenes in solution plotted as a function of the following properties of solvents: (b) surface tension, (c) boiling point, (d) molecular weight, (e) viscosity, (f) dielectric constant and (g) polarity index. Reproduced from ref. 132 with permission from John Wiley & Sons, Copyright 2017.

point, molecular weight, viscosity, polarity index and dielectric constants (Figure 3.12(b)–(g)). For example:

- Water exhibits a surface tension of 72 mN m^{-1} and $Ti_3C_2T_x$ is likely to require solvents with surface tensions over 40 mN m^{-1} to form a stable solution (Figure 3.12(b)).
- Solvents with high-boiling point are more effective in dispersing $Ti_3C_2T_x$ (Figure 3.12(c)).
- The concentrations of stable $Ti_3C_2T_x$ colloidal solutions follow a linear fit correlated with solvent molecular weight, if water (as an example of a 'good' solvent with stable dispersion over at least 96 h) and some poor solvents (less or no dispersion), such as hexane, toluene, and dichlorobenzene are excluded (Figure 3.12(d)).
- High viscosity solvents offer high kinetic dispersion stability of $Ti_3C_2T_x$ (Figure 3.12(e)).
- Solvents with high dielectric constants provide a better dispersion for negatively surface charged $Ti_3C_2T_x$ (Figure 3.12(f)).
- The dielectric constant correlates to the solvent polarity index, suggesting that solvents with a higher polarity index are also good for dispersing $Ti_3C_2T_x$ (Figure 3.12(g)).

In summary, polar solvents with a high surface tension, a high boiling point and possessing a high dielectric constant are beneficial to minimise the solvent–nanosheet and solvent–solvent interactions, thus leading to better dispersion stability for $Ti_3C_2T_x$.[132,133]

The diverse rheological properties (both viscous and viscoelastic characteristics) of MXene dispersions can provide information for various solution-processing techniques. The concentration of MXene suspension and the aspect ratio of MXenes (single-layer and multilayer MXenes) are the most important factors in influencing their rheological behaviours.[134–136] Further, dispersions of single-layer and multilayer MXenes should be treated as two distinct colloidal systems.[135]

Single-layer $Ti_3C_2T_x$ dispersions (with an average particle size less than ~1 μm) exhibit viscosities that increase steeply as the $Ti_3C_2T_x$ concentration (or corresponding volume fraction, φ) is increased, especially at low shear rate; the varying disparity of viscosity with concentration is reduced at higher shear rates (Figure 3.13(a)). By contrast, the multilayer $Ti_3C_2T_x$ dispersions maintain a similar trend in viscosity with concentration, both at low and high shear rates (Figure 3.13(b)). Moreover, single layer $Ti_3C_2T_x$ dispersions specifically show a strong dependency of the low shear rate relative viscosity on concentration at low volume fractions (Figure 3.13(c) and (d)), indicating their processability at very dilute concentrations. Based on this behaviour, single-layer $Ti_3C_2T_x$ dispersions with high relative viscosities obtained at low concentrations could be used for the fabrication of ultra-transparent and conductive films, which are highly important for applications of micro-electronics, energy storage, catalysis, *etc.*[137–140] However, there is an upper concentration limit of stability for single-layer $Ti_3C_2T_x$ dispersions. When a given processing technique requires higher viscosities at high concentrations, the consistency of the single-layer $Ti_3C_2T_x$ dispersions becomes a potential issue. In such instances, multi-layer $Ti_3C_2T_x$ particles, which form stable dispersions with much higher viscosities even at high shear rate, are more suitable.

The dispersions of single-layer flake and multi-layer $Ti_3C_2T_x$ also behave differently with respect to viscoelastic characteristics (G' and G''). Specifically, single-layer $Ti_3C_2T_x$ dispersions exhibit noticeable values of G' ($\varphi < 9.47 \times 10^{-5}$) even at ultra-low concentrations (<0.36 mg mL^{-1}), enabling a high level of control in deposition processes such as spray coating and spin coating; these are both techniques that usually require relatively high elasticity (elasticity of the dispersions helps to eliminate the perturbations associated with conventional viscous fluids during processing).[137] At a moderate concentration of 0.9 mg mL^{-1}, a single-layer $Ti_3C_2T_x$ dispersion exhibits a viscoelastic soft solid behaviour at low frequencies and a power-law scaling behaviour at high frequency, which is suitable for inkjet printing applications (such rheological properties help to maintain the form of the printed substances when laid out on a substrate) (Figure 3.13(e)).[141] When further increasing the concentration to 2.7–3.6 mg mL^{-1}, the responses of G' and G'' with respect to frequency are similar to those of a soft gel,[142] and a processing technique (*e.g.*, wet spinning) suitable for gel-like materials could be applied in fabricating MXene products (Figure 3.13(e)).

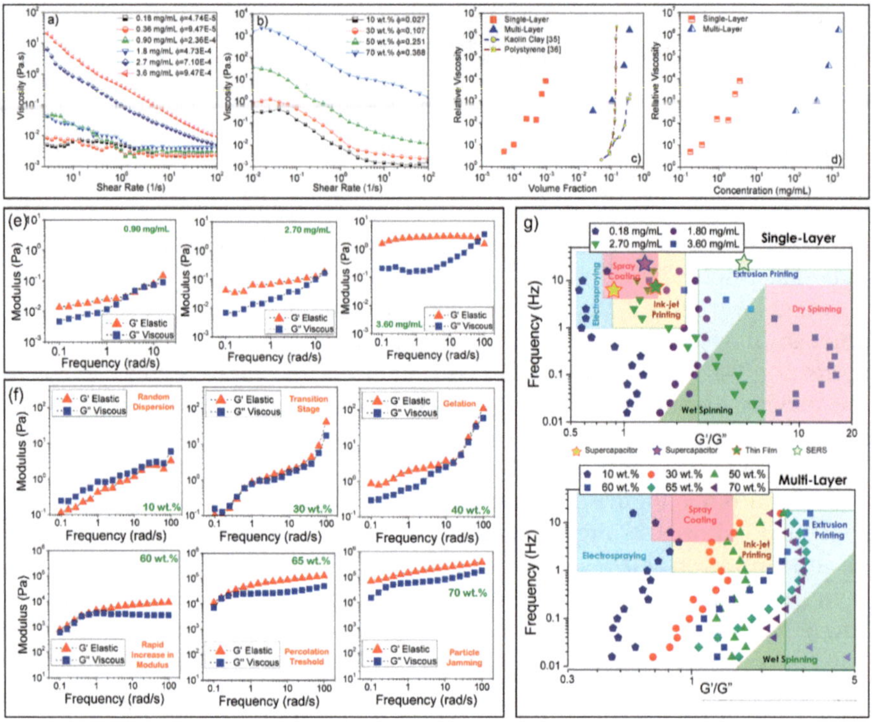

Figure 3.13 Rheological properties of $Ti_3C_2T_x$ dispersions. (a) Viscosity *versus* shear rate for single-layer MXene dispersions. (b) Viscosity *versus* shear rate for multilayer MXene dispersions. (c) Relative viscosity *versus* volume fraction plots for single and multilayer $Ti_3C_2T_x$ dispersions. (d) Viscosity *versus* concentration plots for single and multilayer $Ti_3C_2T_x$ dispersions. (e) Viscoelastic measurements conducted on a single-layer MXene flake suspensions with various concentrations. (f) Viscoelastic measurements conducted on multilayer MXene dispersions with various solid loadings. (g) Frequency dependency of the $G':G''$ ratio (where G' is the elastic modulus and G'' is the viscous modulus) for single-layer and multilayer $Ti_3C_2T_x$ MXenes dispersed in water with various solid loadings (wt%). The favoured MXene film processing methods are indicated. Reproduced from ref. 135 with permission from American Chemical Society, Copyright 2018.

It should be noted that the absolute value of G' for a single-layer MXene dispersion is approximately 2 Pa even when the concentration is 3.6 mg mL^{-1}, which is insufficient for processing routes such as extrusion printing or dry spinning (which require elasticity of >100 Pa for processing).[137,143] The geometrical characteristics of single flakes (high aspect ratio) result in a lower packing density, limiting the interactions of multiparticle and G' accordingly.[144] In contrast, multi-layer MXene particles with a random radial morphology enable higher packing densities. Thus, multilayer dispersions exhibit a weaker electrostatic force than single layer dispersions (surface charge of −28.5 mV for multilayer particles, −45.9 mV for single layer flakes), further promoting higher solid loading for multi-layer dispersions. Moreover, the 'swelling' of multilayer particles results in multiparticle interactions, which also improves the elasticity.[145] As a result, the viscoelastic behaviours for multilayer MXene dispersions over different concentration ranges exhibit similar trends to those of the single-layer dispersions (Figure 3.13(f)) but occur at a much higher concentration regime. Dispersions of multilayer MXenes at concentrations of 40–60 wt% offer high elasticity values of over 100 Pa. When further increasing the concentrations, the high magnitude of G' for multilayer $Ti_3C_2T_x$ dispersions makes it suitable for extrusion printing or dry spinning.[137]

Hence, MXenes dispersions with single-layer or multilayer each have their advantages in terms of specific processing methods (Figure 3.13(g)). The desired properties of an MXene film (*i.e.*, resolution, conductivity, thickness, or transparency) can be prescribed through selecting suitable processing techniques and adjusting MXene dispersions with corresponding rheological properties, which is of great importance to achieve the efficient manufacturing of MXene-based devices for different applications. When depositing MXene materials onto particular substrates, the process also relies heavily on the interactions between the layered MXenes and the hosting substrates. Specifically, surface energies of the dispersions and substrates should match with each other to ensure a good wetting and adhesion. The drying behaviour should also be taken into consideration to optimise the MXene-based functional structures.[130]

3.5 Properties

3.5.1 Oxidation Stability

MXenes suffer from severe oxidative degradation in the presence of aqueous dispersions with dissolved oxygen, especially at elevated temperatures or under sunlight, which drastically deteriorates their characteristics.[146,147] Improving the oxidation stability and prolonging the shelf life of MXenes are essential for their practical applications. Each step from synthesis through storage and usage to post-processing has a strong impact on the resulting features of MXenes. Understanding the fundamentals behind the oxidative degradation is of importance to control the oxidation kinetics.

The defects (for example, the vacancies) present in delaminated MXene flakes (inevitably induced by harsh chemical etching synthesis processes) usually lead to oxidative degradation reactions.[148] The smaller the MXene flakes, the faster the degradation.[148–150] In general, a fresh black $Ti_3C_2T_x$ MXene dispersion gradually degrades to a cloudy-white $Ti_3C_2T_x/TiO_2$ (anatase crystal structure) composite solution after 5 days when placed in an ambient environment (Figure 3.14(a)).[151,152]

The oxidation kinetics of MXene flakes can be altered by the synthesis methods. For instance, the defect density is reduced when using HF with

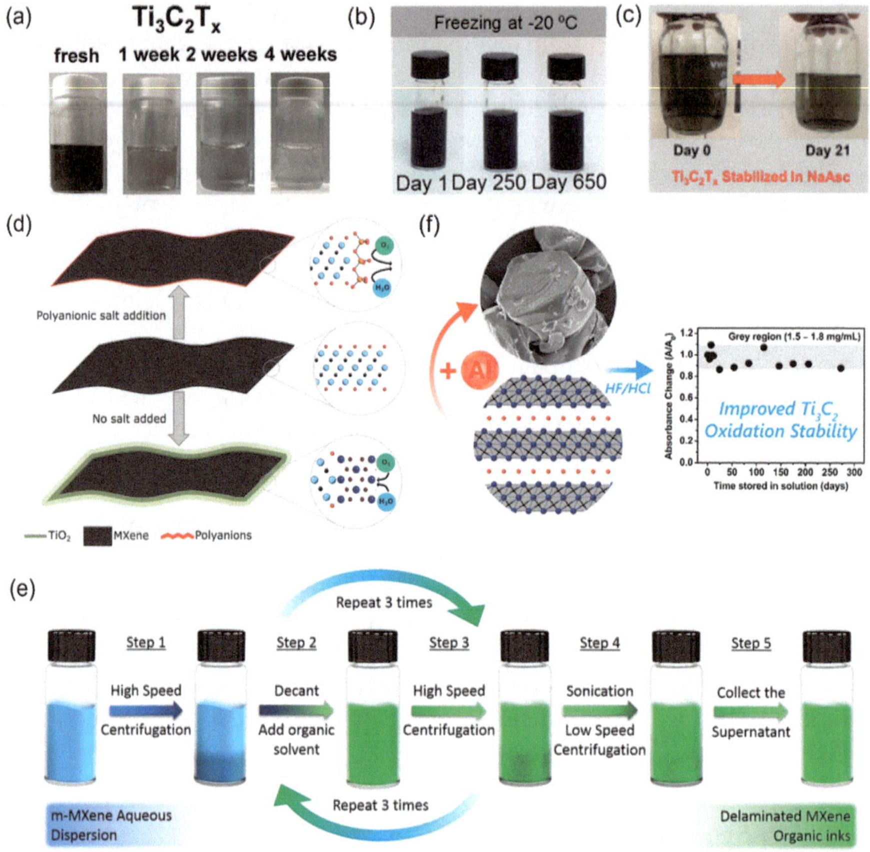

Figure 3.14 (a) Visual appearance of $Ti_3C_2T_x$ in water/O_2 over time. Reproduced from ref. 153 with permission from American Chemical Society, Copyright 2019. (b) Photos of the aqueous MXene dispersion stored at −20 °C for 250 and 650 days, respectively. Reproduced from ref. 154 with permission from American Chemical Society, Copyright 2020. (c) Images of the MXene dispersions at day 0 and day 21 in sodium L-ascorbate (NaAsc). Reproduced from ref. 149 with permission from Elsevier, Copyright 2019. (d) Edge capping of MXene sheets by polyanions. Reproduced from ref. 150 with permission from American Chemical Society, Copyright 2020. (e) Transferring MXenes to organic solvents by repeated centrifugation/redispersion. Reproduced from ref. 129, https://doi.org/10.1038/s41467-019-09398-1, under the terms of the CC BY 4.0 license, http://creativecommons.org/licenses/by/4.0/. (f) Modified MAX precursors with excess aluminium to improve the oxidation stability of the corresponding MXenes. Reproduced from ref. 155 with permission from American Chemical Society, Copyright 2021.

a low etching concentration or a mixture of etchant acids. Storage and post-processing strategies are also efficient in improving the oxygen stability of MXenes. Storing MXenes in an Ar atmosphere and/or retaining samples in a low-temperature environment (5 °C), has the effect of preventing degradation even after 3 weeks.[148] The quality of MXene flakes can be maintained over a period as long as two years by freezing their aqueous dispersions at −20 °C (Figure 3.14(b)).[148,154] The oxidation of MXenes generally starts from the edge sites of a flake, then gradually expands towards the interior.[148] Capping the active edges of MXene flakes with polyanionic salts (*e.g.*, polyphosphates, polysilicates or polyborates) can therefore be effective to mitigate oxidation (Figure 3.14(d)).[150] Adding an antioxidant such as sodium L-ascorbate, NaAsc, to a dispersion is also capable of protecting the edges of the nanosheets by restricting water molecules from reactive sites in aqueous suspensions (Figure 3.14(c)).[149] Besides dissolved oxygen, water is another primary factor in the shortening of the shelf life of MXene dispersions.[152,153] Replacing water with an organic solvent (such as NMP, DMSO, ethanol) through multiple solvent exchange (*e.g.*, three times without sonication) is a powerful technique to minimise the interactions between the MXenes and dissolved oxygen and/or water (Figure 3.14(e)), hence improving the oxidation stability.[129,156,157] Attempts to improve chemical stability have also focused on modifying the MAX precursors in the original MXene synthesis.[158] For example, including excess aluminium during the fabrication of Ti_3AlC_2 MAX precursors has been proven to reduce carbon vacancies in Ti_3AlC_2 grains, thus leading to both improved crystallinity and Ti–C stoichiometry (closer to an ideal ratio of 3 : 2). The associated loss of Ti atoms near carbon vacancies after MAX etching also decreases, which in turn minimise the locations for water hydrolysis to occur and thereby slows the degradation rate of $Ti_3C_2T_x$ MXene flakes. Consequently, the corresponding MXene sheets obtained from such modified MAX precursors demonstrate an increased resistance to oxidation (ten months shelf life, even when stored in ambient conditions) and higher electronic conductivity (Figure 3.14(f)).[155]

3.5.2 Mechanical Properties

The study of the mechanical properties of MXenes is still at an early stage.[120] The elastic modulus of MXenes was predicted to be high in early studies, considering the high elastic modulus of bulk carbides.[105] To date, experimental studies on single layer flakes of $Ti_3C_2T_x$ and $Nb_4C_3T_x$ have been reported, which were performed using static atomic force microscopy (AFM) indentation,[159,160] which is one of the most commonly used methods to investigate the mechanical properties of free-standing ultrathin layered materials.[161] The schematics of AFM indentation measurement is shown in Figure 3.15(a). The Young's modulus, E, of single layer $Ti_3C_2T_x$ was ~330 GPa, with a breaking strength of approximately 17.3 GPa, while bare Ti_3C_2 has an E value of 502 GPa based on molecular dynamics simulations.[162,163] The relatively low E value of the $Ti_3C_2T_x$ film obtained by experiment is caused by the surface terminations and the presence of defects. The experimental measurement of the mechanical properties of $Nb_4C_3T_x$ is similar to the results for $Ti_3C_2T_x$.[159,160] An overall comparison of the effective Young's modulus of monolayer $Nb_4C_3T_x$, $Ti_3C_2T_x$ with other 2D materials studied in similar nanoindentation experiments are shown in Figure 3.15(b). By comparison to free-standing MXene layers, blade coated $Ti_3C_2T_x$ films have shown a tensile strength of ~560 MPa,

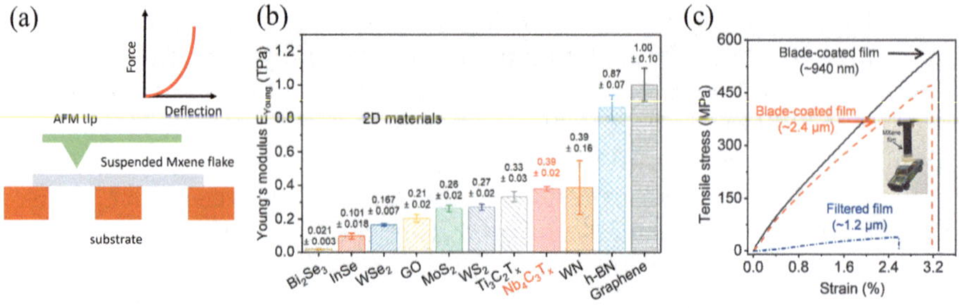

Figure 3.15 Mechanical properties of MXenes. (a) Schematic of AFM tip nanoindentation on a suspended MXene film, and typical signals obtained. (b) Comparison of the effective Young's moduli of monolayer $Nb_4C_3T_x$, $Ti_3C_2T_x$ and other 2D materials studied in similar nanoindentation experiments. Reproduced from ref. 160 with permission from John Wiley & Sons, Copyright 2020. (c) Typical stress–strain curve for blade coated and vacuum-filtered $Ti_3C_2T_x$ films. Reproduced from ref. 164 with permission from John Wiley & Sons, Copyright 2020.

which is associated with a film with a thickness of ~940 nm (Figure 3.15(c)).[164] Thus far, the influence of different constituent elements, crystal and microstructure and surface terminations on the mechanical properties of single layered MXene membranes has not been systematically investigated. There is no doubt that such a systematic study would be of significant interest in order to realise the accurate top-down design of MXenes with bespoke mechanical properties optimised for certain applications.

3.5.3 Electronic and Electrochemical Properties

The electronic properties of MXenes are highly dependent on their compositions and structures as well as on their surface and interlayer chemistry. To date, electronic properties of MXenes, such as the widely tuneable metal-to-semiconductor transition[165,166] ultralow work functions[167] topological insulating behaviour[168] and large anisotropic electronic conduction[169] have drawn intensive interest among many different research communities. Focusing on the principles and flexible tailoring of their electrical properties is essential for the success of applying MXenes to applications such as energy storage, biochemical sensing, electro-magnetic shielding, *etc.*[29–31]

MAX phases are metallic conductors with a typical electrical conductivity of a few tens of kS cm^{-1} at room-temperature[170] where the free electrons of the transition metals act as the charge carriers. Inherited from the MAX precursors, all MXenes without terminations are predicted to be metallic conductors, and normally possess excellent metallic electrical conductivity with reported values up to 20 kS cm^{-1} (Figure 3.16(a))[171] Similar to metals, the electrical conductivity of most MXenes decreases with increasing temperature, since the electron mobility reduces as the thermal energy of the free electrons increases[172,173] Such metallic behaviour can be modified by altering the

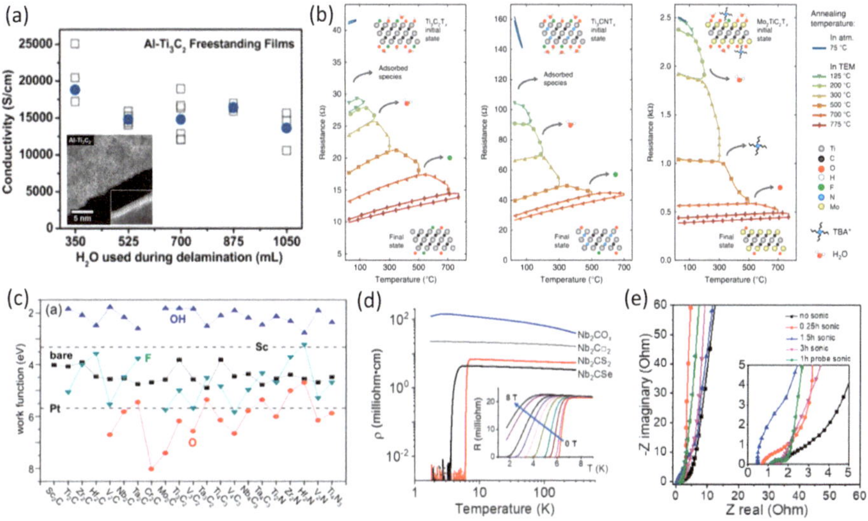

Figure 3.16 Experimental measurement of the electronic properties of selected MXenes, showing: (a) Electronic conductivity of free-standing Al-Ti_3C_2 film suspensions at different stages of delamination, showing a conductivity of *ca.* 20 kS cm^{-1} (squares represent measurement on different samples and blue circles represent average values). Reproduced from ref. 171 with permission from American Chemical Society, Copyright 2021. (b) The evolution of the electronic properties of $Ti_3C_2T_x$, Ti_3CNT_x and $Mo_2TiC_2T_x$ during *in situ* vacuum annealing. Reproduced from ref. 91, https://doi.org/10.1038/s41467-018-08169-8, under the terms of the CC BY 4.0 license, http://creativecommons.org/licenses/by/4.0/. (c) Calculated work functions of various MXenes with different surface terminations where the dashed lines indicate the work function of platinum and scandium as points of reference. Reproduced from ref. 174 with permission from American Chemical Society, Copyright 2016, (d) Temperature-dependent resistivity for Nb_2CT_x MXenes (where □ represents a vacant site), showing the superconductivity of –S and –Se terminated Nb_2C MXenes (the inset shows the resistance *versus* temperature of Nb_2CS_2 under different applied magnetic fields). Reproduced from ref. 103 with permission from American Association for the Advancement of Science, Copyright 2020. (e) Electrochemical impedance spectroscopy (Nyquist plots) for $Ti_3C_2T_x$ electrodes; shorter sonication times produce larger flake sizes, indicating the flake-size dependent electronic properties of MXenes. The inset shows the high-frequency range. Reproduced from ref. 176 with permission from American Chemical Society, Copyright 2018.

M or X constituents and the MXene structure,[91] leading to diverse electronic properties among MXenes. For example, when the outer transition layers of metallic $Ti_3C_2T_x$ are substituted by Mo, the resulting o-MXenes ($Mo_2TiC_2T_x$) become semiconducting.[78]

The surface terminations can change the density of states (DOS) and shift the Fermi-level, which has the effect of tuning the electronic properties of MXenes such as the conductivity (Figure 3.16(b)) and work function (Figure 3.16(c)).[174] Considering electrical behaviour for example, Mo_2CT_x[53] exhibits a negative temperature dependence of resistance as fabricated (with –OH, –O and –F surface groups) but this behaviour can be tuned to a positive temperature dependence when the surface groups are removed. $Ti_3C_2T_x$ as the most studied MXene and with a high concentration of terminations, for example, exhibits a lower charge carrier density when compared with the pristine Ti_3C_2.[91] Additionally, Nb_2CT_x becomes superconducting when the terminating group T, is Se or S, while the O-terminated or non-terminated Nb_2C does not show such a phenomenon (Figure 3.16(d)).[99] In addition to the above examples of metallic and semiconducting MXenes, some oxygen-terminated MXenes (M_2CO_2 where M = W, Mo, and Cr) have been predicted to be topological insulators.[175] These topological insulating phenomena only occur when optimised in materials with certain band gaps. Although this transition mechanism has not been well explained, it does open a new potential area for tailoring MXene-based materials to certain applications. Notably, as the terminations on the MXene surface can be altered or removed through post-processing, there is a feasible approach to engineer MXenes with advantageous electronic properties for applications such as energy storage, sensing, *etc.*

As mentioned in the synthesis section above, the selection of intercalated cations or molecules during the synthesis of MXenes also plays an important role in determining the electrical properties of the respective materials.[120] Small cations such as Li^+ only have a limited effect on the conductivity of MXenes such as highly conductive $Ti_3C_2T_x$. However, organic molecules or large cations such as TMA^+ and TBA^+ can significantly enlarge the interlayer space. A larger interlayer space, in turn, hinders the inter-flake transport of charge carriers and ultimately results in larger electrical resistance[91] The flake size and the point defects, which link to the synthesis process such as the sonication time, also influence the electronic properties of MXenes, where larger flakes prepared with shorter sonication time and less defects give smoother intra-flake electron transport and thus possess higher electrical conductivity (Figure 3.16(e)).[176]

Among the outstanding properties and potential applications of MXene-based materials, MXenes initially drew research interest for electrochemical energy storage and their application in Li-ion batteries and supercapacitors due to their high volumetric/gravimetric capacity/capacitance and ultra-high electron transport rates.[177] Among the MXene family, those compositions with lower formula weights theoretically have higher gravimetric capacity. Therefore, M_2X are expected to be more promising than their M_3X_2, M_4X_3 and M_5X_4 counterparts, which is consistent with all available experiment data. V_2CT_x shows the highest capacity among three tested M_2X-based MXenes (V_2C, Nb_2C and Ti_2C) in lithium-ion batteries at both low (1C) and high (10C) cycling rate (Figure 3.17(a)).[178] As for supercapacitors, 3 μm thick hydrogel electrodes of $Ti_3C_2T_x$ show a volumetric capacitance of up to 1400 F cm^{-3} in protic H_2SO_4 electrolyte. $Ti_3C_2T_x$ electrodes maintained over 50% capacitance at a scanning

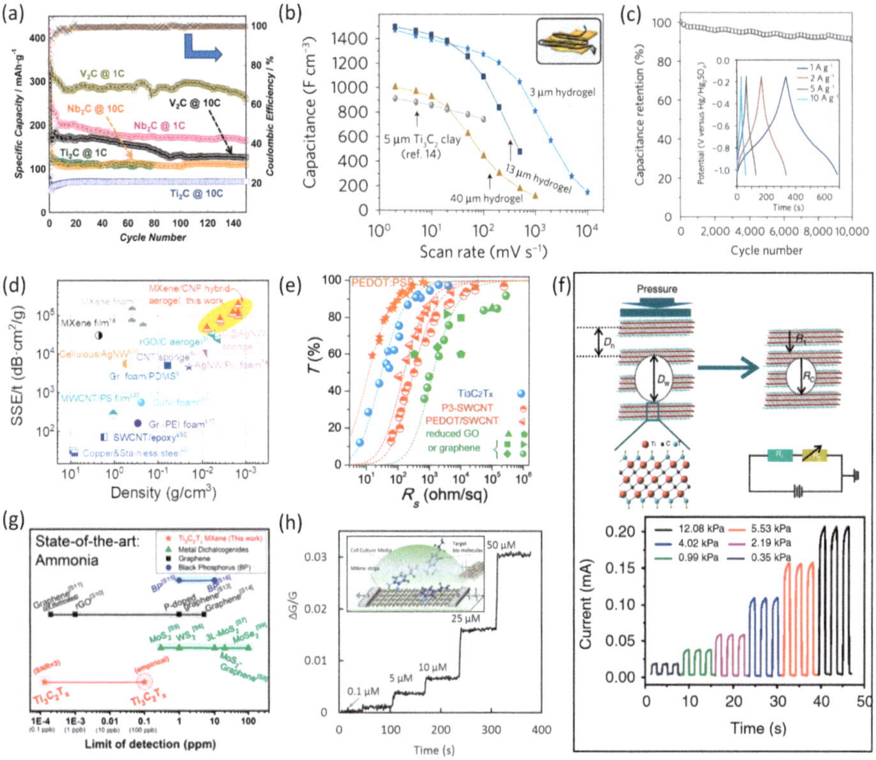

Figure 3.17 Characterisation of distinctive properties of Mxenes for particular applications. (a) Specific lithiation/delithiation gravimetric capacities *versus* cycle number at different scanning rates for several MXene-based electrodes. Reproduced from ref. 178 with permission from American Chemical Society, Copyright 2013. (b) Volumetric capacitance of MXene-based supercapacitors (3 μm, 13 μm, and 40 μm $Ti_3C_2T_y$ hydrogel film compared with 5 μm $Ti_3C_2T_x$ clay) in protic H_2SO_4 electrolyte at different scanning rate and (c) the capacitance retention test of the supercapacitors based on the 13 μm hydrogel films in (b). Reproduced from ref. 179 with permission from Springer Nature, Copyright 2017. (d) Electromagnetic interference (EMI) shielding performance comparison of MXene-based and other typical materials. Reproduced from ref. 186, https://doi.org/10.1002/advs.202000979, under the terms of the CC BY 4.0 license, http://creativecommons.org/licenses/by/4.0/. (e) Transmittance as a function of sheet resistance for MXene films with comparisons taken from other typical literature materials including poly(3,4-ethylenedioxythiophene)polystyrene sulfonate (PEDOT:PSS) and single-wall carbon nanotube (SWCNT). Reproduced from ref. 180 with permission from John Wiley & Sons, Copyright 2017. (f) Working mechanism of an MXene-based piezoresistive sensor and the typical response signal obtained from such a sensor, where a wider distance (D_w) between two interlayers is easier to compress, while a narrower distance (D_n) between two lattices has a smaller compression ratio. Reproduced from ref. 182, https://doi.org/10.1038/s41467-017-01136-9, under the terms of the CC BY

(continued)

rate of 1000 mV s^{-1} (Figure 3.17(b)) and retained over 90% of the capacitance after 10 000 operating cycles (Figure 3.17(c)).[179] The high electrical conductivity of certain MXenes renders them promising candidates for electromagnetic interference (EMI) shielding (Figure 3.17(d)), since EMI shielding effectiveness is theoretically proportional to the electrical conductivity of any given materials. Marvelous electrical conductivity, good mechanical tolerance and facile thin-film processability empower MXenes as promising candidates for flexible transparent electrodes (Figure 3.17(e)), which can then be employed in solar cells, flexible displays and supercapacitors.[180,181] The inverse dependency of MXene conductivity on interlayer spacing can transduce input mechanical signals like pressure and strain into electrical output signals such as current and resistance. These properties are particularly beneficial for MXene-based piezoresistive sensors (Figure 3.17(f))[182] Gas sensors based on MXenes have been reported to show a large signal-to-noise ratio and a highly sensitive response (Figure 3.17(g)) due to the excellent metallic conductivity of the core channels and strong adsorption at the active surface sites of MXenes, which are revealed by noise power spectral density measurements and density functional theory (DFT)[183,184] Moreover, MXenes have also been utilised for bio-sensing applications. Such applications exploit the electronic conductivity, excellent bio-compatibility and stability, and low cytotoxicity of most MXene components. For example, $Ti_3C_2T_x$ based field effect transistors were used for dopamine detection, showing an incremental change in conductivity with higher dopamine concentration in an ambient environment. Stable responses at concentrations down to 0.1 μM can be achieved as shown in Figure 3.17(h).[185]

3.5.4 Optical Properties

MXenes exhibit unique optical properties in a broad wavelength range. The optical responses of MXenes include interband transitions, *i.e.*, strong absorbance in the ultraviolet (UV) range, and plasma excitations, *i.e.*, versatile plasmon modes in the visible and near infrared range (vis and NIR). Like other physiochemical properties, the optical properties of MXenes also vary with structure and composition, such as the type and structure of the metal (M) and carbon/nitrogen (X) sites, as well as the surface terminations.[187–192]

Figure 3.17 *(continued)* 4.0 license, http://creativecommons.org/licenses/by/4.0/. (g) Diagram of the lower limits of detection for state-of-the-art room temperature ammonia gas sensors based on 2D materials including $Ti_3C_2T_x$ MXenes, transition metal dichalcogenides, graphene, and black phosphorus (BP) (note: the *y*-axis has no quantitative meaning). Reproduced from ref. 183 with permission from American Chemical Society, Copyright 2018. (h) Real-time dopamine detection obtained from an MXene-FET device, where the inset shows a schematic of the construction and operation of the MXene-FET biosensing device itself. Reproduced from ref. 185, https://doi.org/10.1002/adma.201504657, under the terms of the CC BY 4.0 license, http://creativecommons.org/licenses/by/4.0/.

MXene colloidal solutions and thin films display different plasmonic colours that span the vis-NIR range as functions of chemical composition and crystal structure (various M sites and *n* values) (Figure 3.18(a)).[191–193] It is found that the excitation peaks in optical spectra shift to higher energies on decreasing the *n* value in $M_{n+1}X_nT_x$ MXenes. For example, the $Ti_3C_2T_x$ MXene has a characteristic absorbance peak centred at 1.6 eV, whereas V_2CT_x has no absorption peak in the UV-vis-NIR spectrum region (Figure 3.18(a)). In MXene solid solutions, for instance, several compositions in the $(Ti, Nb)_2CT_x$ and $(Ti, V)_2CT_x$ systems, changes to M′–M″ ratios result in the variations of plasmonic colours. These colours are also different from monometallic MXenes containing any of the respective constituent metals (Figure 3.18(b)).[190]

Additionally, the absorption features of MXenes change when substituting the carbon atom (*i.e.*, X = C) with a nitrogen atom (X = N), allowing the tuning of the optical properties on demand. The typical absorption peak of $Ti_3C_2T_x$ is located at 780 nm and shifts to 670 nm for Ti_3CNT_x (Figure 3.18(c)). The electrons from nitrogen are distributed in a lower energy band far from the Fermi level, which both influence the optical properties and the electrical conductivity.[189]

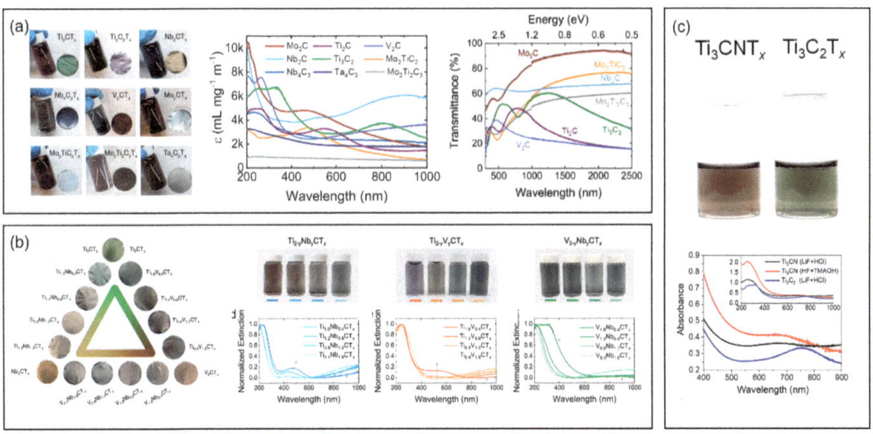

Figure 3.18 (a) Observed colours of MXene colloidal dispersions and corresponding free-standing films (I), UV–vis–NIR optical extinction properties of various MXenes (II) and UV–vis–NIR transmittance spectra of MXene thin films (III). Reproduced from ref. 192 with permission from John Wiley & Sons, Copyright 2020. (b-I) The compositional triangle of $Ti_{2-y}Nb_yCT_x$, $Ti_{2-y}V_yCT_x$, and $V_{2-y}Nb_yCT_x$ showing the continuous solid solutions of M_2C-type MXenes and their corresponding free-standing films, (II–IV) digital images and UV-vis spectra of solid-solution $M'_{2-y}M''_yCT_x$ MXenes, (II) $Ti_{2-y}Nb_yCT_x$, (III) $Ti_{2-y}V_yCT_x$, (IV) $V_{2-y}Nb_yCT_x$. Reproduced from ref. 191 with permission from American Chemical Society, Copyright 2020. (c) Optical images (I) and UV-visible spectra (II) of Ti_3CNT_x and $Ti_3C_2T_x$ solutions. Reproduced from ref. 189 with permission from American Chemical Society, Copyright 2019.

As discussed in Section 3.5.3, different surface terminations lead to a change in the density of states near the Fermi level, which profoundly influence the electronic properties. Specifically, the interband electronic transitions caused by different surface terminations affect the dielectric constant, which then influences the absorption behaviour.[194,195] Taking $Ti_3C_2T_x$ as an example, Ti_3C_2 with surface terminations exhibits a higher absorption than the pristine Ti_3C_2 in the UV range (Figure 3.19(a)).[188] In the visible range, $Ti_3C_2O_2$ shows stronger absorption, while –OH/–F terminated Ti_3C_2 has

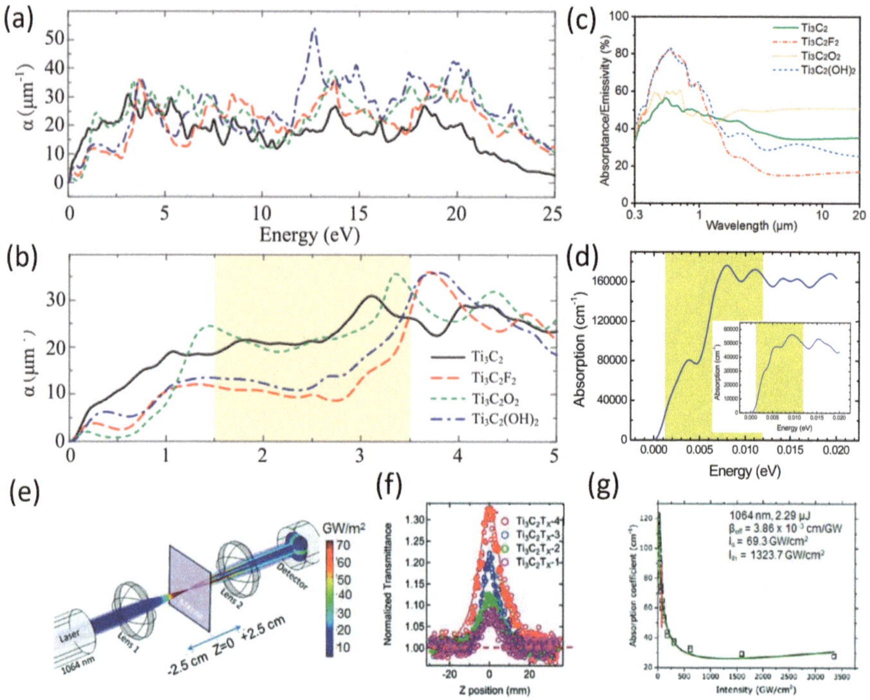

Figure 3.19 (a) Calculated absorption spectra (DFT) of MXene ($Ti_3C_2T_x$) in the UV region, and (b) calculated absorption spectra calculation (DFT) of MXene ($Ti_3C_2T_x$) in the visible range. Both reproduced from ref. 188, https://doi.org/10.1063/1.4948799, under the terms of the CC BY 4.0 license, http://creativecommons.org/licenses/by/4.0/. (c) Calculated absorptance/emissivity spectra (DFT) of $Ti_3C_2T_x$ with different surface terminations in the mid-IR range. Reproduced from ref. 190 with permission from John Wiley & Sons, Copyright 2021. (d) Calculated THz spectra (DFT) of Ti_3C_2 with –F, –OH and/or –O terminations (insert). Reproduced from ref. 187 with permission from Royal Society of Chemistry. (e and f) Schematic Z-scan measurement and saturable absorption properties of $Ti_3C_2T_x$ MXene films. Reproduced from ref. 196 with permission from John Wiley & Sons, Copyright 2018. (g) Dependence of the absorption coefficient of $Ti_3C_2T_x$ MXene films on the illuminating intensity. Reproduced from ref. 197 with permission from John Wiley & Sons., Copyright 2017.

weaker absorption when compared to that of pristine Ti_3C_2 (Figure 3.19(b)).[188] According to the results from DFT calculations, $Ti_3C_2(OH)_2$ and $Ti_3C_2F_2$ also exhibit similar absorption behaviour in the IR range (Figure 3.19(c)).[190] Moreover, optical absorption and extinction coefficients of $Ti_3C_2T_x$ in the terahertz (THz) range are predicted, indicating the potential for MXenes to serve as THz detecting materials (Figure 3.19(d)).[187]

Non-linear optical behaviour is also observed in MXenes.[196,197] As a specific example, $Ti_3C_2T_x$ exhibits nonlinear light absorption (saturable absorption, SA), in which the transmission increases nonlinearly with increasing illuminating intensity (Figure 3.19(e) and (f)).[196] The effective non-linear absorption coefficient decreases with increasing illumination until a threshold intensity is reached (Figure 3.19(g)).[197] This behaviour is explained by a plasmon-induced increase in the ground state absorption, when $Ti_3C_2T_x$ is illuminated by a laser of a wavelength comparable to the plasmon resonance frequency of the MXenes.[196]

Applications based on the optical behaviour of MXenes have been widely demonstrated. Because of their high transmissivity in single- to few-layer form and their simple fabrication processes, MXenes hold potential to be integrated in photonics and microelectronics. Combining high conductivity with metal-like carrier concentrations, transparent MXene films can be applied as electrodes for photodetectors with a detection range from the UV to the NIR regions.[198–202] Compared with traditional metal electrodes, one advantage of the MXene-based electrodes is the fabrication method; various coating routes for MXenes can be utilised in ambient conditions without the requirements of vacuum and high temperature. MXene-based electrodes exhibit responsivities 2–3 times higher than equivalent Au-based devices and quantum efficiencies with sub-nanosecond response times under high biasing voltage (Figure 3.20(a) and (b)).[201] Mo_2CT_x has been applied as a plasmonic photodetector in the visible range with responsivity of 9 A W^{-1} and detectivity of 5×10^{11} J, where the surface plasmon modes in Mo_2CT_x assist in the generation of hot electrons (Figure 3.20(c) and (d)).[203] Moreover, Mo_2CT_x is beneficial for plasmonic photodetection in terms of its reduced production of high dark current. MXenes can also be applied as sensitive SERS substrates due to their plasmonic properties in the NIR range, which is useful for applications including biomedical imaging, high sensitivity environmental analysis and food safety monitoring (Figure 3.20(e))[204] Owing to the flat and negatively charged surface, MXenes enhance the absorption of positively charged dye molecules and promote charge transfer from the MXene to the dye, thus inducing excellent SERS sensitivities and low detection limits.[204,205] The lowest detected Raman signals from $Ti_3C_2T_x$ for Rhodamine 6G dye (Rh6G) could be achieved for a concentration of 10^{-8} M, whereas Ti_2NT_x enhances the Raman signal at 532 nm by a factor of 10^{12} and the MXene composite, $Ti_3C_2T_x$/Au shows detection limits for Rh6G, thiram and diquat of 10^{-12}, 10^{-10} and 10^{-8} M, respectively.[206–208] Inspired by the interaction between $Ti_3C_2T_x$ and electromagnetic waves, where the absorbed wave is negligibly emitted and dissipated in the form of heat within the material,[209] investigations on

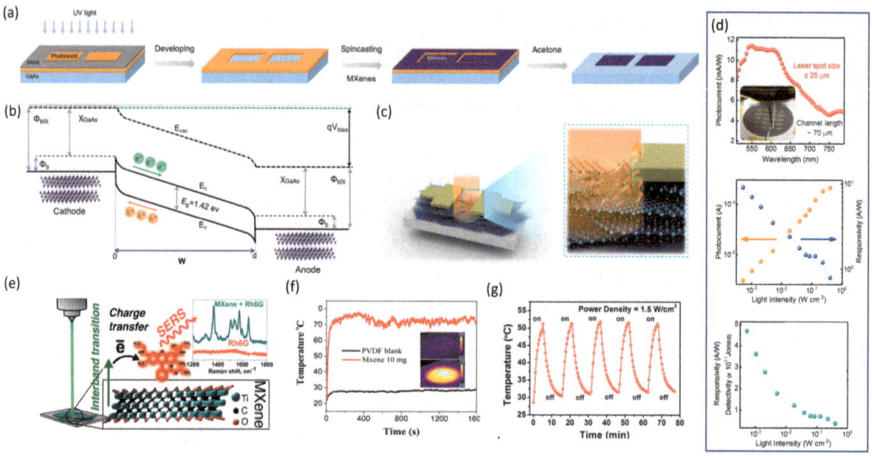

Figure 3.20 (a) The fabrication process and (b) the energy band diagram for MXene–GaAs–MXene photodetector devices. Reproduced from ref. 201 with permission from John Wiley & Sons, Copyright 2019. (c) Schematic of the Mo_2CT_x thin film photodetector showing the migration of the plasmon-assisted hot electrons toward the biased gold electrodes under illumination. (d) The photo-response, photocurrents, responsivity and detectivity of the Mo_2CT_x thin film photodetector. Reproduced from ref. 203 with permission from John Wiley & Sons, Copyright 2019. (e) Schematic of SERS enhancement of Rh6G on $Ti_3C_2T_x$ MXene substrates. Reproduced from ref. 204 with permission from American Chemical Society, Copyright 2017.

MXenes as light-to-heat conversion materials are becoming increasingly significant.[210–213] MXene films absorb sunlight and readily covert it to the thermal energy, reaching temperatures of 70 °C[211] in dry conditions and 50–60 °C[212] in water. Such behaviour suggests that significant progress might be made by utilising MXenes in biomedical applications,[214–216] anticancer therapy/tumour ablation[217–219] and as light-driven actuators.[220–223]

3.5.5 Magnetic Properties

A ground state of ferromagnetic (FM) ordering has been theoretically predicted for the functionalised Cr_2C and Cr_2N MXene systems,[224] while experimentally synthesised MXenes such as $Ti_3C_2T_x$ show no such ordering and are Pauli paramagnetic.[225] MXenes possess a high degree of chemical and structural diversity enabling the tuning of the magnetic ground state, intimating their great promise as 2D magnetic materials with versatile behaviours (Figure 3.21(a)–(c)).[226–228]

In the crystal structures of the MXenes, the transition metal atoms provide electrons in non-bonding d-orbitals which are spin-polarised, and it is these which predominantly contribute to the magnetic behaviour. However, the

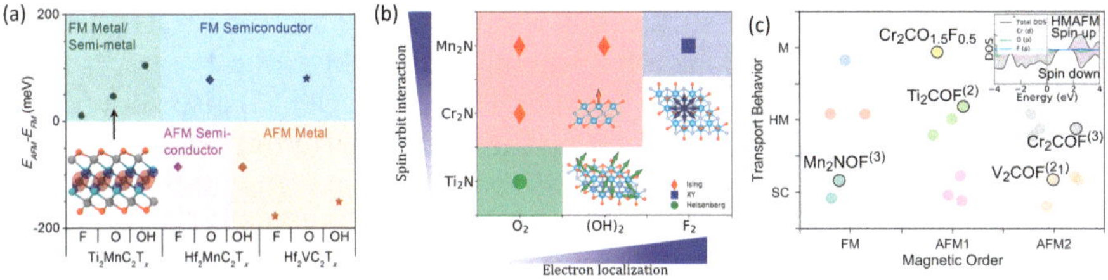

Figure 3.21 (a) Exchange energy and magnetic states in mixed Ti and Hf-based MXenes with various surface termination. Reproduced from ref. 226 with permission from American Chemical Society, Copyright 2017. (b) Accessible spin symmetries of M_2N (M = Ti, Cr, Mn) *via* the tuning of the spin–orbit interaction (a heavier M atom induces a stronger spin–orbit coupling effect) and the electron localisation (which increases with the electronegativity of the surface termination group); the symbols represent Ising (red), XY (blue) and Heisenberg (green) models, respectively. Reproduced from ref. 227 with permission from American Chemical Society, Copyright 2018. (c) Diversity of transport (semiconducting, half-metallic, and metallic) and magnetic behaviours in Janus MXenes, $M_2XO_xF_{2-x}$ (x = 0.5, 1 or 1.5) with asymmetric surface functionalisation. There are three distinct orderings for O and F when x = 1; the composition labels with a superscript $OF^{(3)}$, $OF^{(21)}$ and $OF^{(2)}$ indicate the modes with three-fold rotation, 2_1 screw and two-fold rotation axis structural symmetry, respectively. Reproduced from ref. 228 with permission from American Chemical Society, Copyright 2019.

early transition metals in most MXenes possess small numbers of valence electrons resulting in limited unpaired electron spins and the corresponding non-magnetic nature of many MXenes. Thus, most of the predicted magnetic MXenes are based on magnetic transition metals which have more valence electrons such as Cr, V, Mn, Fe, Co, Ni, Hf, both in single transition metal systems and ordered double transition metal structures.[229,230] The surface functionalisation of MXenes provides an additional degree of freedom with which to tune the magnetic ground states. Moreover, the nitride based MXenes contain an extra electron compared to the equivalent carbides, leading to distinct magnetic characteristics.[227] It is revealed that (–F, –O, –H, –Cl and –OH) functionalised Cr_2C and Cr_2N display long range magnetic ordering with ferromagnetic–antiferromagnetic (FM–AFM) transitions.[224,231,232] For instance, non-terminated Cr_2C is ferromagnetic due to the itinerant Cr d-orbital electrons. The electrons are more localised in surface terminated (–F, –OH, –H or –Cl) Cr_2C, which favour AFM ordering and then induce FM–AFM transitions. The number of electrons in Cr_2N is different from that in Cr_2C, and the bare Cr_2N MXene exhibits AFM characteristics. After surface-passivation with monovalent (–F, –OH) to divalent (–O) functional groups, the electrons with minority spin at the Fermi level are suppressed gradually, and the FM configuration becomes increasingly stable (delocalised electrons with itinerant character are indicated in O-terminated magnetic MXenes, stabilising FM ordering). As a result, the surface-passivated Cr_2NF_2 and $Cr_2N(OH)_2$ are still AFM metals, while Cr_2NO_2 is a half-metal with a FM ground state.[233] Many mixed MXenes, such as $Cr_2M''C_2T_2$ (M'' = Ti, V, T = F, OH, O) also show different magnetic properties depending on the surface terminations; they can be non-magnetic (*e.g.*, $Cr_2TiC_2O_2$, where C ions are weakly magnetised or $Cr_2TiC_2T_2$ where Ti atoms in AFM states are nonmagnetic), FM (*e.g.*, $Cr_2VC_2(OH)_2$, $Cr_2VC_2F_2$, and $Cr_2VC_2O_2$) or AFM (*e.g.*, $Cr_2TiC_2F_2$ and $Cr_2TiC_2(OH)_2$).[234]

3.6 Conclusions and Perspectives

We have summarised some key features of 2D MXenes in this chapter from their chemical synthesis and structures to their fundamental properties. Owing to the tunable atomic-scale thickness, structure, and compositions, MXenes display rich physics and hold promise for a wide range of applications. However, most of the extensive studies have been focused on $Ti_3C_2T_x$, while other examples of the materials have received far less attention.

Of particular interest in the MXene systems is the possibility to tune the properties by atomic arrangement, composition, and surface chemistry engineering, which can be directly realised through the optimisation of top-down synthesis methods. Future investigations should include exploring the synthesis routes for non-Al A elements, developing environmentally friendly/safe synthesis methods, and improving synthesis capability for large-scale production. New synthesis methods are essential for discovering novel MXenes with various M-sites (arrangement of constitutive elements) and X-sites (from C to N and B). Controlling surface chemistry is one of the challenges for the next decade of MXene research. Developing synthesis routes to obtain MXenes with terminations, such as –S, –Se, –Te, –H, –N (other than –O, –OH, or halogens) and the methods to switch between them, engineering defects and/or achieve uniform terminations should all be taken into consideration. Exploration of bottom-up methods to obtain non-terminated 2D carbides or nitrides is another future challenge that needs to be addressed in order to exploit termination-controlled physical properties (*e.g.*, superconductivity) of MXenes. Synthetic progress is at the heart of the continuing development of this large family of 2D materials.

There is plenty of room to discover new MXene phases and almost every new MXene phase possesses unique or perhaps previously unknown physicochemical properties. There are also many predicted properties that have not yet been realised that represent the next frontier for MXenes. Such manifestations lie in the regime of electronic, optical, magnetic, thermal, and thermoelectric properties and include phenomena such as quantum confinement effects, *etc.* Simulation and modelling are guiding much of these developments and also assist in the design of new materials. Considering the substantial influences that precursors have already been shown to exert on properties, potential theoretical initiatives include not only the prediction of new MXene members themselves, but also the identification of new MXene precursors, (both new MAX phases and/or non-MAX phases towards specifically targeted MXenes). To achieve this, the roles of the precursor's characteristics (*e.g.*, structure/defects/stoichiometry) on those of the ensuing MXenes should be further examined. The techniques of machine learning and artificial intelligence are currently in the process of emerging to assist the design of both MAX phases and MXenes.[235] Certainly, based on the accumulation of theoretical results, the limits of MXene properties can be further defined and understood. Thus, it is also of importance to develop theoretical methods that produce outcomes to drive the processes of prediction and materials discovery.

Many exciting applications of these materials are yet to be discovered and demonstrated. MXenes are also *en route* to realisation in industrial applications. Understanding and being able to control the physiochemical properties are the key to extend the possible applications of MXenes. For example, it will be beneficial for practical and industrial applications to find functional MXenes that are resistive to oxidation with improved chemical and high temperature stability. Meanwhile, the study of the health impacts and environmental safety of MXenes will ensure the reliability of MXenes in biomedical applications. Further, understanding the mechanism of electromagnetic wave attenuation in MXenes is essential for electromagnetic interference shielding, wireless communication and antennas, *etc.*

One of the important characteristics among 2D materials is their ability to combine with other nanostructures with additive and programmed properties (mechanical, electronic, plasmonic, electrochemical or catalytic properties, *etc.*). The future design prospects for functional MXenes depends on the establishment of efficient self-assembly techniques with controllable orientation and assembly of MXenes as nanoscale building blocks to obtain hybrid nanoarchitectures. Combined with the discovery of new MXenes and precise control of their structures and compositions, the utilisation of MXenes can thus be further expanded into new technologies, including flexible and wearable devices for energy storage/harvesting, sensors, actuators, optical lenses, artificial memory devices, quantum computing and the components/devices within the Internet of Things (IoT).

Therefore, the landscape of MXenes is still being discovered. The world of 2D MXenes continues to grow at an unprecedented rate, and we expect to see the promise of MXene research being fulfilled in the coming years as manifested by a significant increase in the number of materials, more diversity and more widespread commercialisation.

References

1. D. R. Morris, in *Molten Salt Technology*, Springer US, Boston, MA, 1982, pp. 153–183.
2. F. Thévenot, *J. Eur. Ceram. Soc.*, 1990, **6**, 205.
3. L. S. Ramsdell, *Am. Mineral.*, 1947, **32**, 64.
4. G. S. Upadhyaya, *Mater. Des.*, 2001, **22**, 483.

5. X. Wang, K. Zhu, Y. Ju, Y. Li, W. Li, J. Xu and Y. Hou, *J. Magn. Mater.*, 2019, **489**, 165432.
6. W. S. Williams, *Prog. Solid State Chem.*, 1971, **6**, 57.
7. B. Roebuck, *Powder Metall.*, 2001, **44**, 14.
8. S. T. Oyama, J. C. Schlatter, J. E. Metcalfe and J. M. Lambert, *Ind. Eng. Chem. Res.*, 1988, **27**, 1639.
9. Y. Zhong, X. Xia, F. Shi, J. Zhan, J. Tu and H. J. Fan, *Adv. Sci.*, 2016, **3**, 1500286.
10. R. B. Levy and M. Boudart, *Science*, 1973, **181**, 547.
11. *Materials Science of Carbides, Nitrides and Borides*, ed. Y. G. Gogotsi, and R. A. Andrievski, Springer Netherlands, Dordrecht, 1999.
12. A. K. Geim and K. S. Novoselov, *Nat. Mater.*, 2007, **6**, 183.
13. K. S. Novoselov, A. K. Geim, S. v. Morozov, D. Jiang, M. I. Katsnelson, I. v. Grigorieva, S. v. Dubonos and A. A. Firsov, *Nature*, 2005, **438**, 197.
14. A. G. Kelly, T. Hallam, C. Backes, A. Harvey, A. S. Esmaeily, I. Godwin, J. Coelho, V. Nicolosi, J. Lauth, A. Kulkarni, S. Kinge, L. D. A. Siebbeles, G. S. Duesberg and J. N. Coleman, *Science*, 2017, **356**, 69.
15. T. Lin, I. W. Chen, F. Liu, C. Yang, H. Bi, F. Xu and F. Huang, *Science*, 2015, **350**, 1508.
16. X. Xiao, H. Wang, P. Urbankowski and Y. Gogotsi, *Chem. Soc. Rev.*, 2018, **47**, 8744.
17. D. Zhang, S. Wang, Y. Ma and S. Yang, *J. Energy Chem.*, 2018, **27**, 128.
18. A. Harvey, C. Backes, Z. Gholamvand, D. Hanlon, D. Mcateer, H. C. Nerl, E. Mcguire, A. Andrés Seral-Ascaso, Q. M. Ramasse, N. Mcevoy, S. Winters, N. C. Berner, D. Mccloskey, J. F. Donegan, G. S. Duesberg, V. Nicolosi and J. N. Coleman, *Chem. Mater.*, 2015, **27**, 54.
19. C. J. Zhang, S.-H. Park, O. Ronan, A. Harvey, A. Seral-Ascaso, Z. Lin, N. McEvoy, C. S. Boland, N. C. Berner, G. S. Duesberg, P. Rozier, J. N. Coleman and V. Nicolosi, *Small*, 2017, **13**, 1701677.
20. E. Varrla, C. Backes, K. R. Paton, A. Harvey, Z. Gholamvand, J. McCauley and J. N. Coleman, *Chem. Mater.*, 2015, **27**, 1129.
21. Z. Gholamvand, D. Mcateer, A. Harvey, C. Backes and J. N. Coleman, *Chem. Mater.*, 2016, **28**, 36.
22. C. Zhang, R. Maloney, M. R. Lukatskaya, M. Beidaghi, B. Dyatkin, E. Perre, D. Long, W. Qiao, B. Dunn and Y. Gogotsi, *J. Power Sources*, 2015, **274**, 121.
23. M. Naguib, M. Kurtoglu, V. Presser, J. Lu, J. Niu, M. Heon, L. Hultman, Y. Gogotsi and M. W. Barsoum, *Adv. Mater.*, 2011, **23**, 4248.
24. M. Naguib, O. Mashtalir, J. Carle, V. Presser, J. Lu, L. Hultman, Y. Gogotsi and M. W. Barsoum, *ACS Nano*, 2012, **6**, 1322.
25. M. Naguib and Y. Gogotsi, *Acc. Chem. Res.*, 2015, **48**, 128.
26. B. Anasori, M. R. Lukatskaya and Y. Gogotsi, *Nat. Rev. Mater.*, 2017, **2**, 16098.
27. B. Anasori, Y. Xie, M. Beidaghi, J. Lu, B. C. Hosler, L. Hultman, P. R. C. Kent, Y. Gogotsi and M. W. Barsoum, *ACS Nano*, 2015, **9**, 9507.

28. M. W. Barsoum, *MAX Phases*, Wiley-VCH Verlag GmbH & Co. KGaA, Weinheim, Germany, 2013.
29. M. Naguib, M. W. Barsoum and Y. Gogotsi, *Adv. Mater.*, 2021, **33**, 2103393.
30. Y. Gogotsi and B. Anasori, *ACS Nano*, 2019, **13**, 8491.
31. A. Vahid Mohammadi, J. Rosen and Y. Gogotsi, *Science*, 2021, **372**, eabf1581.
32. B. Anasori, M. R. Lukatskaya and Y. Gogotsi, *2D Metal Carbides and Nitrides (MXenes)*, Springer International Publishing, Cham, 2019, vol. 2.
33. Y. Wei, P. Zhang, R. A. Soomro, Q. Zhu and B. Xu, *Adv. Mater.*, 2021, **33**, 2103148.
34. Q. Tang, Z. Zhou and P. Shen, *J. Am. Chem. Soc.*, 2012, **134**, 16909.
35. X. Xie, S. Chen, W. Ding, Y. Nie and Z. Wei, *Chem. Commun.*, 2013, **49**, 10112.
36. J. Halim, M. R. Lukatskaya, K. M. Cook, J. Lu, C. R. Smith, L.-Å. Näslund, S. J. May, L. Hultman, Y. Gogotsi, P. Eklund and M. W. Barsoum, *Chem. Mater.*, 2014, **26**, 2374.
37. C. E. Ren, K. B. Hatzell, M. Alhabeb, Z. Ling, K. A. Mahmoud and Y. Gogotsi, *J. Phys. Chem. Lett.*, 2015, **6**, 4026.
38. B. Xu, M. Zhu, W. Zhang, X. Zhen, Z. Pei, Q. Xue, C. Zhi, P. Shi, B. Xu, W. Zhang, X. Zhen, P. Shi, M. Zhu, Z. Pei, Q. Xue and C. Zhi, *Adv. Mater.*, 2016, **28**, 3333.
39. F. Shahzad, M. Alhabeb, C. B. Hatter, B. Anasori, S. M. Hong, C. M. Koo and Y. Gogotsi, *Science*, 2016, **353**, 1137.
40. Z. Kang, Y. Ma, X. Tan, M. Zhu, Z. Zheng, N. Liu, L. Li, Z. Zou, X. Jiang, T. Zhai and Y. Gao, *Adv. Electron. Mater.*, 2017, **3**, 1700165.
41. E. Lee, A. Vahid Mohammadi, B. C. Prorok, Y. S. Yoon, M. Beidaghi and D.-J. Kim, *ACS Appl. Mater. Interfaces*, 2017, **9**, 37184.
42. Y. Z. Zhang, K. H. Lee, D. H. Anjum, R. Sougrat, Q. Jiang, H. Kim and H. N. Alshareef, *Sci. Adv.*, 2018, **4**, eaat098.
43. B. Lyu, M. Kim, H. Jing, J. Kang, C. Qian, S. Lee and J. H. Cho, *ACS Nano*, 2019, **13**, 11392.
44. D. Pang, M. Alhabeb, X. Mu, Y. Dall'Agnese, Y. Gogotsi and Y. Gao, *Nano Lett.*, 2019, **19**, 7443.
45. Y. Dong, S. S. K. Mallineni, K. Maleski, H. Behlow, V. N. Mochalin, A. M. Rao, Y. Gogotsi and R. Podila, *Nano Energy*, 2018, **44**, 103.
46. A. Sarycheva, A. Polemi, Y. Liu, K. Dandekar, B. Anasori and Y. Gogotsi, *Sci. Adv.*, 2018, **4**, eaau0920.
47. S. Ahn, T. H. Han, K. Maleski, J. Song, Y. H. Kim, M. H. Park, H. Zhou, S. Yoo, Y. Gogotsi and T. W. Lee, *Adv. Mater.*, 2020, **32**, 2000919.
48. H. Wei, H. Yu, J. Gong, M. Ma, H. Han, Y. Ni, S. Zhang and W. Xu, *Adv. Funct. Mater.*, 2021, **31**, 2007232.
49. M. A. Unal, F. Bayrakdar, L. Fusco, O. Besbinar, C. E. Shuck, S. Yalcin, M. T. Erken, A. Ozkul, C. Gurcan, O. Panatli, G. Y. Summak, C. Gokce, M. Orecchioni, A. Gazzi, F. Vitale, J. Somers, E. Demir, S. S. Yildiz,

H. Nazir, J. C. Grivel, D. Bedognetti, A. Crisanti, K. C. Akcali, Y. Gogotsi, L. G. Delogu and A. Yilmazer, *Nano Today*, 2021, **38**, 101136.
50. L. Verger, C. Xu, V. Natu, H. M. Cheng, W. Ren and M. W. Barsoum, *Curr. Opin. Solid State Mater. Sci.*, 2019, **23**, 149.
51. M. Alhabeb, K. Maleski, T. S. Mathis, A. Sarycheva, C. B. Hatter, S. Uzun, A. Levitt and Y. Gogotsi, *Angew. Chem., Int. Ed.*, 2018, **57**, 5444.
52. J. Zhou, X. Zha, X. Zhou, F. Chen, G. Gao, S. Wang, C. Shen, T. Chen, C. Zhi, P. Eklund, S. Du, J. Xue, W. Shi, Z. Chai and Q. Huang, *ACS Nano*, 2017, **11**, 3841.
53. J. Halim, S. Kota, M. R. Lukatskaya, M. Naguib, M.-Q. Zhao, E. J. Moon, J. Pitock, J. Nanda, S. J. May, Y. Gogotsi and M. W. Barsoum, *Adv. Funct. Mater.*, 2016, **26**, 3118.
54. J. Zhou, X. Zha, F. Y. Chen, Q. Ye, P. Eklund, S. Du and Q. Huang, *Angew. Chem., Int. Ed.*, 2016, **55**, 5008.
55. C. Xu, L. Wang, Z. Liu, L. Chen, J. Guo, N. Kang, X. L. Ma, H. M. Cheng and W. Ren, *Nat. Mater.*, 2015, **14**, 1135.
56. M. Naguib, M. Kurtoglu, V. Presser, J. Lu, J. Niu, M. Heon, L. Hultman, Y. Gogotsi and M. W. Barsoum, *Adv. Mater.*, 2011, **23**, 4248.
57. M. Ghidiu, M. R. Lukatskaya, M. Q. Zhao, Y. Gogotsi and M. W. Barsoum, *Nature*, 2014, **516**, 78.
58. P. Urbankowski, B. Anasori, T. Makaryan, D. Er, S. Kota, P. L. Walsh, M. Zhao, V. B. Shenoy, M. W. Barsoum and Y. Gogotsi, *Nanoscale*, 2016, **8**, 11385.
59. W. Sun, S. A. Shah, Y. Chen, Z. Tan, H. Gao, T. Habib, M. Radovic and M. J. Green, *J. Mater. Chem. A*, 2017, **5**, 21663.
60. T. Li, L. Yao, Q. Liu, J. Gu, R. Luo, J. Li, X. Yan, W. Wang, P. Liu, B. Chen, W. Zhang, W. Abbas, R. Naz and D. Zhang, *Angew. Chem., Int. Ed.*, 2018, **57**, 6115.
61. N. Kurra, M. Alhabeb, K. Maleski, C. H. Wang, H. N. Alshareef and Y. Gogotsi, *ACS Energy Lett.*, 2018, **3**, 2094.
62. M. Li, J. Lu, K. Luo, Y. Li, K. Chang, K. Chen, J. Zhou, J. Rosen, L. Hultman, P. Eklund, P. O. Å. Persson, S. Du, Z. Chai, Z. Huang and Q. Huang, *J. Am. Chem. Soc.*, 2019, **141**, 4730.
63. Y. Li, H. Shao, Z. Lin, J. Lu, L. Liu, B. Duployer, P. O. Å. Persson, P. Eklund, L. Hultman, M. Li, K. Chen, X. H. Zha, S. Du, P. Rozier, Z. Chai, E. Raymundo-Piñero, P. L. Taberna, P. Simon and Q. Huang, *Nat. Mater.*, 2020, **19**, 894.
64. V. Natu, R. Pai, M. Sokol, M. Carey, V. Kalra and M. W. Barsoum, *Chem*, 2020, **6**, 616.
65. A. el Ghazaly, H. Ahmed, A. R. Rezk, J. Halim, O. Å. Persson, L. Y. Yeo and J. Rosen, *ACS Nano*, 2021, **15**, 4293.
66. V. Nicolosi, M. Chhowalla, M. G. Kanatzidis, M. S. Strano and J. N. Coleman, *Science*, 2018, **4**, eaau0920.
67. K. S. Novoselov, A. K. Geim, S. V. Morozov, D. Jiang, Y. Zhang, S. V. Dubonos, I. V. Grigorieva and A. A. Firsov, *Science*, 2004, **306**, 666.
68. M. Naguib, O. Mashtalir, J. Carle, V. Presser, J. Lu, L. Hultman, Y. Gogotsi and M. W. Barsoum, *ACS Nano*, 2012, **6**, 1322.

69. M. Naguib, V. N. Mochalin, M. W. Barsoum, Y. Gogotsi, M. Naguib, V. N. Mochalin, M. W. Barsoum and Y. Gogotsi, *Adv. Mater.*, 2014, **26**, 992.
70. M. Sokol, V. Natu, S. Kota and M. W. Barsoum, *Trends Chem.*, 2019, **1**, 210.
71. C. Zhan, W. Sun, Y. Xie, D. Jiang and P. R. C. Kent, *ACS Appl. Mater. Interfaces*, 2019, **11**, 24885.
72. M. Alhabeb, K. Maleski, B. Anasori, P. Lelyukh, L. Clark, S. Sin and Y. Gogotsi, *Chem. Mater.*, 2017, **29**, 7633.
73. O. Mashtalir, M. Naguib, V. N. Mochalin, Y. Dall'Agnese, M. Heon, M. W. Barsoum and Y. Gogotsi, *Nat. Commun.*, 2013, **4**, 1.
74. M. Naguib, R. R. Unocic, B. L. Armstrong and J. Nanda, *Dalton Trans.*, 2015, **44**, 9353.
75. B. Anasori, Y. Xie, M. Beidaghi, J. Lu, B. C. Hosler, L. Hultman, P. R. C. Kent, Y. Gogotsi and M. W. Barsoum, *ACS Nano*, 2015, **9**, 9507.
76. M. H. Tran, T. Schäfer, A. Shahraei, M. Dürrschnabel, L. Molina-Luna, U. I. Kramm and C. S. Birkel, *ACS Appl. Energy Mater.*, 2018, **1**, 3908.
77. P. Urbankowski, B. Anasori, T. Makaryan, D. Er, S. Kota, P. L. Walsh, M. Zhao, V. B. Shenoy, M. W. Barsoum and Y. Gogotsi, *Nanoscale*, 2016, **8**, 11385.
78. B. Anasori, C. Shi, E. J. Moon, Y. Xie, C. A. Voigt, P. R. C. Kent, S. J. May, S. J. L. Billinge, M. W. Barsoum and Y. Gogotsi, *Nanoscale Horiz.*, 2016, **1**, 227.
79. M. Ghidiu, S. Kota, J. Halim, A. W. Sherwood, N. Nedfors, J. Rosen, V. N. Mochalin and M. W. Barsoum, *Chem. Mater.*, 2017, **29**, 1099.
80. M. Ghidiu, M. R. Lukatskaya, M. Q. Zhao, Y. Gogotsi and M. W. Barsoum, *Nature*, 2015, **516**, 78.
81. I. Persson, A. el Ghazaly, Q. Tao, J. Halim, S. Kota, V. Darakchieva, J. Palisaitis, M. W. Barsoum, J. Rosen and P. O. Å. Persson, *Small*, 2018, **14**, 1703676.
82. F. Liu, A. Zhou, J. Chen, J. Jia, W. Zhou, L. Wang and Q. Hu, *Appl. Surf. Sci.*, 2017, **416**, 781.
83. J. Halim, M. R. Lukatskaya, K. M. Cook, J. Lu, C. R. Smith, L.-Å. Näslund, S. J. May, L. Hultman, Y. Gogotsi, P. Eklund and M. W. Barsoum, *Chem. Mater.*, 2014, **26**, 2374.
84. A. S. Levitt, M. Alhabeb, C. B. Hatter, A. Sarycheva, G. Dion and Y. Gogotsi, *J. Mater. Chem. A*, 2019, **7**, 269.
85. N. Driscoll, A. G. Richardson, K. Maleski, B. Anasori, O. Adewole, P. Lelyukh, L. Escobedo, D. K. Cullen, T. H. Lucas, Y. Gogotsi and F. Vitale, *ACS Nano*, 2018, **12**, 10419.
86. T. L. Tan, H. M. Jin, M. B. Sullivan, B. Anasori and Y. Gogotsi, *ACS Nano*, 2017, **11**, 4407.
87. A. Lipatov, M. Alhabeb, M. R. Lukatskaya, A. Boson, Y. Gogotsi, A. Sinitskii, A. Lipatov, A. Boson, A. Sinitskii, M. Alhabeb, M. R. Lukatskaya, Y. Gogotsi and Y. A. Gogotsi J, *Adv. Electron. Mater.*, 2016, **2**, 1600255.

88. V. Natu, R. Pai, M. Sokol, M. Carey, V. Kalra and M. W. Barsoum, *Chem*, 2020, **6**, 616.
89. M. Anayee, N. Kurra, M. Alhabeb, M. Seredych, M. N. Hedhili, A.-H. Emwas, H. N. Alshareef, B. Anasori, Y. Gogotsi and R. Li, *Chem. Commun.*, 2020, **56**, 6090.
90. N. Kurra, M. Alhabeb, K. Maleski, C. H. Wang, H. N. Alshareef and Y. Gogotsi, *ACS Energy Lett.*, 2018, **3**, 2094.
91. J. L. Hart, K. Hantanasirisakul, A. C. Lang, B. Anasori, D. Pinto, Y. Pivak, J. T. van Omme, S. J. May, Y. Gogotsi and M. L. Taheri, *Nat. Commun.*, 2019, **10**, 1.
92. H. Yu, Y. Wang, Y. Jing, J. Ma, C.-F. Du, Q. Yan, H. Yu, Y. H. Wang, Y. Jing, C. Du, J. M. Ma and Q. Y. Yan, *Small*, 2019, **15**, 1901503.
93. T. Schultz, N. C. Frey, K. Hantanasirisakul, S. Park, S. J. May, V. B. Shenoy, Y. Gogotsi and N. Koch, *Chem. Mater.*, 2019, **31**, 6590.
94. T. Li, L. Yao, Q. Liu, J. Gu, R. Luo, J. Li, X. Yan, W. Wang, P. Liu, B. Chen, W. Zhang, W. Abbas, R. Naz and D. Zhang, *Angew. Chem., Int. Ed.*, 2018, **57**, 6115.
95. C. Wang, H. Shou, S. Chen, S. Wei, Y. Lin, P. Zhang, Z. Liu, K. Zhu, X. Guo, X. Wu, P. M. Ajayan and L. Song, *Adv. Mater.*, 2021, **33**, 2101015.
96. Y. Li, H. Shao, Z. Lin, J. Lu, L. Liu, B. Duployer, P. O. Å. Persson, P. Eklund, L. Hultman, M. Li, K. Chen, X. H. Zha, S. Du, P. Rozier, Z. Chai, E. Raymundo-Piñero, P. L. Taberna, P. Simon and Q. Huang, *Nat. Mater.*, 2020, **19**, 894.
97. W. Sun, S. A. Shah, Y. Chen, Z. Tan, H. Gao, T. Habib, M. Radovic and M. J. Green, *J. Mater. Chem. A*, 2017, **5**, 21663.
98. S. Yang, P. Zhang, F. Wang, A. G. Ricciardulli, M. R. Lohe, P. W. M. Blom and X. Feng, *Angew. Chem., Int. Ed.*, 2018, **57**, 15491.
99. V. Kamysbayev, A. S. Filatov, H. Hu, X. Rui, F. Lagunas, D. Wang, R. F. Klie and D. V. Talapin, *Science*, 2020, **369**, 979.
100. M. Yan, L. Yang, C. Li and Y. Zou, *J. Wuhan Univ. Technol., Mater. Sci. Ed.*, 2019, **34**, 299.
101. H. Dong, P. Xiao, N. Jin, B. Wang, Y. Liu and Z. Lin, *ChemElectroChem*, 2021, **8**, 957.
102. M. Li, J. Lu, K. Luo, Y. Li, K. Chang, K. Chen, J. Zhou, J. Rosen, L. Hultman, P. Eklund, P. O. Å. Persson, S. Du, Z. Chai, Z. Huang and Q. Huang, *J. Am. Chem. Soc.*, 2019, **141**, 4737.
103. V. Kamysbayev, A. S. Filatov, H. Hu, X. Rui, F. Lagunas, D. Wang, R. F. Klie and D. V. Talapin, *Science*, 2020, **369**, 979.
104. S. Y. Pang, Y. T. Wong, S. Yuan, Y. Liu, M. K. Tsang, Z. Yang, H. Huang, W. T. Wong and J. Hao, *J. Am. Chem. Soc.*, 2019, **141**, 9610.
105. M. Naguib, M. Kurtoglu, V. Presser, J. Lu, J. Niu, M. Heon, L. Hultman, Y. Gogotsi and M. W. Barsoum, *Adv. Mater.*, 2011, **23**, 4248.
106. S.-Y. Pang, Y.-T. Wong, S. Yuan, Y. Liu, M.-K. Tsang, Z. Yang, H. Huang, W.-T. Wong and J. Hao, *J. Am. Chem. Soc.*, 2019, **141**, 9610.
107. Y. Gogotsi, *Nat. Mater.*, 2015, **14**, 1079.
108. R. Deng, H. Zhang, Y. Zhang, Z. Chen, Y. Sui, X. Ge, Y. Liang, S. Hu, G. Yu and D. Jiang, *Chin. Phys. B*, 2017, **26**, 067901.

109. P. Urbankowski, B. Anasori, K. Hantanasirisakul, L. Yang, L. Zhang, B. Haines, S. J. May, S. J. L. Billinge and Y. Gogotsi, *Nanoscale*, 2017, **9**, 17722.
110. X. Xiao, H. Yu, H. Jin, M. Wu, Y. Fang, J. Sun, Z. Hu, T. Li, J. Wu, L. Huang, Y. Gogotsi and J. Zhou, *ACS Nano*, 2017, **11**, 2180.
111. X. Xiao, P. Urbankowski, K. Hantanasirisakul, Y. Yang, S. Sasaki, L. Yang, C. Chen, H. Wang, L. Miao, S. H. Tolbert, S. J. L. Billinge, H. D. Abruña, S. J. May and Y. Gogotsi, *Adv. Funct. Mater.*, 2019, **29**, 1809001.
112. X. Sang, Y. Xie, D. E. Yilmaz, R. Lotfi, M. Alhabeb, A. Ostadhossein, B. Anasori, W. Sun, X. Li, K. Xiao, P. R. C. Kent, A. C. T. Van Duin, Y. Gogotsi and R. R. Unocic, *Nat. Commun.*, 2018, **9**, 2266.
113. X. Xiao, H. Yu, H. Jin, M. Wu, Y. Fang, J. Sun, Z. Hu, T. Li, J. Wu, L. Huang, Y. Gogotsi and J. Zhou, *ACS Nano*, 2017, **11**, 2180.
114. C. E. Shuck, A. Sarycheva, M. Anayee, A. Levitt, Y. Zhu, S. Uzun, V. Balitskiy, V. Zahorodna, O. Gogotsi and Y. Gogotsi, *Adv. Eng. Mater.*, 2020, **22**, 1901241.
115. A. El Ghazaly, H. Ahmed, A. R. Rezk, J. Halim, P. O. Å. Persson, L. Y. Yeo and J. Rosen, *ACS Nano*, 2021, **15**, 4287.
116. J. Zhu, J. Zhang, R. Lin, B. Fu, C. Song, W. Shang, P. Tao and T. Deng, *Chem. Commun.*, 2021, **57**, 12611.
117. C. E. Shuck and Y. Gogotsi, *Chem. Eng. J.*, 2020, **401**, 125786.
118. Y. Gogotsi and B. Anasori, *ACS Nano*, 2019, **13**, 8491.
119. Y. Wei, P. Zhang, R. A. Soomro, Q. Zhu and B. Xu, *Adv. Mater.*, 2021, **33**, 2103148.
120. A. Vahid Mohammadi, J. Rosen and Y. Gogotsi, *Science*, 2021, 372.
121. M. Dahlqvist, J. Lu, R. Meshkian, Q. Tao, L. Hultman and J. Rosen, *Sci. Adv.*, 2017, **3**, e1700642.
122. Q. Tao, M. Dahlqvist, J. Lu, S. Kota, R. Meshkian, J. Halim, J. Palisaitis, L. Hultman, M. W. Barsoum, P. O. Å. Persson and J. Rosen, *Nat. Commun.*, 2017, **8**, 1.
123. R. Meshkian, M. Dahlqvist, J. Lu, B. Wickman, J. Halim, J. Thörnberg, Q. Tao, S. Li, S. Intikhab, J. Snyder, M. W. Barsoum, M. Yildizhan, J. Palisaitis, L. Hultman, P. O. Å. Persson and J. Rosen, *Adv. Mater.*, 2018, **30**, 1706409.
124. J. Halim, J. Palisaitis, J. Lu, J. Thörnberg, E. J. Moon, M. Precner, P. Eklund, P. O. Å. Persson, M. W. Barsoum and J. Rosen, *ACS Appl. Nano Mater.*, 2018, **1**, 2455.
125. F. Shahzad, M. Alhabeb, C. B. Hatter, B. Anasori, S. Man Hong, C. M. Koo and Y. Gogotsi, *Science*, 2016, **353**, 1137.
126. M. Aymerich, A. Gómez-Varela, E. Álvarez and M. Flores-Arias, *Materials*, 2016, **9**, 728.
127. M. Zhao, N. Trainor, C. E. Ren, M. Torelli, B. Anasori and Y. Gogotsi, *Adv. Mater. Technol.*, 2019, **4**, 1800639.
128. A. D. Dillon, M. J. Ghidiu, A. L. Krick, J. Griggs, S. J. May, Y. Gogotsi, M. W. Barsoum and A. T. Fafarman, *Adv. Funct. Mater.*, 2016, **26**, 4162.

129. C. Zhang, L. McKeon, M. P. Kremer, S.-H. Park, O. Ronan, A. Seral-Ascaso, S. Barwich, C. Ó. Coileáin, N. McEvoy, H. C. Nerl, B. Anasori, J. N. Coleman, Y. Gogotsi and V. Nicolosi, *Nat. Commun.*, 2019, **10**, 1795.
130. S. Abdolhosseinzadeh, X. Jiang, H. Zhang, J. Qiu and C. Zhang, *Mater. Today*, 2021, **48**, 214.
131. M. Seredych, C. E. Shuck, D. Pinto, M. Alhabeb, E. Precetti, G. Deysher, B. Anasori, N. Kurra and Y. Gogotsi, *Chem. Mater.*, 2019, **31**, 3324.
132. K. Maleski, V. N. Mochalin and Y. Gogotsi, *Chem. Mater.*, 2017, **29**, 1632.
133. Y. Hernandez, M. Lotya, D. Rickard, S. D. Bergin and J. N. Coleman, *Langmuir*, 2010, **26**, 3208.
134. W. Yang, J. Yang, J. J. Byun, F. P. Moissinac, J. Xu, S. J. Haigh, M. Domingos, M. A. Bissett, R. A. W. Dryfe and S. Barg, *Adv. Mater.*, 2019, **31**, 1902725.
135. B. Akuzum, K. Maleski, B. Anasori, P. Lelyukh, N. J. Alvarez, E. C. Kumbur and Y. Gogotsi, *ACS Nano*, 2018, **12**, 2685.
136. S. Abdolhosseinzadeh, R. Schneider, A. Verma, J. Heier, F. Nüesch and C. Zhang, *Adv. Mater.*, 2020, **32**, 2000716.
137. S. Naficy, R. Jalili, S. H. Aboutalebi, R. A. Gorkin III, K. Konstantinov, P. C. Innis, G. M. Spinks, P. Poulin and G. G. Wallace, *Mater. Horiz.*, 2014, **1**, 326.
138. C. J. Zhang, B. Anasori, A. Seral-Ascaso, S.-H. Park, N. McEvoy, A. Shmeliov, G. S. Duesberg, J. N. Coleman, Y. Gogotsi and V. Nicolosi, *Adv. Mater.*, 2017, **29**, 1702678.
139. K. Hantanasirisakul, M. Zhao, P. Urbankowski, J. Halim, B. Anasori, S. Kota, C. E. Ren, M. W. Barsoum and Y. Gogotsi, *Adv. Electron. Mater.*, 2016, **2**, 1600050.
140. M. Mariano, O. Mashtalir, F. Q. Antonio, W.-H. Ryu, B. Deng, F. Xia, Y. Gogotsi and A. D. Taylor, *Nanoscale*, 2016, **8**, 16371.
141. A. P.-Z. Clark, C. Shi, B. C. Ng, J. N. Wilking, A. L. Ayzner, A. Z. Stieg, B. J. Schwartz, T. G. Mason, Y. Rubin and S. H. Tolbert, *ACS Nano*, 2013, **7**, 962.
142. G. Ovarlez, Q. Barral and P. Coussot, *Nat. Mater.*, 2010, **9**, 115.
143. J. H. Y. Chung, S. Naficy, Z. Yue, R. Kapsa, A. Quigley, S. E. Moulton and G. G. Wallace, *Biomater. Sci.*, 2013, **1**, 763.
144. S. Morelly, M. Tang and N. Alvarez, *Polymers*, 2017, **9**, 461.
145. M. Naguib, R. R. Unocic, B. L. Armstrong and J. Nanda, *Dalton Trans.*, 2015, **44**, 9353.
146. C. J. Zhang, S. Pinilla, N. McEvoy, C. P. Cullen, B. Anasori, E. Long, S.-H. Park, A. Seral-Ascaso, A. Shmeliov, D. Krishnan, C. Morant, X. Liu, G. S. Duesberg, Y. Gogotsi and V. Nicolosi, *Chem. Mater.*, 2017, **29**, 4848.
147. O. Mashtalir, K. M. Cook, V. N. Mochalin, M. Crowe, M. W. Barsoum and Y. Gogotsi, *J. Mater. Chem. A*, 2014, **2**, 14334.
148. C. J. Zhang, S. Pinilla, N. McEvoy, C. P. Cullen, B. Anasori, E. Long, S.-H. Park, A. Seral-Ascaso, A. Shmeliov, D. Krishnan, C. Morant, X. Liu, G. S. Duesberg, Y. Gogotsi and V. Nicolosi, *Chem. Mater.*, 2017, **29**, 4848.

149. X. Zhao, A. Vashisth, E. Prehn, W. Sun, S. A. Shah, T. Habib, Y. Chen, Z. Tan, J. L. Lutkenhaus, M. Radovic and M. J. Green, *Matter*, 2019, **1**, 513.
150. V. Natu, J. L. Hart, M. Sokol, H. Chiang, M. L. Taheri and M. W. Barsoum, *Angew. Chem., Int. Ed.*, 2019, **58**, 12655.
151. C. J. Zhang, S. Pinilla, N. McEvoy, C. P. Cullen, B. Anasori, E. Long, S. H. Park, A. Seral-Ascaso, A. Shmeliov, D. Krishnan, C. Morant, X. Liu, G. S. Duesberg, Y. Gogotsi and V. Nicolosi, *Chem. Mater.*, 2017, **29**, 4848.
152. S. Huang and V. N. Mochalin, *ACS Nano*, 2020, **14**, 10251.
153. S. Huang and V. N. Mochalin, *Inorg. Chem.*, 2019, **58**, 1958.
154. J. Zhang, N. Kong, D. Hegh, K. A. S. Usman, G. Guan, S. Qin, I. Jurewicz, W. Yang and J. M. Razal, *ACS Appl. Mater. Interfaces*, 2020, **12**, 34032.
155. T. S. Mathis, K. Maleski, A. Goad, A. Sarycheva, M. Anayee, A. C. Foucher, K. Hantanasirisakul, C. E. Shuck, E. A. Stach and Y. Gogotsi, *ACS Nano*, 2021, **15**, 6420.
156. K. Maleski, V. N. Mochalin and Y. Gogotsi, *Chem. Mater.*, 2017, **29**, 1632.
157. S. Seyedin, J. Zhang, K. A. S. Usman, S. Qin, A. M. Glushenkov, E. R. S. Yanza, R. T. Jones and J. M. Razal, *Global Challenges*, 2019, **3**, 1900037.
158. T. S. Mathis, K. Maleski, A. Goad, A. Sarycheva, M. Anayee, A. C. Foucher, K. Hantanasirisakul, C. E. Shuck, E. A. Stach and Y. Gogotsi, *ACS Nano*, 2021, **15**, 6420.
159. A. Lipatov, H. Lu, M. Alhabeb, B. Anasori, A. Gruverman, Y. Gogotsi and A. Sinitskii, *Sci. Adv.*, 2018, **4**, eaat0491.
160. A. Lipatov, M. Alhabeb, H. Lu, S. Zhao, M. J. Loes, N. S. Vorobeva, Y. Dall'Agnese, Y. Gao, A. Gruverman, Y. Gogotsi and A. Sinitskii, *Adv. Electron. Mater.*, 2020, **6**, 1901382.
161. A. Castellanos-Gomez, V. Singh, H. S. J. Van Der Zant and G. A. Steele, *Ann. Phys.*, 2015, **527**, 27.
162. V. N. Borysiuk, V. N. Mochalin and Y. Gogotsi, *Nanotechnology*, 2015, **26**, 265705.
163. A. Lipatov, H. Lu, M. Alhabeb, B. Anasori, A. Gruverman, Y. Gogotsi and A. Sinitskii, *Sci. Adv.*, 2018, **4**, eaat0491.
164. J. Zhang, N. Kong, S. Uzun, A. Levitt, S. Seyedin, P. A. Lynch, S. Qin, M. Han, W. Yang, J. Liu, X. Wang, Y. Gogotsi and J. M. Razal, *Adv. Mater.*, 2020, **32**, 2001093.
165. M. Khazaei, M. Arai, T. Sasaki, C.-Y. Chung, N. S. Venkataramanan, M. Estili, Y. Sakka and Y. Kawazoe, *Adv. Funct. Mater.*, 2013, **23**, 2185.
166. M. Khazaei, A. Ranjbar, M. Arai, T. Sasaki and S. Yunoki, *J. Mater. Chem. C*, 2017, **5**, 2488.
167. H. A. Tahini, X. Tan and S. C. Smith, *Nanoscale*, 2017, **9**, 7016.
168. M. Khazaei, A. Ranjbar, M. Arai and S. Yunoki, *Phys. Rev. B*, 2016, **94**, 125152.
169. T. Hu, H. Zhang, J. Wang, Z. Li, M. Hu, J. Tan, P. Hou, F. Li and X. Wang, *Sci. Rep.*, 2015, **5**, 1.
170. B. Anasori and Y. Gogotsi, *2D Metal Carbides and Nitrides (MXenes)*, Springer, Cham, 2019.

171. T. S. Mathis, K. Maleski, A. Goad, A. Sarycheva, M. Anayee, A. C. Foucher, K. Hantanasirisakul, C. E. Shuck, E. A. Stach and Y. Gogotsi, *ACS Nano*, 2021, **15**, 6420.
172. K. Hantanasirisakul and Y. Gogotsi, *Adv. Mater.*, 2018, **30**, 1804779.
173. M. Khazaei, A. Ranjbar, M. Arai, T. Sasaki and S. Yunoki, *J. Mater. Chem. C*, 2017, **5**, 2488.
174. Y. Liu, H. Xiao and W. A. Goddard, *J. Am. Chem. Soc.*, 2016, **138**, 15853.
175. H. Weng, A. Ranjbar, Y. Liang, Z. Song, M. Khazaei, S. Yunoki, M. Arai, Y. Kawazoe, Z. Fang and X. Dai, *Phys. Rev. B*, 2015, **92**, 75436.
176. K. Maleski, C. E. Ren, M. Q. Zhao, B. Anasori and Y. Gogotsi, *ACS Appl. Mater. Interfaces*, 2018, **10**, 24491.
177. M. R. Lukatskaya, O. Mashtalir, C. E. Ren, Y. Dall'Agnese, P. Rozier, P. L. Taberna, M. Naguib, P. Simon, M. W. Barsoum and Y. Gogotsi, *Science*, 2013, **341**, 1502.
178. M. Naguib, J. Halim, J. Lu, K. M. Cook, L. Hultman, Y. Gogotsi and M. W. Barsoum, *J. Am. Chem. Soc.*, 2013, **135**, 15966.
179. M. R. Lukatskaya, S. Kota, Z. Lin, M. Q. Zhao, N. Shpigel, M. D. Levi, J. Halim, P. L. Taberna, M. W. Barsoum, P. Simon and Y. Gogotsi, *Nat. Energy*, 2017, **2**, 1.
180. C. Zhang, B. Anasori, A. Seral-Ascaso, S.-H. Park, N. McEvoy, A. Shmeliov, G. S. Duesberg, J. N. Coleman, Y. Gogotsi and V. Nicolosi, *Adv. Mater.*, 2017, **29**, 1702678.
181. C. Zhang, *J. Energy Chem.*, 2021, **60**, 417.
182. Y. Ma, N. Liu, L. Li, X. Hu, Z. Zou, J. Wang, S. Luo and Y. Gao, *Nat. Commun.*, 2017, **8**, 1.
183. S. J. Kim, H. J. Koh, C. E. Ren, O. Kwon, K. Maleski, S. Y. Cho, B. Anasori, C. K. Kim, Y. K. Choi, J. Kim, Y. Gogotsi and H. T. Jung, *ACS Nano*, 2018, **12**, 986.
184. H.-J. Koh, S. J. Kim, K. Maleski, S.-Y. Cho, Y.-J. Kim, C. W. Ahn, Y. Gogotsi and H.-T. Jung, *ACS Sens.*, 2019, **4**, 1365.
185. B. Xu, M. Zhu, W. Zhang, X. Zhen, Z. Pei, Q. Xue, C. Zhi, P. Shi, B. Xu, W. Zhang, X. Zhen, P. Shi, M. Zhu, Z. Pei, Q. Xue and C. Zhi, *Adv. Mater.*, 2016, **28**, 3333.
186. Z. Zeng, C. Wang, G. Siqueira, D. Han, A. Huch, S. Abdolhosseinzadeh, J. Heier, F. Nüesch, C. Zhang and G. Nyström, *Adv. Sci.*, 2020, **7**, 2000979.
187. Y. I. Jhon, M. Seo and Y. M. Jhon, *Nanoscale*, 2018, **10**, 69.
188. G. R. Berdiyorov, *AIP Adv.*, 2016, **6**, 055105.
189. K. Hantanasirisakul, M. Alhabeb, A. Lipatov, K. Maleski, B. Anasori, P. Salles, C. Ieosakulrat, P. Pakawatpanurut, A. Sinitskii, S. J. May and Y. Gogotsi, *Chem. Mater.*, 2019, **31**, 2941.
190. Y. Li, C. Xiong, H. Huang, X. Peng, D. Mei, M. Li, G. Liu, M. Wu, T. Zhao and B. Huang, *Adv. Mater.*, 2021, **33**, 2103054.
191. M. Han, K. Maleski, C. E. Shuck, Y. Yang, J. T. Glazar, A. C. Foucher, K. Hantanasirisakul, A. Sarycheva, N. C. Frey, S. J. May, V. B. Shenoy, E. A. Stach and Y. Gogotsi, *J. Am. Chem. Soc.*, 2020, **142**, 19110.

192. K. Maleski, C. E. Shuck, A. T. Fafarman and Y. Gogotsi, *Adv. Opt. Mater.*, 2021, **9**, 2001563.
193. K. Hantanasirisakul and Y. Gogotsi, *Adv. Mater.*, 2018, **30**, 1804779.
194. X. Jiang, A. v. Kuklin, A. Baev, Y. Ge, H. Ågren, H. Zhang and P. N. Prasad, *Phys. Rep.*, 2020, **848**, 1.
195. K. Chaudhuri, Z. Wang, M. Alhabeb, K. Maleski, Y. Gogotsi, V. Shalaev and A. Boltasseva, in *2D Metal Carbides and Nitrides (MXenes): Structure, Properties and Applications*, ed. B. Anasori, and Y. Gogotsi, Springer International Publishing, Cham, 2019, pp. 327–346.
196. Y. Dong, S. Chertopalov, K. Maleski, B. Anasori, L. Hu, S. Bhattacharya, A. M. Rao, Y. Gogotsi, V. N. Mochalin and R. Podila, *Adv. Mater.*, 2018, **30**, 1705714.
197. X. Jiang, S. Liu, W. Liang, S. Luo, Z. He, Y. Ge, H. Wang, R. Cao, F. Zhang, Q. Wen, J. Li, Q. Bao, D. Fan and H. Zhang, *Laser Photonics Rev.*, 2018, **12**, 1700229.
198. J. Xu, J. Shim, J. H. Park and S. Lee, *Adv. Funct. Mater.*, 2016, **26**, 5328.
199. C. Yi, Y. Chen, Z. Kang, Y. Ma, Y. Yue, W. Liu, M. Zhu and Y. Gao, *Adv. Electron. Mater.*, 2021, 7, 2000955.
200. W. Song, Q. Liu, J. Chen, Z. Chen, X. He, Q. Zeng, S. Li, L. He, Z. Chen and X. Fang, *Small*, 2021, **17**, 2100439.
201. K. Montazeri, M. Currie, L. Verger, P. Dianat, M. W. Barsoum and B. Nabet, *Adv. Mater.*, 2019, **31**, 1903271.
202. Z. Kang, Y. Ma, X. Tan, M. Zhu, Z. Zheng, N. Liu, L. Li, Z. Zou, X. Jiang, T. Zhai and Y. Gao, *Adv. Electron. Mater.*, 2017, **3**, 1700165.
203. D. B. Velusamy, J. K. El-Demellawi, A. M. El-Zohry, A. Giugni, S. Lopatin, M. N. Hedhili, A. E. Mansour, E. di Fabrizio, O. F. Mohammed and H. N. Alshareef, *Adv. Mater.*, 2019, **31**, 1807658.
204. A. Sarycheva, T. Makaryan, K. Maleski, E. Satheeshkumar, A. Melikyan, H. Minassian, M. Yoshimura and Y. Gogotsi, *J. Phys. Chem. C*, 2017, **121**, 19983.
205. Y. Peng, P. Cai, L. Yang, Y. Liu, L. Zhu, Q. Zhang, J. Liu, Z. Huang and Y. Yang, *ACS Omega*, 2020, **5**, 26486.
206. B. Soundiraraju and B. K. George, *ACS Nano*, 2017, **11**, 8892.
207. H. Xie, P. Li, J. Shao, H. Huang, Y. Chen, Z. Jiang, P. K. Chu and X.-F. Yu, *ACS Sens.*, 2019, **4**, 2303.
208. R. Liu, L. Jiang, C. Lu, Z. Yu, F. Li, X. Jing, R. Xu, W. Zhou and S. Jin, *Spectrochim. Acta, Part A*, 2020, **236**, 118336.
209. F. Shahzad, M. Alhabeb, C. B. Hatter, B. Anasori, S. Man Hong, C. M. Koo and Y. Gogotsi, *Science*, 2016, **353**, 1137.
210. J. Xuan, Z. Wang, Y. Chen, D. Liang, L. Cheng, X. Yang, Z. Liu, R. Ma, T. Sasaki and F. Geng, *Angew. Chem., Int. Ed.*, 2016, **128**, 14789.
211. R. Li, L. Zhang, L. Shi and P. Wang, *ACS Nano*, 2017, **11**, 3752.
212. H. Lin, X. Wang, L. Yu, Y. Chen and J. Shi, *Nano Lett.*, 2017, **17**, 384.
213. D. Xu, Z. Li, L. Li and J. Wang, *Adv. Funct. Mater.*, 2020, **30**, 2000712.
214. K. Huang, Z. Li, J. Lin, G. Han and P. Huang, *Chem. Soc. Rev.*, 2018, **47**, 5109.

215. M. Soleymaniha, M.-A. Shahbazi, A. R. Rafieerad, A. Maleki and A. Amiri, *Adv. Healthcare Mater.*, 2019, **8**, 1801137.
216. H. Lin, Y. Chen and J. Shi, *Adv. Sci.*, 2018, **5**, 1800518.
217. C. Dai, Y. Chen, X. Jing, L. Xiang, D. Yang, H. Lin, Z. Liu, X. Han and R. Wu, *ACS Nano*, 2017, **11**, 12696.
218. H. Lin, S. Gao, C. Dai, Y. Chen and J. Shi, *J. Am. Chem. Soc.*, 2017, **139**, 16235.
219. X. Han, J. Huang, H. Lin, Z. Wang, P. Li and Y. Chen, *Adv. Healthcare Mater.*, 2018, **7**, 1701394.
220. G. Cai, J.-H. Ciou, Y. Liu, Y. Jiang and P. S. Lee, *Sci. Adv.*, 2019, **5**, aaw7956.
221. X. Fan, Y. Ding, Y. Liu, J. Liang and Y. Chen, *ACS Nano*, 2019, **13**, 8124.
222. N. Tao, D. Zhang, X. Li, D. Lou, X. Sun, C. Wei, J. Li, J. Yang and Y.-N. Liu, *Chem. Sci.*, 2019, **10**, 10765.
223. C. Yang, D. Xu, W. Peng, Y. Li, G. Zhang, F. Zhang and X. Fan, *Nanoscale*, 2018, **10**, 15387.
224. M. Khazaei, M. Arai, T. Sasaki, C.-Y. Chung, N. S. Venkataramanan, M. Estili, Y. Sakka and Y. Kawazoe, *Adv. Funct. Mater.*, 2013, **23**, 2185.
225. Y. Yoon, T. A. Le, A. P. Tiwari, I. Kim, M. W. Barsoum and H. Lee, *Nanoscale*, 2018, **10**, 22429.
226. L. Dong, H. Kumar, B. Anasori, Y. Gogotsi and V. B. Shenoy, *J. Phys. Chem. Lett.*, 2017, **8**, 422.
227. N. C. Frey, H. Kumar, B. Anasori, Y. Gogotsi and V. B. Shenoy, *ACS Nano*, 2018, **12**, 6319.
228. N. C. Frey, A. Bandyopadhyay, H. Kumar, B. Anasori, Y. Gogotsi and V. B. Shenoy, *ACS Nano*, 2019, **13**, 2831.
229. Z. Zhang, X. Liu, J. Yu, Y. Hang, Y. Li, Y. Guo, Y. Xu, X. Sun, J. Zhou and W. Guo, *Wiley Interdiscip. Rev.:Comput. Mol. Sci.*, 2016, **6**, 324.
230. Y. Zhang and F. Li, *J. Magn. Magn. Mater.*, 2017, **433**, 222.
231. M. Je, Y. Lee and Y.-C. Chung, *Thin Solid Films*, 2016, **619**, 131.
232. J. He, P. Lyu, L. Z. Sun, Á. Morales García and P. Nachtigall, *J. Mater. Chem. C*, 2016, **4**, 6500.
233. G. Wang, *J. Phys. Chem. C*, 2016, **120**, 18850.
234. J. Yang, X. Zhou, X. Luo, S. Zhang and L. Chen, *Appl. Phys. Lett.*, 2016, **109**, 203109.
235. N. C. Frey, J. Wang, G. Iván, I. Vega Bellido, B. Anasori, Y. Gogotsi and V. B. Shenoy, *ACS Nano*, 2019, **13**, 3031.

CHAPTER 4

Hydrides and Halides

JOSHUA W. MAKEPEACE

School of Chemistry, University of Birmingham, United Kingdom
E-mail: j.w.makepeace@bham.ac.uk

4.1 Introduction

Sodium chloride and its 'rock salt' structure (Figure 4.1) may well be the most common first taste of materials and solid-state chemistry for school students around the world. In this way, and because of the focus on the health effects of 'salt', halide materials are among the most well-known inorganic materials, to chemists and the general public alike.

Indeed, metal halides hold a special place in our understanding of the ordered structures of solids. The diffraction patterns and proposed atomic structures of sodium chloride (Figure 4.2), potassium chloride, potassium bromide and calcium fluoride were among those presented in W. L. Bragg's landmark 1913 paper, communicated to the Royal Society of London by his father W. H. Bragg, as the first recorded crystal structures solved through X-ray diffraction.[1]

While less well-connected to people's daily lives, hydrides occupy a similarly fundamental position in our understanding of astrochemistry, having been among the first molecules to have been detected outside of our Solar System.[2] Here, we aim to take a contemporary view of the chemistry and applications of halide and hydride materials, which find themselves at the forefront of a range of energy-focused materials chemistry research. We will explore the relationship between the fundamental chemistry of these materials—their structures and reactivity—and these functional properties. Halide compounds are known for almost every metal, even very unreactive

Inorganic Materials Series
Non-oxide and Heteroanionic Materials
Edited by Duncan H. Gregory
A catalogue record for this book is available from the British Library

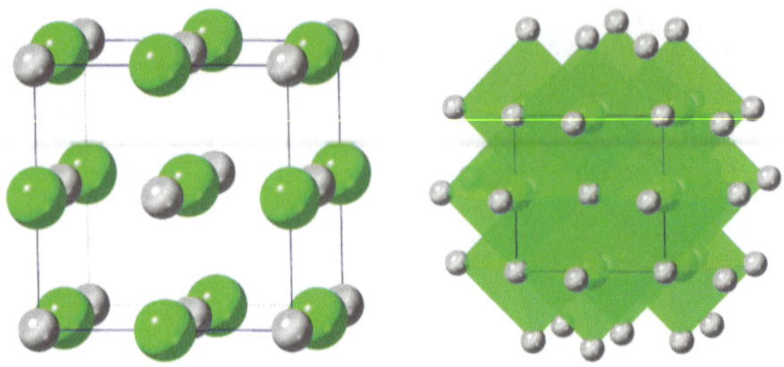

Figure 4.1 The crystal structure of sodium chloride, with sodium ions shown in grey and chloride ions in green, with a polyhedral representation of the structure as $ClNa_6$ octahedra shown to the right.

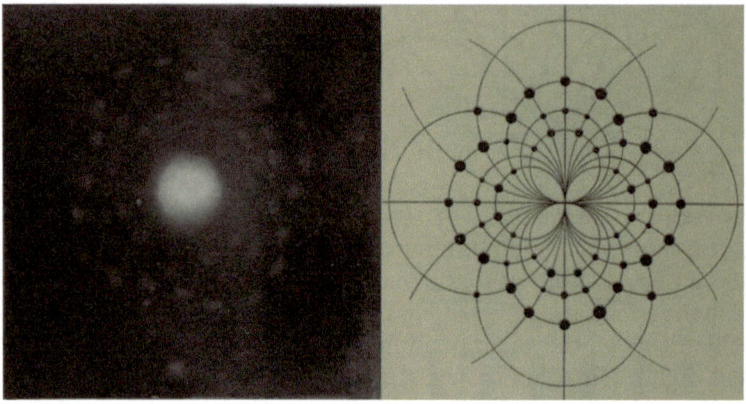

Figure 4.2 The original diffraction pattern and reconstructed schematic of sodium chloride presented in W. L. Bragg's manuscript on the first crystal structures solved by X-ray diffraction. Reproduced from ref. 1 with permission from The Royal Society.

metals like gold and platinum, and are generally accessible from reactions of metals with halogen acids. Hydrides are also widespread, with binary hydrides known for Group I and II metals, early transition metals, f-block metals and p-block metalloids (Figure 4.3), largely formed *via* direct reaction between the element and gaseous hydrogen. An unusual 'hydride gap' is found in the mid and later d-block metals, aside from palladium, though they can be found in alloy-based hydrides.[3]

Given the breadth of the halide and hydride materials, the survey presented in this chapter will by necessity not be exhaustive. For comprehensive descriptions of the basic chemistry of halides and hydrides, their occurrence and synthesis, the reader is pointed to numerous seminal textbooks on the subject.[4–6] However, it is hoped that it will provide a snapshot of the current relevance of these materials, and whet the appetite for further investigations.

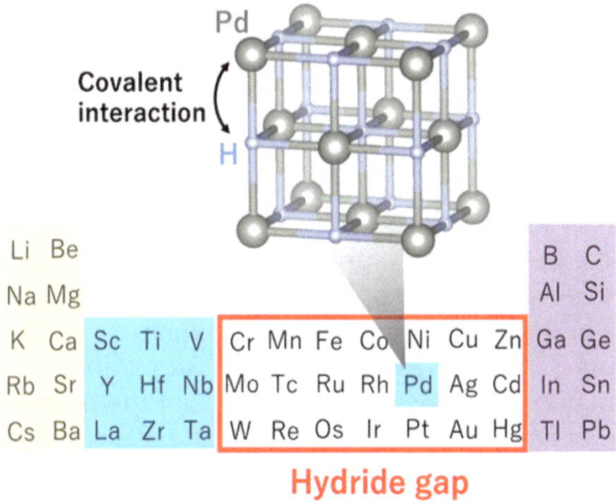

Figure 4.3 Hydride-forming elements, with ionic hydrides indicated in yellow, metallic in blue, and molecular in purple. Reprinted from ref. 3 with permission from American Chemical Society, Copyright 2021.

4.2 Metal Hydrides

4.2.1 Hydrogen Storage, Purification and Compression Materials

Since the earliest conceptions of hydrogen as a clean-burning fuel, the challenge of storing hydrogen gas more densely has been identified as a key barrier to its application. Nowhere is this challenge more starkly defined than in the area of hydrogen-powered transportation, where storing in the order of 5 kg of hydrogen (with equivalent energy to around 15 kg or 20 L of petrol) requires fitting around 11 000 L of hydrogen gas into a space ideally measured in the tens of litres. The most intuitive, physical approaches to solving this challenge of compression and liquefaction of hydrogen are limited in their effectiveness. This is largely due to the extreme conditions required: less than 20 K to liquefy hydrogen and pressures of around 700 bar. This requires either elaborate cryogenic equipment or sophisticated carbon composite pressure vessels, and still only achieves volumetric energy densities in the region of one-quarter that of petrol. These limitations have spurred materials chemists to explore metal hydride materials as a potential solution to this challenge, and in doing so have developed an extensive understanding of the structural, thermodynamic and kinetic properties of a wide range of metal hydrides.

The essential principle of a reversible hydrogen store based on a hydride is to apply an elevated pressure to cause the formation of a metal hydride, and then when hydrogen is required, to apply heat (ideally waste heat from the engine/fuel cell) to the hydride to destabilise it and release hydrogen. In many ways, transition metal and intermetallic hydrides served as the inspiration for the practical application of reversible hydrogen storage. Their volumetric hydrogen densities often exceed that of liquid hydrogen, and they can store and release hydrogen close to room temperature and with modest application of hydrogen pressure. Achieving such high hydrogen density without the need for high pressure also offers significant safety advantages compared to pressurised tanks. The principal drawback of these materials for mobile applications such as hydrogen fuel cell cars is the high mass of the transition metal components. Often the proportion of hydrogen by mass is less than 2%. However, for stationary storage applications and transport methods where the mass of the hydrogen store is less critical—such as river/canal barges or forklifts, these materials offer many attractive qualities.

Some examples of the most common types of metallic hydrides are shown in Table 4.1, encompassing many of the common alloy ratios of known hydride materials.[7] For many of the presented materials, the conditions for reversible hydrogen storage are very mild. This can be understood conceptually from consideration of the structures of the materials in their hydride and metal states. Given the desire for low-temperature and pressure hydrogenation reactions, it is reasonable to state that the hydride phases which form reversibly under these conditions will not have required the heavy metal atoms to significantly change configuration. Indeed, this is the case for many of these materials, where the array of metals is conserved, while hydrogen is incorporated into the interstices of the crystal structure. This is demonstrated in Figure 4.4, which shows the crystal structures of $LaNi_5$ and its corresponding hydride, $LaNi_5H_6$. Hydrogen is incorporated into the octahedral and tetrahedral holes in the metal array, partially occupying these sites in a disordered fashion.[8] Other facile hydride formation reactions require only modest distortions of the metal array.

The formation reactions of these hydrides follow a similar scheme, whereby initial hydrogen incorporation is by random occupancy of interstitial sites (forming a solid solution of hydride and metal, sometimes referred to as an

Table 4.1 Some common metallic hydrides, where p_{pl} refers to the equilibrium pressure of the hydride-forming reaction at the temperature stated. Adapted from ref. 7 with permission from Springer Nature, Copyright 2001.

Type	Metal	Hydride	wt%	p_{pl}/bar	T/K
A	Pd	$PdH_{0.6}$	0.56	0.02	298
AB_5	$LaNi_5$	$LaNi_5H_6$	1.37	2	298
AB_2	ZrV_2	$ZrV_2H_{5.5}$	3.01	10^{-8}	323
AB	FeTi	$FeTiH_2$	1.89	5	303
A_2B	Mg_2Ni	Mg_2NiH_4	3.59	1	555
bcc	TiV_2	TiV_2H_4	2.6	10	313

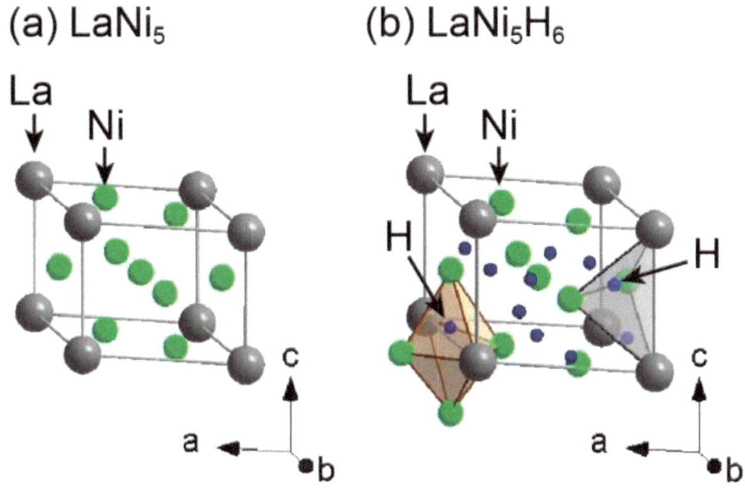

Figure 4.4 The crystal structures of (a) $LaNi_5$ and (b) $LaNi_5H_6$, with octahedral and tetrahedral holes in the structure indicated in brown and grey, respectively. Reproduced from ref. 8 https://doi.org/10.3390/molecules28031256, under the terms of the CC-BY 4.0 license, https://creativecommons.org/licenses/by/4.0/.

α-phase) until a critical hydrogen concentration where the formation of the ordered metal hydride structure occurs (termed the β-phase). Once the material is converted into the ordered hydride, further incorporation of hydrogen becomes very unfavourable and would occur in a disordered fashion. This hydrogenation behaviour is reflected in pressure–composition isotherm (PCI) data, whereby the metal is exposed to controlled quantities of hydrogen and allowed to react until an equilibrium pressure is reached, from which the degree of hydrogen incorporation can be inferred. An example of these data is shown in Figure 4.5.[9] The pressure plateau, which occurs during β-phase formation, is technologically useful since the hydride can be used to deliver a stable hydrogen pressure. In reality, many hydrides exhibit sloped plateaus, which reflect stable deviations from the precise ordered stoichiometry, *i.e.* a non-stoichiometric hydrogenation reaction.

The measurement of PCI data across multiple temperatures allows for the determination of the thermodynamic parameters of the reaction to be established through the Van't Hoff relation (eqn (4.1)), where p/p_o is the partial hydrogen pressure, ΔH is the enthalpy change for the hydrogenation reaction, R is the universal gas constant, T is the temperature and ΔS is the entropy change for the hydrogenation reaction:

$$\ln\frac{p}{p_o}=\frac{\Delta H}{RT}-\frac{\Delta S}{R} \tag{4.1}$$

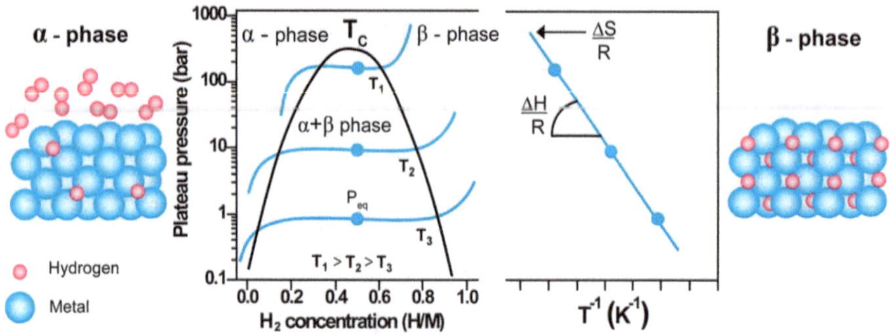

Figure 4.5 Hydrogenation reactions as followed by pressure–composition isotherms. Reproduced from ref. 9 with permission from The Royal Society of Chemistry.

The entropy change during hydrogenation is fairly constant between different metal hydride systems because it is dominated by the entropy of gaseous hydrogen, 130 J mol^{-1} K^{-1}. The enthalpy of the reaction in simple metal hydrides is a measure of the average M–H bond strength, and as such is governed by the identity of the metals involved and the structure of the metal (in that it determines the geometry and therefore electrostatic interactions associated with the interstitial sites which hydrogen occupies). Research efforts have therefore focused on the tailoring of structure and composition in order to achieve practical combinations of hydrogen plateau pressure and temperature. Alloying is the principal tool with which this is easily achieved, given the significant range of metal substitutions which can be achieved in alloy formulations. As a heuristic rule, for an alloy to form a stable hydride, at least one of the metals used in the alloy must be capable of forming a binary hydride.[10]

Metal substitutions within well-known hydrides such as $LaNi_5$ have been successfully used to modify the plateau pressure,[11] and can also influence the hydrogen storage capacity of the material if substitutions have less favourable metal–hydrogen interactions and therefore destabilise interstitial sites.[12] One issue in the repeated hydrogen storage cycling of alloy materials can sometimes result in the formation of binary hydrides if there is a thermodynamic instability of the ternary hydride with respect to the binary phases, and has been recorded in $LaNi_5$ with the formation of LaH_3 and Ni.[13,14] While the use of lower hydrogenation pressure can be effective in avoiding regions of instability towards disproportionation,[15] substitution is also effective, with the formation of $LaNi_{5-x}Sn_x$ alloys significantly improving the cycling stability, which was attributed to the presence of tin retarding the mobility of metal atoms towards nucleation of the new phases[16] or preventing defects which might serve as nucleation sites.[15]

While the ability to occupy interstitial sites in the metal array is one of the reasons why these metallic hydrides are so readily reversible as hydrogen stores, it also imposes some fundamental limitations on the gravimetric hydrogen content. Closer examination of the data shown in Table 4.1 shows that the hydrogen to metal ratio in these materials is always less than 2 : 1, and often closer to 1 : 1. The hexagonal and cubic close-packed metal arrays on which most of these materials are based each have one octahedral and two tetrahedral interstitial sites per metal atom. While this might imply a maximum 3 : 1 hydrogen to metal ratio, it is often energetically unfavourable for hydrogen to occupy adjacent sites, such as where an octahedral and tetrahedral site share a face, that would result in close approach of the hydrogens occupying the two sites. Efforts to quantify this effect have suggested a minimum hydrogen–hydrogen distance of 2.1 Å is possible. The size of the interstitial site will also influence its favourability for hydride occupancy and similar quantification efforts have suggested a minimum hole radius of 0.40 Å.[17]

While conventional alloying approaches have remained limited by the upper limit of hydrogen content discussed above, recent efforts to exceed the 2 : 1 ratio have involved the alloying of five or more different metals in similar molar quantities to produce so-called high-entropy alloys. Highly disordered arrangements of the metals are observed within the different alloy structure types, as summarised in Figure 4.6,[18] and are generally synthesised by arc-melting and/or high-energy ball-milling approaches. While still a relatively new area of research, there have been a number of promising studies on the beneficial effects of high-entropy alloys for hydrogen storage.[19] In theory, the ability to mix and match a larger number of metals in the alloy allows for greater control over the hydrogen affinity of the material, with the incorporation of non-hydriding elements used to promote higher plateau pressures.[19] The use of a large number of different radius metals is also expected to result in the formation of highly-strained crystal structures with more irregularly-shaped interstitial sites. This argument has been proposed to explain the unusual hydrogen storage properties of a TiVZrNbHf alloy, which showed a hydrogen to metal ratio of approximately 2.5 : 1.[20] A detailed powder diffraction experiment on the material demonstrated hydrogen occupancy of both the octahedral and tetrahedral sites, in support of this argument.

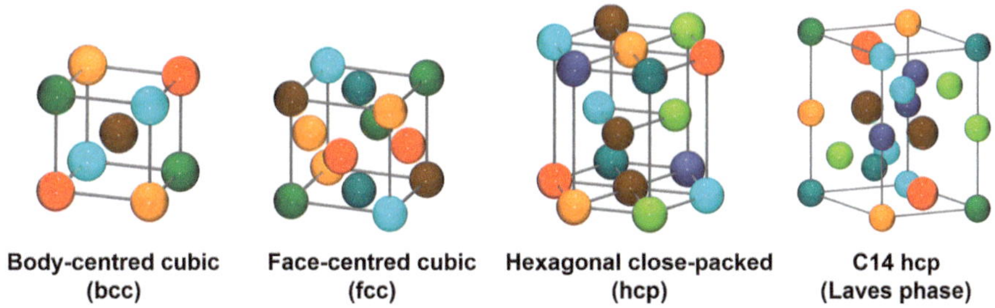

Figure 4.6 High entropy alloy structure types, with different coloured balls representing different metals. Reproduced from ref. 18 with permission from The Royal Society of Chemistry.

Aside from the thermodynamics of the hydrogen storage reactions, kinetic optimisation has also been explored for many transition metal systems. The titanium iron alloy is a useful example of this, being a cheap alloy it is very attractive for larger-scale stationary hydrogen storage. However, the titanium readily oxidises in air, retarding the hydrogen absorption reaction. As such, the system typically requires 'hard activation', consisting of high-pressure, high-temperature hydrogen absorption and release cycling, before a stable cycling regime can be established, and then must be kept scrupulously clear of oxidisers. Alternative mechanical activation treatments change grain sizes or break up oxide layers using milling, torsion or deformation approaches.[21–23] Partial substitutions with a wide range of elements have also been shown to improve the kinetics of the hydrogen storage reactions and reduce the need for hard activation. Possible explanations for these effects include substitutional elements forming sacrificial oxygen 'getters', the formation of hydrogen-conductive secondary phases or cracking of the oxide layer through differential expansion of minor phases.[24]

Moving away from transition metal metallic hydrides, rare earth (RE) metal hydrides also form readily, and many such REH_2 structures have been known since the 1950s, forming fluorite-type crystal structures.[25] Furthermore, rare earth metals are often able to form structures where the hydrogen to metal ratio exceeds two (*e.g.*, UH_3, LaH_{2+x}, CeH_3, GdH_3). While significantly limited by the low hydrogen weight percentage which can be stored in the materials and the expense of rare-earth metals, this has not prevented such materials finding practical use. Uranium hydride has very high hydrogen density by volume and a very low plateau pressure at room temperature (<0.1 Pa). Given the availability of uranium within the nuclear power industry and the need to safely store hydrogen, and in particular the radioactive tritium isotope which is critical for nuclear fusion processes, uranium hydride beds have been used as a hydrogen store in this context for many decades.[26–28]

Ultimately, the targets set by agencies such as the United States Department of Energy for the gravimetric hydrogen storage capacity required for mobile applications (currently at 6.5 wt% H_2 considered as a proportion of the system weight including tank, *etc.*[29]) have driven an extensive search for other, lightweight hydrides which might be suitable for on-board storage. A logical extension of the transition metal hydrides is to look at hydrides of lighter metals; lithium hydride (LiH, 12.6 wt%), beryllium hydride (BeH_2, 18.2 wt%), magnesium hydride (MgH_2, 7.6 wt%) and aluminium hydride (AlH_3, 10.0 wt%) all exceed the target level. With the bonding in these hydrides being more ionic in nature than for transition metal hydrides, these materials can be considered more like inorganic salts, with metal cations and hydride ions. The decomposition of the hydride requires reduction of the metal cation, and the temperature of decomposition has been shown to be strongly correlated with the reduction potential of the cation,[30] leading to a very high hydrogen release temperature of over 700 °C for lithium hydride. Discounting beryllium hydride due to its high toxicity, it is magnesium and aluminium hydrides which have had the most extensive attention for practical application.

Magnesium hydride releases hydrogen above 300 °C and also suffers from slow hydrogen absorption and release kinetics. As such, it has been the subject of significant efforts to tune its kinetic and thermodynamic properties. The purported explanations for slow hydrogen uptake and release have included poor hydrogen diffusion through MgO, $Mg(OH)_2$ or MgH_2 layers formed on the surface of magnesium, particle agglomeration, and the slow nucleation of Mg.[31] During the reaction between magnesium and hydrogen, high-energy ball milling has been shown to greatly enhance the reaction kinetics, presumably by breaking up agglomerates and surface passivation layers. A wide range of additives also have beneficial effects on the hydrogen storage and release kinetics, including a variety of carbonaceous materials, metal halides and oxides,[32] with Nb_2O_5 being among the most widely studied.[33] These additives are thought to play a number of key roles, including acting as dispersants to prevent particle agglomeration as well as enhancing the activation and/or transport of hydrogen, either individually or through the formation of conductive intermediates. In an extension of the dispersant strategy, many attempts have also been made to produce nanoconfined magnesium hydride, maximising the surface area to volume ratio in order to produce the most favourable reaction kinetics.[34] By their nature, using supports to do this reduces the hydrogen storage capacity of the composite, and so other efforts have focussed on preparing MgH_2 in more sophisticated nanostructured forms such as nanowires and ribbons.[31,35]

In addition to improving the kinetics of the hydrogen storage and release reactions, the thermodynamic stability of MgH_2 has been tailored to reduce the temperature at which hydrogen release becomes favourable. Akin to the strategy outlined for the transition metal hydrides, alloying-based approaches offer significant scope for achieving this. Some of the most successful examples of this are with Group IV elements like Si/Ge/Sn, which form alloys with Mg, reducing the enthalpy from 75 kJ mol_{H2}^{-1} to below 20 kJ mol_{H2}^{-1} in the case of the Ge system.[31] Alloying with transition metals and rare earth metals opens the possibility of forming ternary hydrides like Mg_2NiH_4 and Mg_2CoH_5, and although these systems often do not have significantly reduced reaction enthalpies, their kinetics can be enhanced by the transition metal's ability to activate hydrogen. Examples of these reactions are given below:

$$2MgH_2 + Si \rightarrow Mg_2Si + 2H_2 \tag{4.2}$$

$$2MgH_2 + Ge \rightarrow Mg_2Ge + 2H_2 \tag{4.3}$$

$$2MgH_2 + Sn \rightarrow Mg_2Sn + 2H_2 \tag{4.4}$$

$$Mg_2NiH_4 \rightarrow Mg_2Ni + 2H_2 \tag{4.5}$$

$$Mg_2CoH_5 \rightarrow 2MgH_2 + Co + ½H_2 \tag{4.6}$$

As for nanoconfinement approaches, the introduction of heavier metals to improve the thermodynamics of the hydrogen storage and release reactions has a significant impact on the hydrogen storage content of the material. This illustrates the significant challenge in tailoring the properties of metal hydrides to meet the diverse requirements of a practical system.

Aluminium hydride, or alane, has been the other lightweight binary hydride to receive significant focus, due to its very high hydrogen content, which can be released in a single step desorption below 150 °C.[36] Also, despite being pyrophoric when freshly prepared, it has been shown to be safely stored for long periods under ambient conditions without significant degradation, presumably due to the formation of a surface oxide layer. AlH_3 also has interesting structural properties, with seven known polymorphs observed depending on the synthesis conditions.[37] Only a few of those polymorphs have been well studied, and their structures are based on corner-sharing AlH_6 octahedra, as shown in Figure 4.7. The alpha phase is understood to be the most stable phase, with other polymorphs converting to it under thermal treatment. However, it was found that at low temperatures, direct hydrogen loss from the beta and gamma phases is possible, and indeed is more rapid than from the alpha phase.[38] The same study also showed the importance of particle size effects, with significantly higher apparent activation energy observed for a larger particle size sample of the alpha phase.

While its high hydrogen content and low decomposition temperature make AlH_3 very attractive, the hydrogen release reaction is only very mildly endothermic (~7 kJ mol_{H2}^{-1})[39] and involves a positive entropy change due to the release of hydrogen gas. As such, the reverse reaction is highly unfavourable and direct hydrogenation of aluminium metal requires incredibly harsh reaction conditions which would never be used in a practical setting. Synthesis methods usually involve reactions between lithium aluminium hydride and aluminium chloride in ethereal solvents:

$$3LiAlH_4 + AlCl_3 \rightarrow 4AlH_3 + 3LiCl \tag{4.7}$$

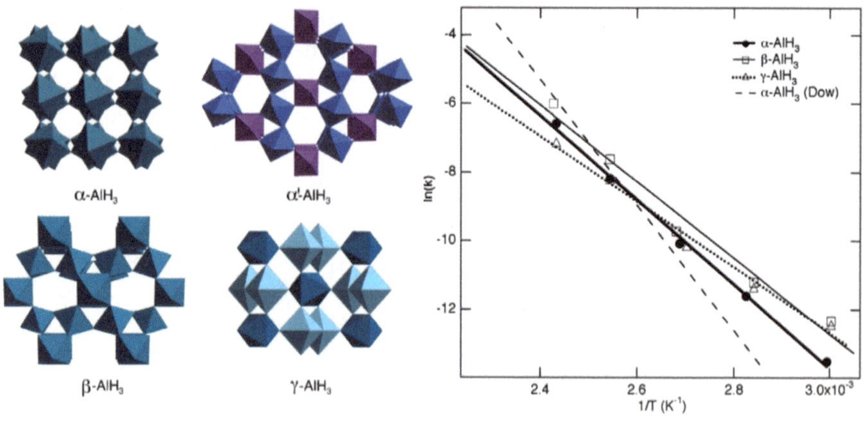

Figure 4.7 Polymorphs of aluminium hydrides and associated Arrhenius plots for hydrogen release from α-, β- and γ-AlH_3, including a specimen of γ-AlH_3 with large particle size (Dow). Polymorphs reproduced from ref. 36 with permission from Elsevier, Copyright 2011 and kinetic data reprinted from ref. 37 with permission from American Chemical Society, Copyright 2005.

However, in reality, to obtain a pure product and remove all solvent and LiCl by-product, a complex set of reaction steps are required involving excess $Li[AlH_4]$ and $Li[BH_4]$.[36] The lack of a straightforward means of regenerating Al produced from the decomposition of AlH_3 severely limits the practical application prospects of the material as a hydrogen store.

Aluminium-based hydrides have had significantly more detailed exploration and led to a new wave of research into hydrogen storage materials at the turn of the 21st Century. In a logical progression from increasing the hydrogen storage density by looking to lighter metal hydrides, the next tranche of materials we will examine incorporate additional hydrogen through the use of complex anions with multiple hydrogen atoms. This was initially spurred through sodium aluminium hydride, or sodium alanate ($NaAlH_4$), which can release hydrogen in two steps, releasing 3.7 wt% and 1.9 wt% H_2 as a proportion of the starting mass, respectively:

$$3Na[AlH_4] \rightarrow Na_3[AlH_6] + 2Al + 3H_2 \tag{4.8}$$

$$Na_3[AlH_6] \rightarrow 3NaH + Al + 3/2H_2 \tag{4.9}$$

Bogdanovic and Schwickardi demonstrated that, with addition of 2 mol% titanium(III) chloride, the system could reversibly store hydrogen at 3–4 wt% under dehydrogenation conditions of 270 °C, and hydrogenation at 170 °C and 150 bar.[40] Whilst still lower than the target values, this represented a very positive starting point for further development. Two further classes of complex anion materials soon emerged as hydrogen storage candidates: metal borohydrides based on the $[BH_4]^-$ anion, and metal amides/imides, based on the $[NH_2]^-$ and $[NH]^{2-}$ ions. These materials have been reviewed extensively,[41–48] and we will attempt only to give a general overview of the most promising systems and draw out some general attributes of the different classes.

Group I metal alanates generally decompose *via* the mechanism outlined for $Na[AlH_4]$ above, with the $[AlH_4]^-$ species converting into the $[AlH_6]^{3-}$ species, before decomposition to the metal hydride. Group II metal alanates have differing decomposition pathways, with $Mg[AlH_4]_2$ decomposing straight to MgH_2 and Al, while $Ca[AlH_4]_2$ decomposes *via* a $Ca[AlH_5]$ phase.[44] Among the alanates, only the sodium example is reversible under moderate conditions, although this material has been successfully scaled up into a multi-kg hydrogen storage system.[49,50]

The borohydrides show similar differences between the Group I and II metal derivatives. For lithium and sodium borohydride, thermal decomposition occurs at high temperatures and results in the formation of the metal or metal hydride and elemental boron:

$$Li[BH_4] \rightarrow LiH + B + 3/2H_2 \tag{4.10}$$

$$Na[BH_4] \rightarrow Na + B + 2H_2 \tag{4.11}$$

The Group II borohydrides appear to have more complex dehydrogenation pathways which have been the subject of some discussion in the literature, and appear to be quite sensitive to the reaction conditions.[41,51,52] Dehydrogenation products often include metal borides (*e.g.*, MgB_2/CaB_6). The controlled decomposition of magnesium borohydride, $Mg[BH_4]_2$, which contains 15 wt% hydrogen, was able to produce $Mg[B_3H_8]_2$ at 200 °C, and this material was able to be re-hydrogenated at 250 °C/120 bar, giving a reversible hydrogen content of 2.5 wt%.[53] In contrast, partial hydrogenation of MgB_2 to $Mg[BH_4]_2$ was found to require 800–950 bar of hydrogen gas at 400 °C.[54,55] Aside from lack of reversibility, the other key issue with the use of metal borohydrides is the potential for diborane release as a minor impurity in the hydrogen gas stream. Not only does this deplete the hydrogen store through boron loss, but diborane is toxic and highly detrimental to the Polymer Electrolyte Membrane (PEM) fuel cells in which the hydrogen may be used.

The use of complex anions is taken to its extreme in ammonium borohydride ($[NH_4][BH_4]$), a 24.5 wt% H_2 material where the attraction between protic and hydridic hydrogen in close proximity is such a driving force for hydrogen release that the material is unstable at room temperature, decomposing to produce the molecular material, ammonia borane (eqn (4.12)). Ammonia borane retains the ability to form dihydrogen bonds, in part explaining its stable solid form at room temperature[56] (Figure 4.8). Ammonia borane itself can sequentially lose its stored hydrogen, forming polyamidoborane materials and eventually boron nitride (eqn (4.13)–(4.15)). Eqn (4.12) and (4.13) are complete by around 150 °C, releasing around 13 wt% H_2. Combined with its air stability, this has attracted significant attention for practical applications.

$$[NH_4][BH_4] \rightarrow NH_3BH_3 + H_2 \tag{4.12}$$

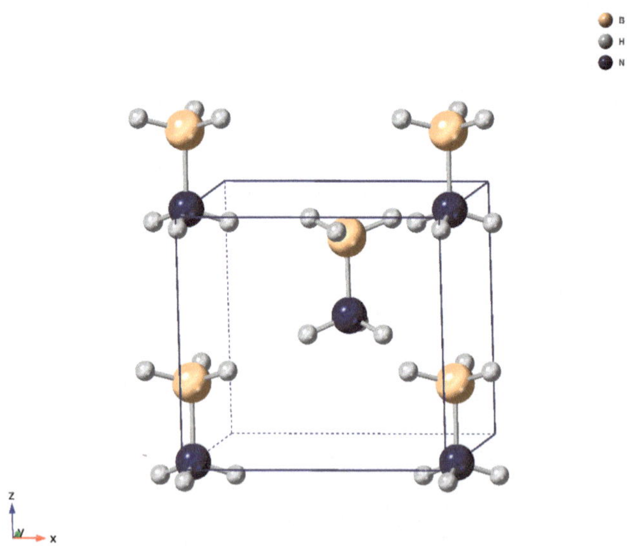

Figure 4.8 Crystal structure of ammonia borane, with nitrogen shown in blue, boron in orange and hydrogen in grey.

$$nNH_3BH_3 \rightarrow [NH_2BH_2]_n + nH_2 \quad (4.13)$$

$$[NH_2BH_2]_n \rightarrow [NHBH]_n + nH_2 \quad (4.14)$$

$$[NHBH]_n \rightarrow [NB]_n + nH_2 \quad (4.15)$$

Ammonia borane's hydrogen release occurs over a broad temperature range, and also results in release of small quantities of undesirable diborane. Reaction with lithium or sodium hydride results in the formation of an amidoborane ($M[NH_2BH_3]$), which has been shown to release hydrogen over a much narrower temperature range below 100 °C, and also suppresses diborane release.[57]

Unfortunately, the protic–hydridic attraction which drives hydrogen release at low temperatures in these materials also results in stable polymeric structures which are very difficult to rehydrogenate. Direct efforts have proved largely unsuccessful, with the most effective rehydrogenation being through reaction with hydrazine and ammonia.[58,59]

The metal amides differ from the other complex anion materials in that the hydrogen atoms have partial positive charge. Only the Group I metal amides heavier than lithium can release hydrogen directly upon heating, decomposing to the elements:

$$Na[NH_2] \rightarrow Na + \tfrac{1}{2}N_2 + H_2 \quad (4.16)$$

Whilst this reaction releases 5.2 wt% H_2, the loss of nitrogen from the system makes the reaction difficult to reverse. However, lithium amide and the Group II metal amides decompose to produce the relevant metal imide with ammonia release:

$$2Li[NH_2] \rightarrow Li_2[NH] + NH_3 \quad (4.17)$$

$$Mg[NH_2]_2 \rightarrow Mg[NH] + NH_3 \quad (4.18)$$

To make a useful hydrogen store, amides must be formed in a composite with a metal hydride, which can react with ammonia to form an amide and hydrogen gas:

$$LiH + NH_3 \rightarrow Li[NH_2] + H_2 \quad (4.19)$$

Combining eqn (4.17) and (4.19) gives a 6.5 wt% hydrogen storage and release reaction (4.20), which has shown reversible operation at around 200–250 °C and less than 10 bar pressure.[60,61] There has been some debate as to whether the hydrogen release is by the sequential reactions implied by eqn (4.17) and (4.19), or directly by interaction of hydride with protic hydrogen on the amide groups.[57,62–64]

$$Li[NH_2] + LiH \rightarrow Li_2[NH] + H_2 \quad (4.20)$$

Mixed lithium–magnesium systems have also been devised in response to the observation that magnesium amide loses ammonia at a lower temperature than lithium amide, while the reaction of LiH with ammonia is faster.[65,66] Lithium–magnesium amide–hydride composite systems in various ratios have delivered reversible hydrogen content in excess of 7 wt% below 250 °C.[67–69] Amide-based systems have the potential to release small quantities of ammonia which would poison PEM fuel cells and deplete the capacity of the material. This can be reduced by using excess hydride[70] or additives.[71]

The remarkable reversibility of the amide-hydride composite systems compared with borohydrides and alanates can be traced to a structural analogy with transition metal hydrides. Examining the structures of lithium amide and lithium imide, it can be seen that the nitrogen array remains in a face-centred configuration in both structures.[72] The change in the hydrogen content upon conversion of amide to imide groups is facilitated by the filling of lithium vacancies in the structure. In this way, only lithium ions are required to be mobile for the hydrogen storage to be facilitated. This phenomenon can be observed in the formation of solid solutions of lithium amide and imide during hydrogen storage and release by *in situ* X-ray powder diffraction (Figure 4.9).[73]

Similar strategies have been employed to improve the hydrogen storage properties of complex anion materials as were highlighted for the transition metal hydrides, including the use of metal halide and oxide dopants,[71,74–76] mixed metal systems to tune thermodynamic parameters,[77,78] and microstructure engineering and nanoscaffolding to improve kinetics and suppress impurity gas release.[9,34,35,79–81]

Through our survey of the application of metal hydrides, a number of emerging trends with respect to achieving reversible, high-density hydrogen storage under moderate conditions can be observed, namely:

- Hydrides are an incredibly effective way of storing hydrogen at high gravimetric and volumetric density, though achieving this whilst also

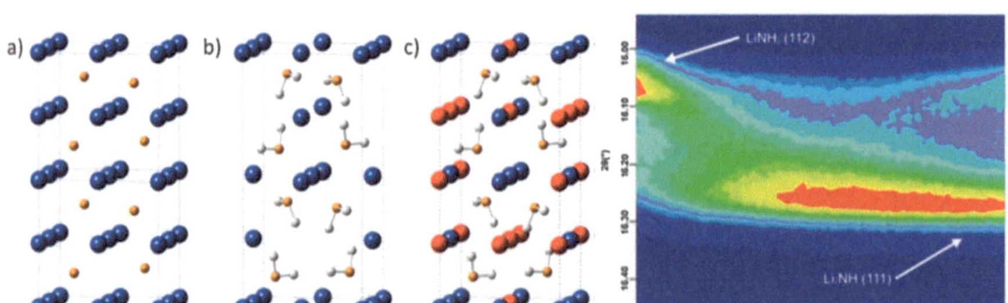

Figure 4.9 Crystal structures of (a) lithium imide (b) lithium amide and (c) lithium amide with nitrogen shown in orange, hydrogen in grey, lithium in blue and lithium vacant sites indicated in red, alongside a contour plot of X-ray diffraction data during dehydrogenation of lithium amide, showing the formation of a solid solution. Reproduced from ref. 73 with permission from the PCCP Owner Societies.

facilitating straightforward reversible storage of that hydrogen remains very challenging.

- Surface area to volume ratios and favourable microstructures have a strong influence on the kinetics of hydrogen storage and release, generated both through mechanical treatments and nanostructured composites.
- Conserving the topology of the metal/non-hydrogen array in hydrogenated and dehydrogenated structures significantly promotes reversible hydrogen storage reactions.
- A range of additives/dopants improve hydrogen storage reactions, though their precise function is not clear. Theories include: additives forming hydrogen activating or conductive intermediates, additives taking a direct role in hydrogen activation, and additives physically dispersing the active material to create higher surface areas.
- The use of alloying and reactive composite approaches can successfully modify the thermodynamics of hydrogen storage materials to achieve more favourable hydrogen release conditions.

A comprehensive (but not exhaustive) figure of the hydrogen storage densities of metal hydrides is shown in Figure 4.10.[82] Whilst it can be argued that hydrides have not yet fulfilled the promise of replacing compressed hydrogen as the technology of choice for mobile hydrogen storage systems, it cannot be denied that an impressive breadth of understanding of the structure

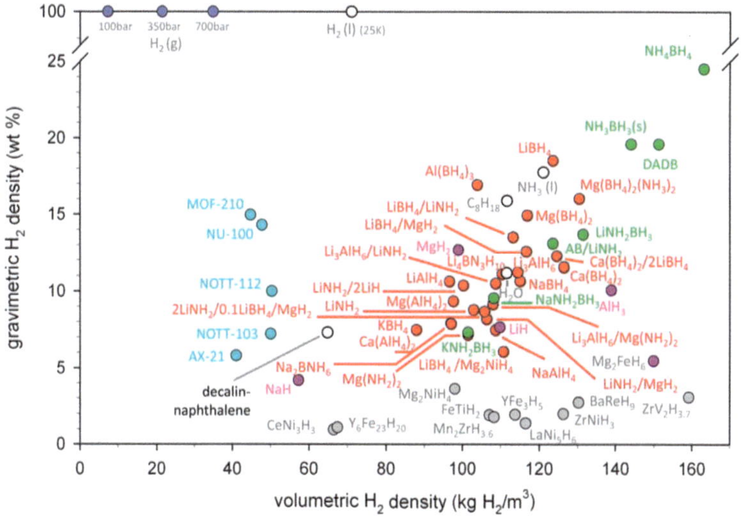

Figure 4.10 Hydrogen storage densities of a range of metal hydride materials, compared with compressed and liquid hydrogen. Pink = ionic hydrides; grey = metallic hydrides; red = complex anion materials; green = molecular hydrides. Reproduced from ref. 82 with permission from The Royal Society of Chemistry.

and properties of metal hydrides has emerged through the pursuit of that goal. There are also a number of applications where hydrides have been employed in practical hydrogen storage systems, at least at the demonstration stage. These include the University of Birmingham canal boat, the 'Ross Barlow', which incorporates a $Ti_{0.93}Zr_{0.05}(Mn_{0.73}V_{0.22}Fe_{0.04})_2$ hydrogen store,[83] hydrogen-powered forklifts,[84,85] and a number of stationary energy storage applications.[86] Hydrides have also been used for decades to help power submarines achieving vastly superior range to battery-powered equivalents.[87]

The extensive cataloguing of the hydrogen storage and release reactions of metal hydrides has prompted interest in exploring other applications. Many uses of hydrogen require the delivery of compressed hydrogen, including in the refuelling of current commercial hydrogen powered vehicles which store hydrogen at 700 bar.[88–90] For fast refuelling, this can require compression of hydrogen up to 1000 bar. Most existing compressors are conventional mechanical systems, which have relatively high failure rates and raise concerns relating to the purity of the hydrogen delivered because of the lubricants which must be used.[87]

Given their high volumetric hydrogen density and reversible hydrogen storage reactions, transition metal hydrides are interesting candidates to act as hydrogen compressors, particularly given the high enthalpy changes associated with the reaction. In this way, metal hydride compressors would be thermally-driven compression systems, removing the need for many moving parts and thus reducing the chances of mechanical failure.[91] The basic principle of operation of such a system can be understood by returning to the pressure–composition isotherm, and considering two temperatures, T_{high} and T_{low}, giving the red(2) and blue(1) isotherm data shown in Figure 4.11. The sequence of operation can be given as follows.

1. The hydride is charged with hydrogen at T_{low} to a pressure of P_{L}, reaching a hydrogen capacity of b1 at point B.
2. The hydride is heated to T_{high} (Q in), driving hydrogen off the hydride at P_{H} (Process D). The volume of the system is calibrated to store or draw off hydrogen at P_{H} until the hydride reaches the lower end of the hydride plateau, a hydrogen capacity of a2 (Point C). This delivers ΔC of hydrogen capacity at P_{H}.
3. The system is cooled to T_{low} (Q out) and re-charged with hydrogen to reach b1, P_{L} (Process A).

Using quite reasonable temperature swings (*e.g.*, 25–150 °C), metal hydrides can theoretically achieve compression ratios ($P_{\text{H}}/P_{\text{L}}$) of 10–100, and even 5–10 when considering very high hydrogen delivery pressures.[91] The highly-tailorable thermodynamics of alloy systems means plateau pressures are possible from very low pressure to kbar at temperatures in this range. However, the real behaviour of hydrides impacts the ability to achieve this. As described earlier, real hydrides do not have perfectly flat absorption plateaus, meaning the real pressure difference between P_{H} and P_{L} is diminished.

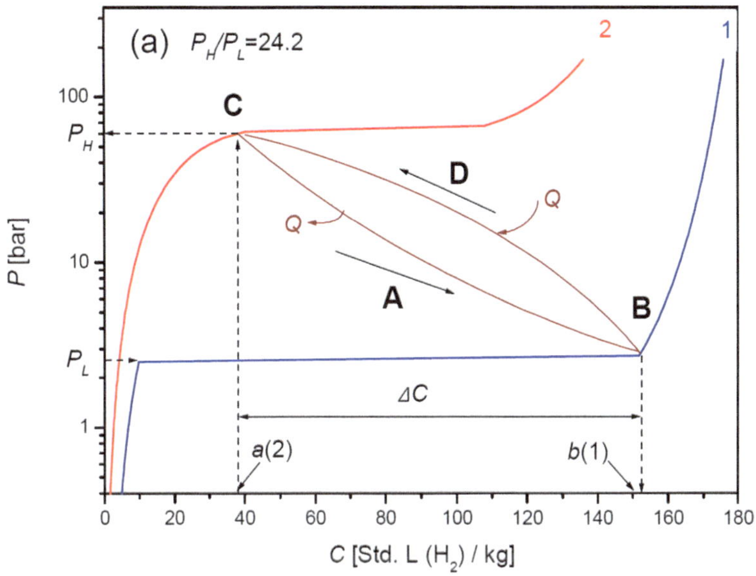

Figure 4.11 Idealised operation of a metal hydride compressor, showing isotherms at the low (blue) and high (red) temperatures. Reproduced from ref. 91 with permission from Elsevier, Copyright 2014.

The difference can be further diminished by systems which show hysteresis between absorption and desorption curves, which can be exacerbated by rapid operation. One way to address this is with multi-stage compressor systems, which involve matched hydride beds which sequentially increase the hydrogen pressure. These have been demonstrated to achieve high compression ratios in two-three steps, including for delivery pressure in the 1 kbar range.[92,93]

The need to cycle for many thousands of cycles also puts high demands on the material to have stable morphology and resistance to trace impurities that may be present in the hydrogen gas supply. Tests of AB_5 and AB_2 alloys for 60 000–90 000 hydrogen storage and release cycles showed that while slow progressive diminishment of the capacity was observed, material capacity could be easily regenerated through higher temperature treatment, suggesting that a regular maintenance interval would result in very long lifetimes.[94,95]

The final aspect of hydrogen handling to which metal hydrides can be applied is the purification of hydrogen. Some applications, and in particular the PEM fuel cells used in mobile applications, have very stringent hydrogen purity requirements due to the sensitivity of the proton-conducting membranes which are used.[96] Because many hydride-forming metals do not readily react with other gases under anywhere near as mild conditions, metal hydrides can be used as a completely selective method of purifying hydrogen. The same principle can be applied when considering the selective absorption of hydrogen from mixtures of gases. There are two key approaches to the use of metal hydrides in hydrogen separation and purification: the formation of thin membranes from hydride-forming metals, and the use of fluorine-coated, hydride-forming metals. These methods offer potential advantages over traditional gas separation techniques such as pressure-swing adsorption where smaller scaler hydrogen purification is required.[97]

The basic principle of the hydrogen separation membrane is that the metal is held at a temperature where the hydride stability can be controlled by a technologically-useful pressure differential across the membrane, hydrogen mobility in the metal is high, and the effects of hydrogen embrittlement are minimised. Palladium is the most mature metal membrane technology, with palladium alloys (with *e.g.*, copper or silver) being found less prone to embrittlement and having higher hydrogen mobility.[98,99] Of course, the hydrogen flux will be dependent on the surface area of the membrane, and a minimum thickness of 10s–100s of microns is required for the membrane to have sufficient mechanical strength, and so efforts to reduce the amount of expensive palladium have been explored. These have included coating thin palladium layers onto porous supports.[100] Interesting recent developments have centred on the use of vanadium, which has a body-centred cubic structure. Not being a close-packed structure, the metal array has more open interstitial sites which results in higher hydrogen permeability; however, its use has been limited by surface oxidation effects which limit hydrogen uptake kinetics. Coating vanadium membranes with a nanometre-thick layer of palladium on each side prevents this oxidation and simultaneously minimises palladium use. Tubular membranes of this type were recently applied to the purification of hydrogen produced from the decomposition of ammonia, and were successfully used for the refuelling of a PEM fuel cell hydrogen car.[101,102]

Metal hydride beds or slurries can also be used to purify hydrogen, in that they will selectively absorb hydrogen, leaving impurities free to be removed prior to desorbing the now pure hydrogen stream. The principal challenge with this methodology is the sensitivity of various metal hydrides to impurities like O_2, H_2O, H_2S, CH_3SH, CO and CO_2.[103] While individual alloys might be tailored for well-characterised impurity compositions, a more general approach has involved the fluorination of the metal hydride surfaces, which impedes the interaction of impurities with the metal hydride, but does not prevent the diffusion of hydrogen to and from the metal.[104–106]

4.2.2 Thermal Energy Storage Materials

While the previous applications of hydrogen storage, purification and compression are all concerned with hydrogen specifically, there is another application of hydrogenation and dehydrogenation reactions that is simply focused on the heat stored and released during the processes: thermal energy storage. Thermal energy storage is a broad field of energy research, given the wide variety of potential heat cycles available. Materials approaches commonly employed include sensible heat stores, which rely simply on the heat capacity of the material and favourable storage properties, and phase change materials, which store thermal energy *via* melt–freeze cycles. These 'physical' heat storage mechanisms can be employed using a wide variety of organic and inorganic materials and typically store heat at a density of around 100–500 kJ kg^{-1}, rising as high as around 1000 kJ kg^{-1} for some fluorides and metal alloys.[107] However, the heat per unit volume and mass obtained during many chemical reactions can significantly exceed these values, leading to a long-standing interest in thermochemical energy storage. Hydrides feature prominently in this area, where energy storage densities rise to several thousand kilojoules per kilogram in the case of lightweight hydrides.[108] Together with the extensive ability to tailor the temperature of the reactions through metal choice from sub-zero to over 1000 °C (making them applicable to a variety of waste heat sources) this has made metal hydrides very attractive for thermal energy storage. While there are a number of technological applications of thermal energy storage (refrigeration, actuators, heat engines), much attention has focused on the potential to store renewable energy, particularly given the advent of concentrated solar power stations.[108] The working principle of a hydride-based thermal energy store is shown in Figure 4.12.[12] Many of the challenges with development of thermal energy storage systems mirror those of hydrogen storage materials: excellent stability, reversibility, cheap materials and fast kinetics. The key difference is that for concentrated solar applications, hydrogen storage and release reactions are required to work at very high temperatures (300–1000 °C). This results in new challenges, where some materials like Li/LiH are extremely corrosive at high temperature, or

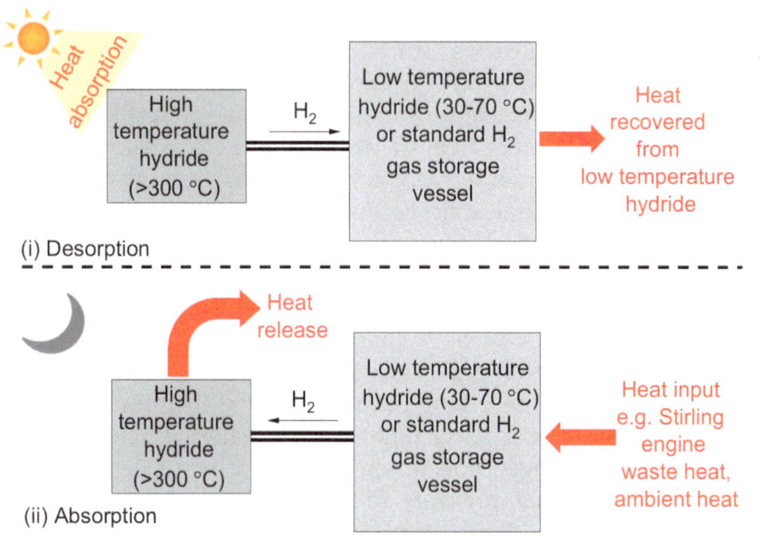

Figure 4.12 Schematic of thermal energy (i) storage and (ii) release by metal hydrides. Reproduced from ref. 12, https://doi.org/10.3389/fenrg.2021.616115, under the terms of the CC-BY 4.0 licence, http://creativecommons.org/licenses/by/4.0/.

involve liquid phases. Heat conduction is also a key factor to be considered for practical systems. Other types of thermochemical energy storage based on metal carbonates, oxides, and hydroxides have also been considered and are in some cases also reaching pilot scale demonstrations. While these systems often consist of cheaper materials than hydrides, they have challenges associated with their reversibility and potential side reactions.[107]

There are a number of materials which are promising for high-temperature thermal energy storage, including MgH_2 (and associated ternary alloys), CaH_2, TiH_x, and ZrH_x.[108,109] Complex anion hydrides have attractive thermal energy capacity, but would need to overcome issues with reversibility and side reactions if they were to be considered for practical application.[110]

4.2.3 Catalysts

The use of transition metal hydrides like PdH_x are among the most well-known catalysts for hydrogenation reactions, taking advantage of the metal's ability to dissociate molecular hydrogen and transfer it to the molecule of interest. More recently, however, hydride materials have been revealed to have promising catalytic function in another field: the catalytic synthesis of ammonia. While ammonia synthesis *via* the Haber–Bosch (Fe-catalysed) or Kellogg (Ru-catalysed) processes is very well established at large scale, interest in smaller-scale ammonia production has intensified with the prospect of coupling it to renewable energy installations and so potentially harnessing ammonia as a sustainable fuel and hydrogen store.[111–113] Unfortunately, the catalysts used for large-scale industrial processes lose their effectiveness at lower temperatures and pressures, which has spurred research into the development of ammonia synthesis catalysts for use under milder conditions.[114]

Typical catalyst formulations for ammonia synthesis look like the following: Cs–Ru/MgO-CNT. These formulations may be described by three key components, the active metal centre (Ru) distributed on a structural support (MgO-CNT) with an electronic promoter (Cs salt) to promote stronger interactions with H_2 and N_2. Catalyst design has evolved to consider the most active formulations as those which permit the reaction to be effectively spread over multiple parts of the catalyst mixture. This design principle seeks to overcome two key limitations to the reaction kinetics: hydrogen inhibition and kinetic scaling relations. In both contexts, hydrides have been at the forefront of new catalyst developments.

Ruthenium catalysts have long been considered the most active catalysts for ammonia synthesis at lower temperatures. However, their activity is in part hindered by hydrogen poisoning: the saturation of the ruthenium surface with hydrogen that prevents further reaction.[115,116] A calcium-based electride material, $[Ca_{24}Al_{28}O_{64}]^{4+}(e^-)_4$, a cage-like material with electrons localised within cages in the structure, showed significant ammonia synthesis activity as a support for Ru.[117,118] It's impressive activity—an order of magnitude higher than state-of-the-art Ru formulations—was attributed to the very low work function of the material enhancing its electron donation (akin

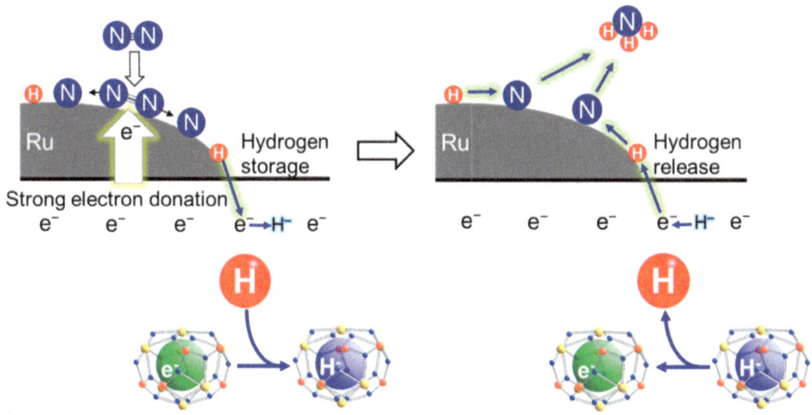

Figure 4.13 Proposed ammonia synthesis mechanism for Ru–Ca-electride composite, with delocalised cage electrons shown in green. Reproduced from ref. 118 with permission from American Chemical Society, Copyright 2017.

to a traditional electronic promoter), and the ability of the free electrons to reversibly react with hydrogen from the surface of Ru (Figure 4.13):

$$H_{ads} + e^- \rightarrow H^- \tag{4.21}$$

This work was extended to a number of other electrides, including Ca_2N, which consists of Ca_2N layers separated by delocalised electron layers.[119] Under ammonia synthesis conditions, it was found that the material existed as an electride–hydride solid solution, $Ca_2NH_{1-x}e_x$, facilitated by the fact that Ca_2N and calcium nitride hydride (Ca_2NH) have very similar crystal structures.[120] Both it and calcium hydride were found to display very high ammonia synthesis activity in composites with ruthenium. It has been proposed that interaction with the transition metal promotes the release of hydrogen from the hydride/nitride–hydride, dramatically lowering the work function of the material from 4.0 eV to 2.3 eV in the case of CaH_2. This is thought to promote the cleavage of nitrogen, while the hydride vacancies facilitate exchange of hydrogen with the metal surface, reducing poisoning effects. Considering the hypothesis that the formation of hydride vacancies is critical to high ammonia synthesis activity, a calcium fluoride–hydride phase was synthesised, with the use of fluorine to destabilise Ca–H bonds in the structure. Ru–CaFH catalysts recorded appreciable ammonia synthesis activity at temperatures as low as 50 °C. A series of oxyhydrides have also recently been reported to demonstrate favourable ammonia synthesis properties,[121–123] cementing hydrides as a new family of catalyst components for ammonia synthesis under mild conditions.

The second series of hydride ammonia synthesis catalysts places an arguably even more prominent position on the hydride component: that of a second catalytic centre. The search for multi-site catalysts is derived from an understanding of kinetic limitations of catalytic reactions on surfaces which has developed over a long time. One of the first articulations of this limitation is in the Sabatier principle, which states that an active catalyst will have an intermediate binding strength for the reactants, binding them strongly enough to activate them, but not so strongly as to not be able to release the products. This principle is illustrated in the many examples of 'volcano plots' of catalytic activity, such as one for ammonia synthesis activity *versus* nitrogen binding energy for a range of transition metals shown in Figure 4.14.[124]

A more detailed explanation of the chemistry behind this intuitive principle is found in scaling relations, which illustrate that the enthalpy of reactant adsorption is proportional to the activation energy of that adsorption. That is, that a surface which reduces the activation energy for reactant binding will also bind those reactants more strongly. Furthermore, a proportional relationship exists for the adsorption enthalpies of hydrogenated versions of the reactant (*i.e.*, N, NH, NH_2 and NH_3).[125] This relation implies that, on a single metal surface, it is not possible to design a catalyst that both activates the reactants strongly and will also release products easily. As such, there is motivation to search for catalysts which have multiple sites which might break this scaling relation. This is in part inspired by biological systems for nitrogen activation, which have been proposed to change coordination environments during different stages in the reaction.[126]

Composites of lithium and barium hydride with transition metals were proposed to break the scaling relation for ammonia synthesis.[127,128] In the proposed mechanism, the nitrogen is adsorbed and dissociated on the transition metal surface, then adsorbed nitrogen atoms are transferred to the hydride, hydrogenated to form amide/imide groups which then react with hydrogen to produce ammonia and re-form the hydride (Figure 4.15). Evidence for the breaking of scaling relations came from a much 'flatter' volcano

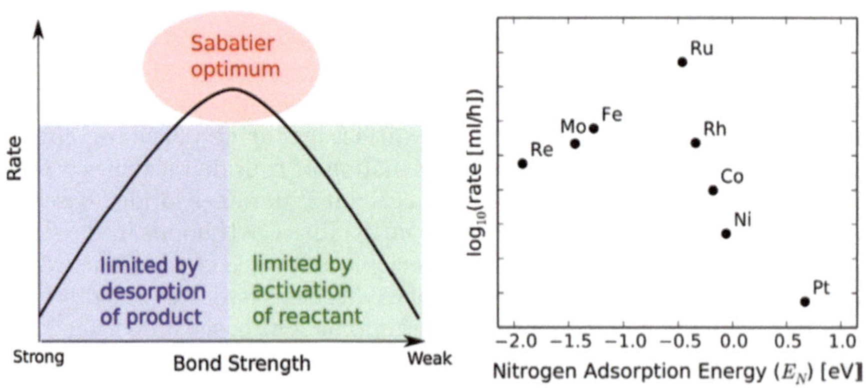

Figure 4.14 A schematic illustration of the Sabatier principle, and an example of a volcano plot for ammonia synthesis activity. Adapted from ref. 124 with permission from Elsevier, Copyright 2015.

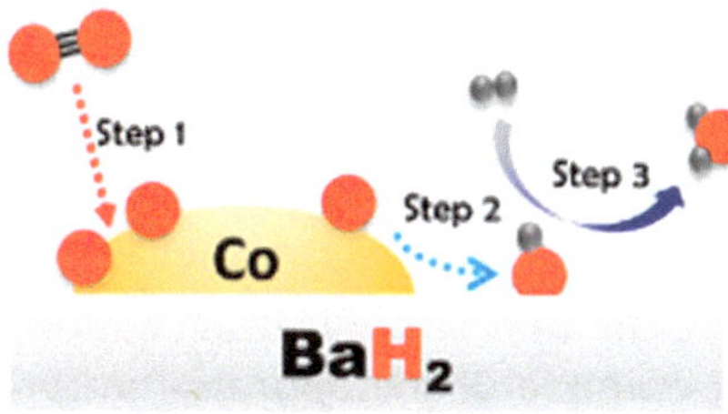

Figure 4.15 Two-site ammonia synthesis catalysis using a BaH_2–Co composite, showing (Step 1) dissociative adsorption of nitrogen on the transition metal, (Step 2) reaction of adsorbed nitrogen with hydride from barium hydride to form an imide group, and (Step 3) reaction between imide and hydrogen to form ammonia. Adapted from ref. 128 with permission from American Chemical Society, Copyright 2017.

curve, including high activity rates for metals previously considered inactive for ammonia synthesis, such as vanadium and manganese.

These reactions have also been demonstrated in a chemical looping ammonia synthesis protocol, where the nitrogenation and hydrogenation steps are separated, enabling the individual optimisation of the two steps. The effectiveness of the protocol relies on bulk conversion of the hydride into amide/imide groups. These reactions have been demonstrated for lithium and barium hydride transition metal composites.[129] Lithium nitride hydride (Li_4NH) is able to undergo bulk conversion to lithium imide and lithium nitride *via* imide–nitride–hydride solid solutions, and has been demonstrated to give high ammonia production by chemical looping both with and without the addition of a transition metal.[130]

The attention given to hydride materials in the context of ammonia synthesis is perhaps the first step of a broader consideration of the role in which they might play as hydrogen activators and/or reducing agents. Indeed, some recent reports highlight activity in new areas such as methanol synthesis[131] and selective hydrogenation.[132]

4.2.4 Battery and Fuel Cell Materials

Hydrides have been a part of the development of rechargeable battery technology since the 1960s, with the development of the nickel–metal hydride (Ni–MH) battery as the key precursor to lithium-ion batteries. Arguably this is the most commercially-successful application of metal hydrides, with millions of Ni–MH batteries used in a range of portable electronics and hybrid vehicles. The energy storage reaction is based on the following generalised redox processes:

$$H_2O + M + e^- \rightarrow OH^- + MH \tag{4.22}$$

$$Ni(OH)_2 + OH^- \rightarrow NiO(OH) + H_2O + e^- \tag{4.23}$$

The metal hydride is generally an AB_5 or AB_2 alloy material, itself a mixture of numerous metals. The key factors which concern the development of metal hydrides for the Ni–MH battery are familiar from our consideration of hydrogen storage: the stability of hydride formation over thousands of charge and discharge cycles, the hydrogen content (which ultimately determines the capacity of the battery), the rate of (de)hydrogenation, and the cost of the elements involved. An example of the optimisation process is considering the $LaNi_5$ alloy. In order to reduce the costs associated with the lanthanum, mischmetal could be substituted, which is a mixture of lanthanoid metals derived from minerals with less purification required. The nickel may then be partially substituted with cobalt, nickel or aluminium to help reduce the oxidation of the alloy surface under the alkaline environment of the battery.[133] In this way, more complex alloys are tailored to the precise requirements of the battery system. While commercial hydrides for Ni–MH batteries are well established, efforts to explore alternative alloys that might offer higher specific energies are the subject of ongoing research, particularly to focus on the substitution of lanthanides in the AB_5 alloys due to the geopolitical factors that concern the supply of lanthanides.[134]

Lithium-ion batteries (LIBs) have gradually supplanted Ni–MH because of their increased energy density; however, this does not spell the end of the role of metal hydrides in batteries, which have recently come to prominence in efforts to develop the next generation of LIBs. The commercial anode in an LIB is graphite based, with lithium intercalating between graphite layers to achieve a maximum theoretical capacity of 372 mA h g^{-1}. Conversion anodes, where lithiation results in bulk conversion to a new phase rather than simple intercalation, have received attention due to the higher capacities which can be achieved. Of particular interest have been magnesium hydride anodes, where reaction to produce LiH and Mg has a theoretical capacity of 2038 mA h g^{-1}, and a higher theoretical volumetric capacity than lithium metal itself. The actual reversible capacity observed is significantly lower, but still around four times the capacity recorded for graphite electrodes and occurs at a lower voltage compared to Li^+/Li than many other anodes.[135]

$$C_6 + Li^+ + e^- \rightarrow LiC_6 \tag{4.24}$$

$$MgH_2 + 2Li^+ + 2e^- \rightarrow 2LiH + Mg \tag{4.25}$$

Analogous reactions have also been explored for a range of ternary magnesium hydrides as well as titanium hydride, which also has favourable electronic conductivity, a requirement for anode materials. All of these hydrides show attractive gravimetric and volumetric capacity compared with graphite, as illustrated in Figure 4.16.[136]

Conversion electrodes also often suffer from voltage hysteresis, where the lithiation voltage differs significantly from the delithiation voltage, which reflects kinetic barriers to the diffusion of the solid materials as they convert. The hydride conversion reaction relies on diffusion of H^-, a more polarisable anion than F^-, O^{2-}, and S^{2-}, upon which other conversion anodes are based.

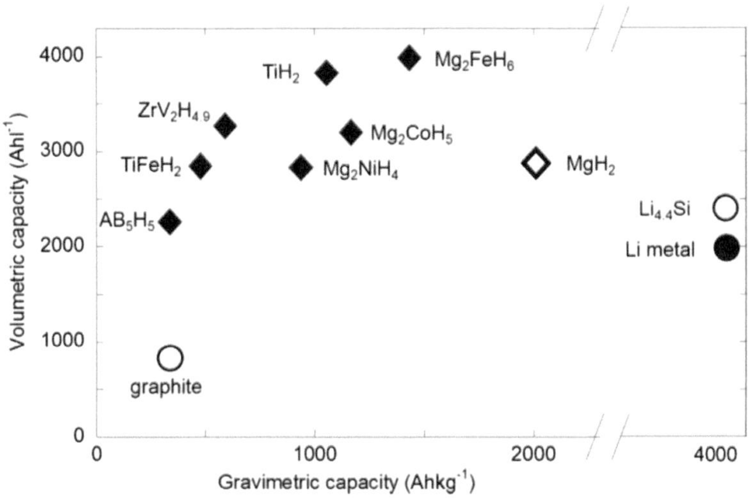

Figure 4.16 Capacity of various anode candidates for LIBs. Reproduced from ref. 136, https://doi.org/10.3762/bjnano.6.186, under the terms of the CC BY 2.0 license, http://creativecommons.org/licenses/by/2.0.

As such, the transport properties of hydride are more favourable and are reflected in some of the lowest voltage hysteresis values.[135]

Unsurprisingly, given our knowledge gained from exploring MgH_2 for hydrogen storage, the repeated conversion between Mg and MgH_2 is not without challenges. The kinetics of the reaction remain very slow, with slow cycling rates required to achieve anything close to the expected capacity. The high voltage range is limited by the irreversible formation of Li–Mg alloys, limiting the total capacity. The rate and extent of the conversion reaction is also dependent on the intimate contact of the LiH, Mg and the electronically-conductive carbon additive. Given that there is a 83% change in the volume of the active components between Mg + 2LiH and MgH_2,[136] the loss of contact through void formation during cycling presents a significant challenge to overcome, and a loss of capacity is observed. Efforts to improve performance follow a familiar list from hydrogen storage optimisation efforts: additives and alloying to improve kinetics and conductivity, and nanostructuring of the hydride, with additional work focusing on the use of binders to minimise the mechanical stresses associated with the volume changes.[110,136,137] These efforts have showed some promise, with reversible capacity of around 500 mA h g^{-1} demonstrated over more than 20 cycles.[138,139] While still some distance from commercial performance, the high capacities will undoubtedly spur further investigation and optimisation. Metal alanates and borohydrides also have high theoretical capacities as anodes, however these systems are less advanced in terms of avoiding irreversible capacity loss.[110]

Among anode materials for LIBs, lithium metal itself is something of a holy grail, with unmatched gravimetric capacity that is highly attractive for mobile energy storage. However, the susceptibility of lithium to the formation of dendritic metal structures during cycling presents significant safety concerns. These dendrites can propagate through the liquid electrolyte and short circuit the battery, which is especially dangerous with flammable liquid electrolytes. Solid ionic conductors are under consideration for a new generation of battery electrolytes which have the potential to minimise this concern and thus enable higher energy densities through the use of lithium anodes. The demands on solid electrolytes are significant: high ionic conductivity (ideally matching that of existing liquid electrolytes), robust electrochemical stability and chemical stability when in contact with both lithium metal and cathode materials.

Lithium borohydride was the first hydride material to be considered for solid electrolyte applications, after it was shown that upon a phase change from an orthorhombic phase to a higher-symmetry hexagonal phase at around 110 °C, the lithium-ion conductivity jumps by three orders of magnitude to form a fast ion conductor.[140] As can be seen in Figure 4.17 the high temperature phase has well-aligned lithium positions which facilitates rapid diffusion compared to the low temperature phase. The formation of mixed halide–borohydride phases was found to stabilise the high temperature phase, to room temperature in the case of the iodide.[141] Lithium borohydride contains a number of favourable characteristics for use in batteries, including low density, a wide electrochemical stability window against lithium metal, mechanical deformability, and low-resistance interfaces with lithium metal. It has been successfully tested (at elevated temperature) in batteries with a number of different configurations, showing promising performance.[137]

Building on the success of $Li[BH_4]$ as a solid electrolyte, more complex cage-like B–H anions have been developed, including $[B_{10}H_{10}]^{2-}$, $[B_{12}H_{12}]^{2-}$ and $[CB_9H_{10}]^-$. These large and low charge density anions provide wide conduction pathways for lithium and sodium ions, showing superionic or liquid-like conductivity.[142,143] Hydrides also offer promise for magnesium ion

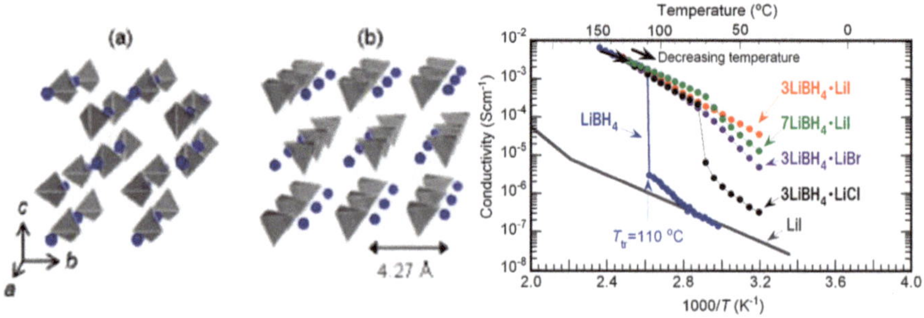

Figure 4.17 The (a) low temperature and (b) high temperature structures of $LiBH_4$, with Li shown in blue and BH_4 as grey tetrahedra, and variable-temperature conductivity data for $LiBH_4$ and halide composites. Reprinted from ref. 140 and 141 with permission from American Chemical Society, Copyright 2007, and 2009, respectively.

conduction, with $Mg[BH_4][NH_2]$ showing one of the highest reported Mg-ion conductivities facilitated by zig-zag Mg chains in the crystal structure.[144,145]

While hydrides have demonstrated significant promise in the conduction of cations, the conduction of hydride ions themselves offers fascinating potential to harness the strong reducing power of hydride to be used in electrochemical ammonia production, among other possibilities. Precious few hydride ion conductors have been characterised with barium hydride, calcium nitride–hydride, and oxygen-doped lanthanum hydride being the most promising materials to date.[146–148]

4.2.5 Superconductivity

The phenomenon of superconductivity, the resistance-free conduction of charge, and the search for room-temperature superconductors, has captured the imagination of scientists since its discovery in 1911.[149] In conventional superconductors such as mercury and magnesium diboride, the emergence of superconductivity is limited to below 40 K,[150] which severely restricts its potential uses. More recently, cuprate and iron pnictide materials greatly advanced the temperature range at which superconductivity is exhibited to liquid nitrogen temperatures, exceeding 160 K for a cuprate material.[151]

Metallic hydrogen was theorised to be a viable room-temperature superconductor,[152] however the formation of metallised hydrogen is expected to require pressures in the region of 400 GPa/4 $\times$ 10^6 bar,[153] which is a different kind of practical limitation. Given the challenges associated with forming metallic hydrogen, it was proposed that hydrogen-rich materials such as metal hydrides might become superconducting at lower pressures, with the hydride providing a 'chemical compression' due to the high volumetric hydrogen densities found in the materials.[154] While superconductivity is observed in hydrides of metals such as palladium, silver, gold and silicon, it is at less than 20 K.[155–157] The lack of high-temperature superconductivity in conventional hydride materials has prompted the search for super-stoichiometric hydrides, theorised to possess very high temperature superconductivity at elevated pressure.[153] Synthesis of these materials is extremely challenging, and often occurs by reactions of metal and hydrogen at high pressures in diamond anvil cells, shown schematically in Figure 4.18.[158] H_3S was the first such material experimentally reported, with a watershed finding that it showed superconductivity at 203 K and 155 GPa.[159] The crystal structures of these superhydrides are clathrate-like,[160] with a selection shown in Figure 4.19 to illustrate the dense packing of hydrogen in the materials, aiding metallisation.

A series of rare-earth and actinide superhydrides have been successfully synthesised and shown to achieve high-temperature superconductivity, as shown in Table 4.2. Recently, the group was expanded through the synthesis of CaH_6, the first main group metal superhydride in this category of high temperature superconductors.[161] Given the intense focus on achieving room-temperature superconductivity, reports of such behaviour are not

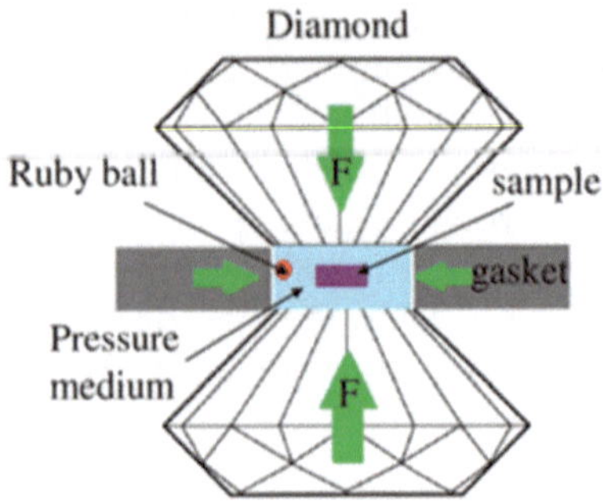

Figure 4.18 Schematic of a diamond anvil cell. Adapted from ref. 158 with permission from the PCCP Owner Societies.

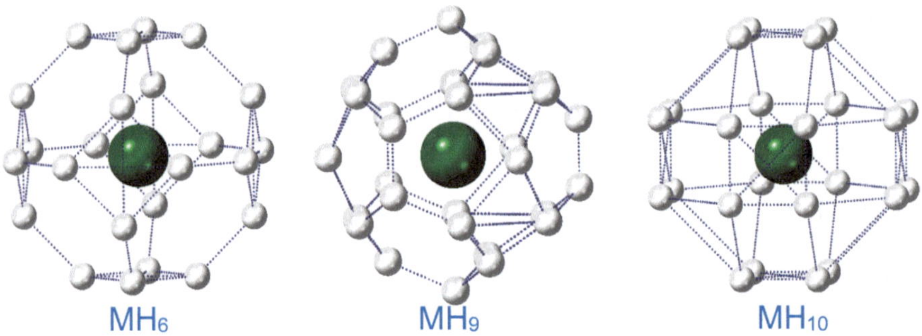

Figure 4.19 Arrangements of hydrogen around metal ions shown for selected superhydride structures. Metal ions are shown in green and hydrogen positions in grey.

Table 4.2 Reported superconductivity in metal superhydrides.

Material	T_c/K	p/GPa	References
YH_6	220	183	167
YH_9	243/256	201/177	167 and 168
CeH_9	100	130	169
CeH_{10}	115	95	169
ThH_9	146	170	170
ThH_{10}	161	170	170
LaH_{10}	250	170	171
$LaYH_{10}$	253	183	172
CaH_6	215	172	161

without controversy; a study of C-doped H_3S has since been retracted,[162] while a very recent report of superconductive N-doped lutetium hydride near room temperature and at a more modest pressure of 1 GPa[163] has been quickly disputed by a number of groups.[164–166] Undoubtedly the search for hydride superconductors has reinvigorated the search for the phenomenon under mild conditions, and will likely deliver a number of fascinating advances in the years to come.

4.3 Metal Halides

4.3.1 Metal Halide Perovskites

Photovoltaic materials are at the heart of humanity's attempt to shift away from a fossil-fuel powered society. While the plummeting cost of silicon solar cells has undoubtedly been the most consequential development in the photovoltaic field in recent decades,[173] the search for new materials which might offer advantages in efficiency, cost and resource footprint continue to motivate significant research efforts in the field. Among those efforts, the development of metal halide perovskite solar cells has captured headlines, in part due to their rapid advance in recorded device power conversion efficiency, rising from 3.8% in the first reported device in 2009[174] to 25.8% in most recent reports,[175] or 33.2% in tandem cells with silicon.[176] This rate of improvement in the efficiency significantly outstrips other photovoltaic classes, as shown in Figure 4.20.[177]

The perovskite structure, based on the perovskite mineral $CaTiO_3$, is one of the canonical crystal structure types. Perovskite materials have found wide application including in microwave technology,[178] ionic conductivity,[179–181] superconductivity[182] and capacitors.[183] With the generalised formula ABX_3, the basic structure has a cubic unit cell (Figure 4.21) with corner-sharing BX_6 octahedra, and the large A cation located in the 12-coordinate void created by the octahedra. The stability of this cubic structure is governed by the relative

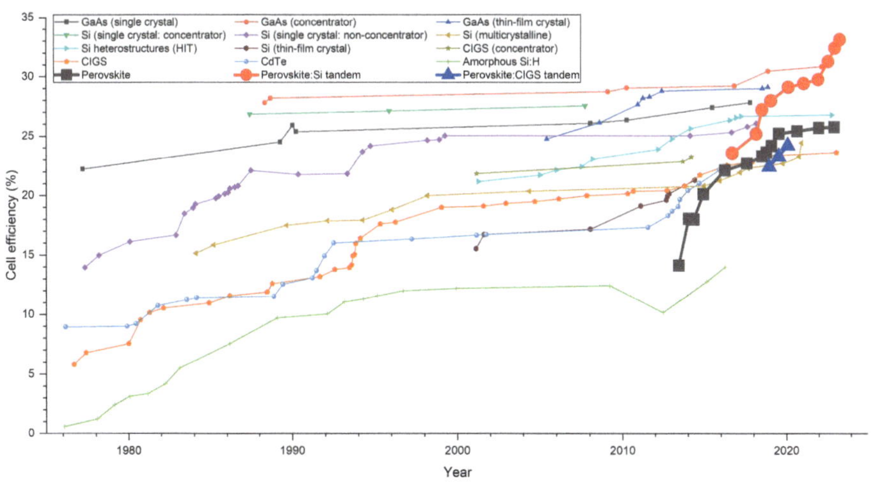

Figure 4.20 Chart of power conversion efficiencies for a range of solar photovoltaic technologies, with perovskite cell technologies shown in bold. Data collected from ref. 177.

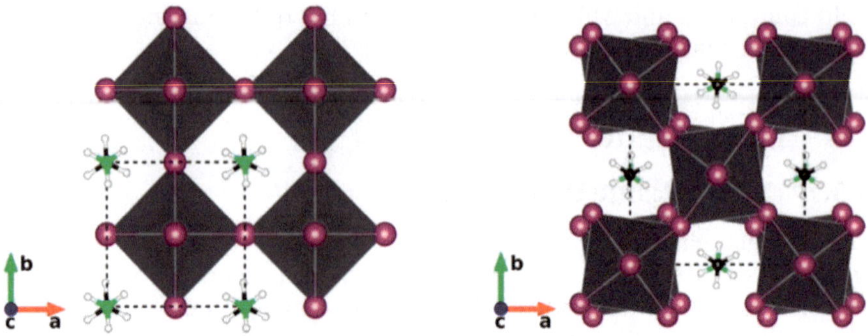

Figure 4.21 The idealised cubic perovskite structure of MAPI (left) and the distorted tetragonal room temperature structure (right). Iodine shown in purple, carbon in black, nitrogen in green and hydrogen in white. Reproduced from ref. 186 with permission from American Chemical Society, Copyright 2014.

ionic radii of the three ions A, B and X. The Goldschmidt tolerance factor, t,[184] (eqn (4.26)) is often used to assess the likelihood of perovskite formation, where r refers to the ionic radius.

$$t = \frac{r_A + r_X}{\sqrt{2}(r_B + r_X)} \tag{4.26}$$

A value of 0.9–1.0 is generally considered to give a stable cubic structure. This prediction offers reasonable accuracy, although alternative geometric conditions have been recently proposed with improved accuracy.[185] Deviations from this idealised structure are common, where size mismatch of the ions is accommodated by tilting of the octahedra, forming distorted structures which typically display tetragonal (also shown in Figure 4.21), orthorhombic or rhombohedral crystal symmetry, depending on the nature of the distortion.

Lead halide perovskites are of particular interest for photovoltaics, with the foundational material being methylammonium lead iodide: $[CH_3NH_3]PbI_3$, often shortened to MAPI. Its cubic and tetragonal crystal structures are shown in Figure 4.21.[186] In addition to its impressive basic performance, the material is attractive due to its relatively easy synthesis by simple techniques such as spin coating for thin film formation, and can be formulated at around 100 °C rather than at temperatures in excess of 1400 °C as is required for silicon cells. These factors have led to an explosion of research into the development of MAPI and related materials. The key drawbacks of MAPI are its low stability with increasing temperature and prolonged irradiation, and the toxicity of lead. Here, we will aim to illustrate a few key ways in which varying the composition and structure of the perovskites can influence their performance and address some of these perceived shortcomings.

The function of metal halide perovskites as photovoltaics is dependent on the visible light band gap. In MAPI, the methylammonium cation has very minimal contribution to either the valence band or the conduction band of the material. In fact, the band structure of MAPI can be well approximated to that of lead iodide, resulting from the mixing of the iodine 5p and lead 6s/6p orbitals (Figure 4.22).[187] As such, the band gap can be tuned by altering the composition of the X anion, with a number of halide partial and complete substitutions having been investigated.

As an example, the formation of Br–I solid solutions based on the MAPI material results in a significant modification of the UV-vis absorbance spectrum, as demonstrated by the spectra recorded in Figure 4.23.[188] The substitution of the smaller bromide ion results in shorter Pb–X bonds and consequently greater orbital overlap, which ultimately widens the band gap of the material. In this sense the iodide is most desirable in terms of the optoelectronic properties for a solar cell, as it can capture the widest range of the solar spectrum. However, the facile tunability of the band gap has allowed for the recent development of promising LEDs based on halide perovskites, often with mixed halide compositions. LEDs with emission in the near-IR, red and green regions have been developed with good performance metrics, with promising signs of development in the blue region, also.[189–191]

Variation in the X anion is not only motivated by tuning the band gap of the resultant material; anion migration is behind one of the principal instabilities of metal halide perovskites under illumination and/or elevated temperatures.[190] Given the increasing electronegativity and smaller size of the halides as the group is ascended, stronger bonding is expected for the smaller halides, reducing their mobility. However, the segregation of mixed halides into single halide ordered domains during operation can lead to instabilities in the material performance.[192]

Concerns regarding the toxicity of lead have also resulted in research efforts to explore other metals which could be used in perovskite solar cells. Only tin-based formulations have shown significant promise, with device

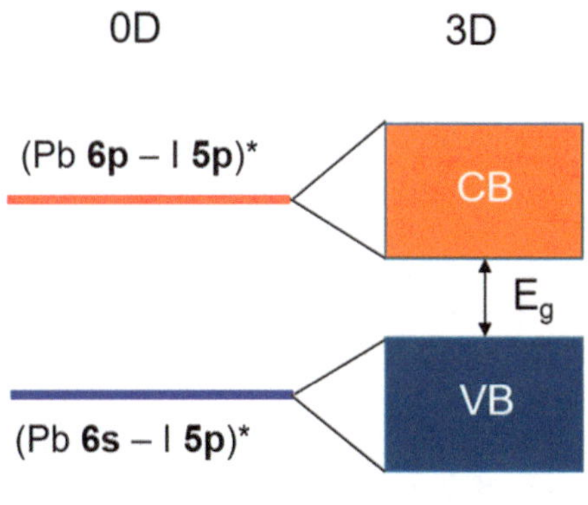

Figure 4.22 The electronic structure of PbI_2. Reproduced from ref. 187 with permission from American Chemical Society, Copyright 2020.

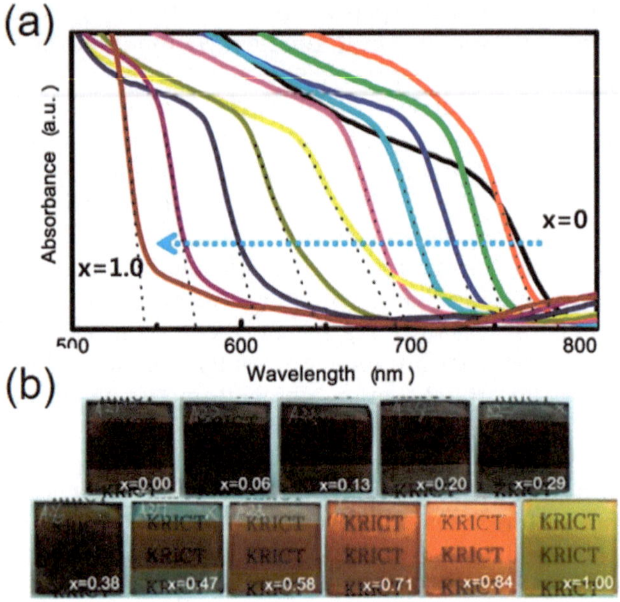

Figure 4.23 (a) UV-vis absorbance spectra and (b) resultant film appearance of $MAPbI_{1-x}Br_x$. Reproduced from ref. 188 with permission from American Chemical Society, Copyright 2013.

efficiencies of around 13% recorded.[193] Tin-based perovskites possess slightly smaller band gaps than their lead counterparts, resulting from the smaller radius of tin compared with lead. However, an additional instability in the materials results from the fact that tin is more easily oxidised than lead, making tin materials more susceptible to defect formation which may trap charge carriers.[187] Various additives to inhibit tin oxidation have been used to improve the stability of tin halide perovskites.[194]

Although the A-site cation (methylammonium in the starting case) has limited direct involvement in the band gap energy levels, it does have an indirect influence given its inevitable role in the distortion of the perovskite structure described above. The rotation of methylammonium on its site is thought to stabilise the separation of charges formed during light absorption, allowing the charge carriers to travel unusually long distances in the structure,[195,196] which is critical to device performance, since the electron and hole must stay separated if the absorbed energy is to be used productively. However, this ease of rotation reflects the modest bond strength between the cation and anion, and methylammonium volatility significantly limits the thermal stability of the material.[187]

In response to the instability of MAPI, alternative cations have been investigated, with significant effort devoted to formamidinium (FA; $[CN_2H_5]^+$) and caesium cations, as well as mixed cation systems. While both FAPI and $CsPbI_3$ show enhanced stability compared with MAPI, their perovskite structures are not particularly stable, with both forming wide band gap phases. Annealing at elevated temperature or the induction of crystal strain through particle size control or synthesis additives is required to stabilise the cubic perovskite phase.[197–199] Alternatively, mixed cation phases have been explored in order to stabilise a cubic average structure through disorder, although local heterogeneity is again a source of instability. Recent work has demonstrated the use of surface modifications to induce the required strain to stabilise the photoactive perovskite phase of FAPI.[200]

4.3.2 Battery Materials

The use of halides in batteries spans all components, from liquid electrolytes to electrode coatings. Fluoride doping of lithium-ion battery cathodes such as $LiNiO_2$ has been demonstrated to improve the cycling stability of the materials through maintaining higher transition metal oxidation states to prevent cation disordering.[201] Coatings of electrodes with fluoride layers has been deployed as a means to minimise the formation of reactive interfaces between electrodes and electrolytes.[202]

Progressing from these examples of the use of halides as an additive to oxide-based battery systems, halide salts have been considered as potential cathodes for LIBs in their own right. Fluorides have been the principal focus given the higher chemical potential of the (de)lithiation reactions, with transition metal salts such as CuF_2, NiF_2, BiF_3, FeF_2, and FeF_3 as key examples.[201] Part of the attraction of these materials is the potential to access multiple equivalents of lithium through the following general reaction:

$$x\mathrm{Li}^{+} + \mathrm{MF}_{x} \rightarrow x\mathrm{LiF} + \mathrm{M} \tag{4.27}$$

The potential for multiple equivalents of lithium stored results in higher potential capacity of these fluorides compared with conventional transition metal oxide cathodes, sometimes by a factor of more than two. In a similar vein to the work described for hydrides, these materials can often be described as conversion electrodes, although there has been recent discussion of the mechanisms more accurately being described through a series of displacive transitions.[203] Nevertheless, they face many of the same challenges observed in hydride conversion electrodes, namely poor diffusion kinetics, significant volume changes during cycling, and poor reversibility. For example, the CuF_2 system exhibits poor cyclability due to the formation of Cu^{+} during lithiation, which can dissolve in the electrolyte.[204] The strongly ionic nature of the metal–fluorine bonding interaction also leads to wide band gaps in the fluoride materials, limiting the electronic conductivity which is required for an electrode. As was observed in the case of the hydrides, research efforts to overcome these issues through nano-composites (*e.g.*, with graphite) and morphological control have led to significant improvements in the electrochemical performance of such electrodes.[205]

While not having resulted in the same commercial development as other battery types, there has also been a significant history of research into the

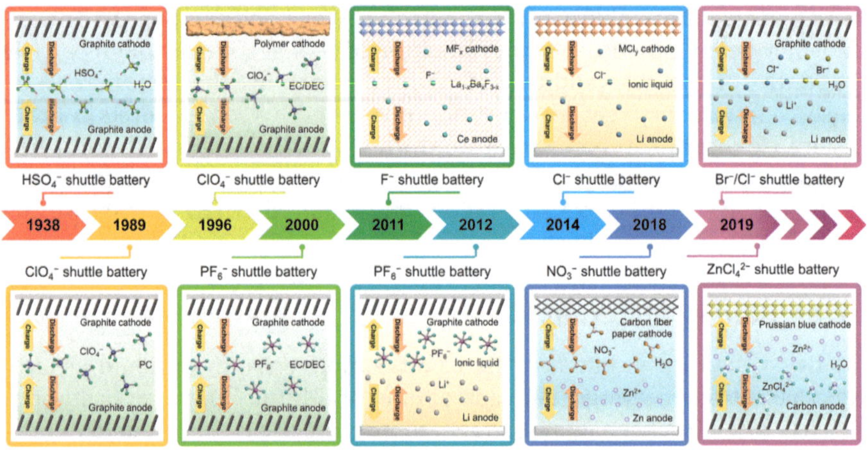

Figure 4.24 A schematic diagram of the timeline of development of anion shuttle batteries. Reproduced from ref. 207 with permission from Elsevier, Copyright 2015.

development of batteries where anions (often halides) are the transported species, as summarised in Figure 4.24. While many such batteries use halide-containing ions in solution, recent developments in these chemistries have involved the use of metal halide electrode materials, and so are of interest to our survey of halide materials chemistry.

In this area, fluoride and chloride ion batteries have seen the most extensive research given their high volumetric energy content, potentially far in excess of LIBs.[206] Fluorine is also much more naturally abundant than lithium, which may relieve resource pressures in the development of high energy density batteries. Fluoride batteries typically consist of a variable oxidation state metal fluoride cathode such as BiF_3, a metal anode such as cerium, and a solid electrolyte fluoride ion conductor. Transition metal fluorides typically operate as conversion electrodes, with similar cyclability issues as described above for their use in LIBs. Transition metal oxides are also used as intercalation cathodes through the formation of oxyfluoride materials, though typically with lower capacities.

The solid fluoride ion conductors which have been employed in fluoride batteries avoid the use of flammable and hazardous fluorine-containing liquid electrolytes. However, achieving high ionic conductivities remains challenging. The vast majority of fluoride ion conductors are based on one of two structure types: tysonite (LaF_3) and fluorite (CaF_2), structures shown in Figure 4.25.

Tysonite is a trigonal structure which can be described as slabs of LaF interspersed by fluoride layers, which are the primary source of fluoride conduction.[207] Fluoride conductivity is enhanced in these materials by doping with divalent cations in order to induce fluoride vacancies, in materials such as $La_{0.95}Ba_{0.05}F_{2.95}$. In contrast, the fluorite structure is cubic, based on a face

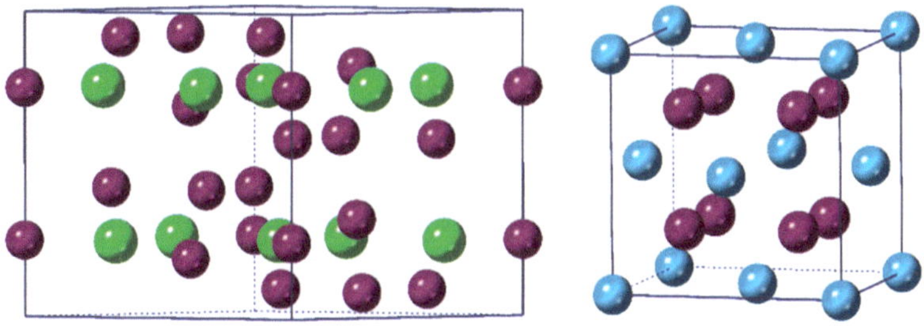

Figure 4.25 The crystal structures of tysonite (left) and fluorite (right), with lanthanum shown in green, calcium in blue and fluorine in purple.

centred cubic arrangement of calcium ions with fluorides occupying all of the tetrahedral holes in the structure. Fluorite is a well-known structure in ionic conductors, with many oxide ion conductors used in solid oxide fuel cells having the fluorite structure. The structure possesses a large number of interstitial sites and high-symmetry conduction pathways. The ion conduction properties are usually enhanced through doping of the cation site with an aliovalent metal ion. Doping with a monovalent cation produces fluoride vacancies, which improves conductivity. However, most fluorite-based fluoride conductors are based on trivalent substitutions (*e.g.* $Ba_{0.6}La_{0.4}F_{2.4}$, $Ba_{0.7}Sb_{0.3}F_{2.3}$), where conduction is promoted by the formation of F^- interstitials.[208]

While both structure types have produced promising conductors, the conductivities are typically low ($10^{-6}/10^{-7}$ S cm^{-1} at room temperature), necessitating elevated temperature operation. The most promising room temperature results have been obtained using the tetragonally-distorted fluorite material $BaSnF_4$, which, with a room temperature fluoride ion conductivity of around 10^{-4} S cm^{-1}, was able to operate in a full electrochemical cell.[209,210] However, fluoride batteries continue to be plagued by cyclability issues as a result of dissolution of the electrodes within electrolyte materials.

Chloride batteries typically use an electropositive metal such as lithium as the anode, with chlorides or oxychlorides of bismuth, iron, cobalt or vanadium as typical cathodes.[206] While there are many examples of chloride batteries which operate using mixed ionic liquid electrolytes, there are significant challenges associated with the dissolution of the electrodes within the electrolyte, which has spurred some effort to develop solid electrolytes. Among the most promising is the perovskite $CsSnCl_3$.[211]

4.4 Conclusions

In this chapter, we have observed the ways in which fundamental chemical concepts including crystal structures, diffusion, electronic structure and thermodynamic properties underpin the functions of metal halide and hydride materials. We have also seen the myriad ways in which chemists attempt to manipulate those properties. This body of work has positioned hydride materials at the forefront of energy storage technology development, while halides have been central to some of the most exciting recent advancements in solar energy technologies.

These applications offer an interesting symmetry in strategies to tailoring key performance characteristics, where cation and/or anion mixing has proved successful in altering hydrogen desorption and light absorption and emission properties alike. A more nuanced understanding of the different degrees of disorder in those materials and the way in which those disordered structures evolve over varying length scales will likely yield even greater control over the materials properties.

In key applications, environmental instabilities continue to hamper application of halides and hydrides, whether it be through water sensitivity or corrosion effects. While these phenomena are in some ways very well understood, a more complete documentation of degrees of air/water sensitivity, for example, could well highlight valuable compromises which could be made between those sensitivities and other material properties, or indeed point towards strategies for moderating them through alloying/substitutions, encapsulation or other approaches.

It is clear that despite constituting some of our most well-known materials, the chemistry of halides and hydrides continues to contribute to a diverse range of contemporary science, including some of the most pressing challenges faced by humanity. Continued research into these fascinating materials will undoubtedly uncover further functional materials which will contribute to the next generation of hydrogen stores, catalysts, batteries and solar cells, to name but a few of the ways in which they may be used. Frontier developments such as the use of high-pressure hydrides as superconductors and the development of anion batteries further offer a window into the new applications which likely await hydride and halide materials.

References

1. W. L. Bragg, The Structure of Some Crystals an Indicated by Their Diffraction of X-Rays, *Proc. R. Soc. A*, 1913, **89**(610), 248–277.
2. M. Gerin, D. A. Neufeld and J. R. Goicoechea, Interstellar Hydrides, *Annu. Rev. Astron. Astrophys.*, 2016, **54**, 181.
3. H. Mizoguchi, S. W. Park and H. Hosono, A View on Formation Gap in Transition Metal Hydrides and Its Collapse, *J. Am. Chem. Soc.*, 2021, **143**(30), 11345.
4. F. Cotton, G. Wilkinson, C. Murillo and M. Bochmann, *Advanced Inorganic Chemistry*, Wiley-Interscience, New York, 6th edn, 1999.
5. A. Wells, *Structural Inorganic Chemistry*, Oxford University Press, Oxford, 1st edn, 2012.
6. N. Greenwood and A. Earnshaw, *Chemistry of the Elements*, Butterworth-Heinemann, Oxford, 2nd edn, 1997.

7. L. Schlapbach and A. Züttel, Hydrogen Storage for Mobile Applications, *Nature*, 2001, **414**, 353.
8. T. Sato, H. Saitoh, R. Utsumi, J. Ito, Y. Nakahira, K. Obana, S. Takagi and S. I. Orimo, Hydrogen Absorption Reactions of Hydrogen Storage Alloy $LaNi_5$ under High Pressure, *Molecules*, 2023, **28**, 1256.
9. R. Bardhan, A. M. Ruminski, A. Brand and J. J. Urban, Magnesium Nanocrystal-Polymer Composites: A New Platform for Designer Hydrogen Storage Materials, *Energy Environ. Sci.*, 2011, **4**(12), 4882.
10. J. J. Reilly, in *Proc. Int. Symp. Hydrides for Energy Storage*, ed. A. F. Andresen and A. J. Maeland, Pergamon Press, Oxford, 1977.
11. J. M. Joubert, V. Paul-Boncour, F. Cuevas, J. Zhang and M. Latroche, $LaNi_5$ Related AB5 Compounds: Structure, Properties and Applications, *J. Alloys Compd.*, 2021, **862**, 158163.
12. P. Modi and K. F. Aguey-Zinsou, Room Temperature Metal Hydrides for Stationary and Heat Storage Applications: A Review, *Front. Energy Res.*, 2021, **9**, 616115.
13. G. Sandrock, P. Goodell, E. Huston and P. Golben, On the Disproportionation of Intermetallic Hydrides, *Z. Phys. Chem.*, 1989, **164**, 1285.
14. B. P. Tarasov, M. S. Bocharnikov, Y. B. Yanenko, P. V. Fursikov and M. V. Lototskyy, Cycling Stability of RNi_5 (R = La, La+Ce) Hydrides during the Operation of Metal Hydride Hydrogen Compressor, *Int. J. Hydrogen Energy*, 2018, **43**(9), 4415.
15. E. M. Borzone, M. V. Blanco, A. Baruj and G. O. Meyer, Stability of $LaNi_{5-X}Sn_x$ Cycled in Hydrogen, *Int. J. Hydrogen Energy*, 2014, **39**(16), 8791.
16. R. C. Bowman, C. H. Luo, C. C. Ahn, C. K. Witham and B. Fultz, The Effect of Tin on the Degradation of $LaNi_{5-Y}Sn_y$ Metal Hydrides during Thermal Cycling, *J. Alloys Compd.*, 1995, **217**(2), 185.
17. D. G. Westlake, Site Occupancies and Stoichiometries in Hydrides of Intermetallic Compounds: Geometric Considerations, *J. Less-Common Met.*, 1983, **90**(2), 251.
18. Y. Ma, Y. Ma, Q. Wang, S. Schweidler, M. Botros, T. Fu, H. Hahn, T. Brezesinski and B. Breitung, High-Entropy Energy Materials: Challenges and New Opportunities, *Energy Environ. Sci.*, 2021, **14**(5), 2883.
19. F. Marques, M. Balcerzak, F. Winkelmann, G. Zepon and M. Felderhoff, Review and Outlook on High-Entropy Alloys for Hydrogen Storage, *Energy Environ. Sci.*, 2021, **14**(10), 5191.
20. M. Sahlberg, D. Karlsson, C. Zlotea and U. Jansson, Superior Hydrogen Storage in High Entropy Alloys, *Sci. Rep.*, 2016, **6**, 1.
21. H. Emami, K. Edalati, J. Matsuda, E. Akiba and Z. Horita, Hydrogen Storage Performance of TiFe after Processing by Ball Milling, *Acta Mater.*, 2015, **88**, 190.
22. K. Edalati, J. Matsuda, A. Yanagida, E. Akiba and Z. Horita, Activation of TiFe for Hydrogen Storage by Plastic Deformation Using Groove Rolling and High-Pressure Torsion: Similarities and Differences, *Int. J. Hydrogen Energy*, 2014, **39**(28), 15589.

23. K. Edalati, J. Matsuda, H. Iwaoka, S. Toh, E. Akiba and Z. Horita, High-Pressure Torsion of TiFe Intermetallics for Activation of Hydrogen Storage at Room Temperature with Heterogeneous Nanostructure, *Int. J. Hydrogen Energy*, 2013, **38**(11), 4622.
24. E. M. Dematteis, N. Berti, F. Cuevas, M. Latroche and M. Baricco, Substitutional Effects in TiFe for Hydrogen Storage: A Comprehensive Review, *Mater. Adv.*, 2021, **2**(8), 2524.
25. W. L. Korst and J. C. Warf, Rare Earth-Hydrogen Systems, *Inorg. Chem.*, 1966, **5**(10), 1719–1726.
26. A. D. Shugard, C. R. Tewell, D. F. Cowgill and R. D. Kolasinski, *Uranium for Hydrogen Storage Applications : A Materials Science Perspective*, 2010.
27. G. R. Longhurst, R. A. Anderl, R. J. Pawelko and C. J. Stoots, Storage and Assay of Tritium in STAR, *Fusion Sci. Technol.*, 2005, **48**(1), 332.
28. R. S. Carlson, The Uranium-Tritium System - the Storage of Tritium, in *Radiation Effects and Tritium Technology for Fusion Reactors*, ed. J. S. Watson, and F. W. Wiffen, 1976, pp. 36–52.
29. U.S. Department of Energy: Office of Energy Efficiency & Renewable Energy, *DOE Technical Targets for Onboard Hydrogen Storage for Light-Duty Vehicles*, https://www.energy.gov/eere/fuelcells/doe-technical-targets-onboard-hydrogen-storage-light-duty-vehicles, accessed on August 8, 2018.
30. W. Grochala and P. P. Edwards, Thermal Decomposition of the Non-Interstitial Hydrides for the Storage and Production of Hydrogen, *Chem. Rev.*, 2004, **104**(3), 1283.
31. Y. Shang, C. Pistidda, G. Gizer, T. Klassen and M. Dornheim, Mg-Based Materials for Hydrogen Storage, *J. Magnesium Alloys*, 2021, **9**(6), 1837.
32. Y. Wang and Y. Wang, Recent Advances in Additive-Enhanced Magnesium Hydride for Hydrogen Storage, *Prog. Nat. Sci.:Mater. Int.*, 2017, **27**(1), 41.
33. J. C. Crivello, B. Dam, R. V. Denys, M. Dornheim, D. M. Grant, J. Huot, T. R. Jensen, P. de Jongh, M. Latroche, C. Milanese, D. Milčius, G. S. Walker, C. J. Webb, C. Zlotea and V. A. Yartys, Review of Magnesium Hydride-Based Materials: Development and Optimisation, *Appl. Phys. A:Mater. Sci. Process.*, 2016, **122**(2), 1.
34. P. E. de Jongh, M. Allendorf, J. J. Vajo and C. Zlotea, Nanoconfined Light Metal Hydrides for Reversible Hydrogen Storage, *MRS Bull.*, 2013, **38**(06), 488.
35. T. Sadhasivam, H. T. Kim, S. Jung, S. H. Roh, J. H. Park and H. Y. Jung, Dimensional Effects of Nanostructured Mg/MgH_2 for Hydrogen Storage Applications: A Review, *Renewable Sustainable Energy Rev.*, 2017, **72**, 523.
36. J. Graetz, J. J. Reilly, V. A. Yartys, J. P. Maehlen, B. M. Bulychev, V. E. Antonov, B. P. Tarasov and I. E. Gabis, Aluminum Hydride as a Hydrogen and Energy Storage Material: Past, Present and Future, *J. Alloys Compd.*, 2011, **509**(Suppl. 2), S517.

37. F. Brower, N. Matzek, P. Reigler, H. Rinn, C. Roberts, S. DL, J. Snover and K. Terada, Preparation and Properties of Aluminum Hydride, *J. Am. Chem. Soc.*, 1976, **98**(9), 2450.
38. J. Graetz and J. J. Reilly, Decomposition Kinetics of the AlH_3 Polymorphs, *J. Phys. Chem. B*, 2005, **109**(47), 22181.
39. J. Graetz, New Approaches to Hydrogen Storage, *Chem. Soc. Rev.*, 2009, **38**(1), 73.
40. M. Schwickardi and B. Bogdanovic, Ti-Doped Alkali Metal Aluminium Hydrides as Potential Novel Reversible Hydrogen Storage Materials, *J. Alloys Compd.*, 1997, **254**, 1.
41. M. Paskevicius, L. H. Jepsen, P. Schouwink, R. Černý, D. B. Ravnsbæk, Y. Filinchuk, M. Dornheim, F. Besenbacher and T. R. Jensen, Metal Borohydrides and Derivatives-Synthesis, Structure and Properties, *Chem. Soc. Rev.*, 2017, **46**, 1565.
42. J. Yang and S. Hirano, Improving the Hydrogen Reaction Kinetics of Complex Hydrides, *Adv. Mater.*, 2009, **21**(29), 3023.
43. I. P. Jain, P. Jain and A. Jain, Novel Hydrogen Storage Materials: A Review of Lightweight Complex Hydrides, *J. Alloys Compd.*, 2010, **503**(2), 303.
44. S.-I. Orimo, Y. Nakamori, J. R. Eliseo, A. Züttel and C. M. Jensen, Complex Hydrides for Hydrogen Storage, *Chem. Rev.*, 2007, **107**(10), 4111.
45. P. Chen, Z. Xiong, G. Wu, Y. Liu, J. Hu and W. Luo, Metal–N–H Systems for the Hydrogen Storage, *Scr. Mater.*, 2007, **56**(10), 817.
46. D. H. Gregory, Lithium Nitrides, Imides and Amides as Lightweight, Reversible Hydrogen Stores, *J. Mater. Chem.*, 2008, **18**(20), 2321.
47. H.-W. Li, Y. Yan, S. Orimo, A. Züttel and C. M. Jensen, Recent Progress in Metal Borohydrides for Hydrogen Storage, *Energies*, 2011, **4**(12), 185.
48. S. Garroni, A. Santoru, H. Cao, M. Dornheim, T. Klassen, C. Milanese, F. Gennari and C. Pistidda, Recent Progress and New Perspectives on Metal Amide and Imide Systems for Solid-State Hydrogen Storage, *Energies*, 2018, **11**, 1027.
49. J. M. Bellosta von Colbe, O. Metz, G. a. Lozano, P. K. Pranzas, H. W. Schmitz, F. Beckmann, A. Schreyer, T. Klassen and M. Dornheim, Behavior of Scaled-up Sodium Alanate Hydrogen Storage Tanks during Sorption, *Int. J. Hydrogen Energy*, 2012, **37**(3), 2807.
50. T. A. Johnson, D. E. Dedrick and R. Behrens, *Durability Study of a Vehicle-Scale Hydrogen Storage System*, 2010.
51. W. I. F. David, S. K. Callear, M. O. Jones, P. C. Aeberhard, S. D. Culligan, a H. Pohl, S. R. Johnson, K. R. Ryan, J. E. Parker, P. P. Edwards, C. J. Nuttall and a. Amieiro-Fonseca, The Structure, Thermal Properties and Phase Transformations of the Cubic Polymorph of Magnesium Tetrahydroborate, *Phys. Chem. Chem. Phys.*, 2012, **14**(33), 11800.
52. T. Matsunaga, F. Buchter, K. Miwa, S. Towata, S. Orimo and A. Züttel, Magnesium Borohydride: A New Hydrogen Storage Material, *Renewable Energy*, 2008, **33**(2), 193.
53. M. Chong, A. Karkamkar, T. Autrey, S. Orimo, S. Jalisatgi and C. M. Jensen, Reversible Dehydrogenation of Magnesium Borohydride to Magnesium

Triborane in the Solid State under Moderate Conditions, *Chem. Commun.*, 2011, **47**(4), 1330.
54. G. Severa, E. Rönnebro and C. M. Jensen, Direct Hydrogenation of Magnesium Boride to Magnesium Borohydride: Demonstration of >11 Weight Percent Reversible Hydrogen Storage, *Chem. Commun.*, 2010, **46**(3), 421.
55. M. P. Pitt, C. J. Webb, M. Paskevicius, D. Sheptyakov, C. E. Buckley and E. M. A. Gray, In Situ Neutron Diffraction Study of the Deuteration of Isotopic $Mg^{11}B_2$, *J. Phys. Chem. C*, 2011, **115**(45), 22669.
56. J. Wang, C. Freysoldt, Y. Du and L. Sun, First-Principles Study of Intrinsic Defects in Ammonia Borane, *J. Phys. Chem. C*, 2017, **121**(41), 22680.
57. Z. Xiong, C. K. Yong, G. Wu, P. Chen, W. Shaw, A. Karkamkar, T. Autrey, M. O. Jones, S. R. Johnson, P. P. Edwards and W. I. F. David, High-Capacity Hydrogen Storage in Lithium and Sodium Amidoboranes, *Nat. Mater.*, 2008, **7**(2), 138.
58. A. D. Sutton, A. K. Burrell, D. A. Dixon, E. B. Garner, J. C. Gordon, T. Nakagawa, K. C. Ott, J. P. Robinson and M. Vasiliu, Regeneration of Ammonia Borane Spent Fuel by Direct Reaction with Hydrazine and Liquid Ammonia, *Science*, 2011, **331**(6023), 1426.
59. Z. Tang, Y. Tan, X. Chen and X. Yu, Regenerable Hydrogen Storage in Lithium Amidoborane, *Chem. Commun.*, 2012, **48**(74), 9296.
60. P. Chen, Z. Xiong, J. Luo, J. Lin and K. L. Tan, Interaction of Hydrogen with Metal Nitrides and Imides, *Nature*, 2002, **420**(6913), 302.
61. P. Chen, Z. Xiong, J. Luo, J. Lin and K. L. Tan, Interaction between Lithium Amide and Lithium Hydride, *J. Phys. Chem. B*, 2003, **107**, 10967.
62. T. Ichikawa, N. Hanada, S. Isobe, H. Leng and H. Fujii, Mechanism of Novel Reaction from $LiNH_2$ and LiH to Li_2NH and H_2 as a Promising Hydrogen Storage System, *J. Phys. Chem. B*, 2004, **108**, 7887.
63. J. Z. Hu, J. H. Kwak, Z. Yang, W. Osborn, T. Markmaitree and L. L. Shaw, Probing the Reaction Pathway of Dehydrogenation of the $LiNH_2$+LiH Mixture Using in Situ ^1H NMR Spectroscopy, *J. Power Sources*, 2008, **181**(1), 116.
64. L. L. Shaw, W. Osborn, T. Markmaitree and X. Wan, The Reaction Pathway and Rate-Limiting Step of Dehydrogenation of the $LiHN_2$+LiH Mixture, *J. Power Sources*, 2008, **177**(2), 500.
65. H. Y. Leng, T. Ichikawa, S. Hino and N. Hanada, New Metal-N-H System Composed of $Mg(NH_2)_2$ and LiH for Hydrogen Storage, *J. Phys. Chem. B*, 2004, **108**, 8763.
66. T. Markmaitree, W. Osborn and L. Shaw, Comparisons between MgH_2- and LiH-Containing Systems for Hydrogen Storage Applications, *Int. J. Hydrogen Energy*, 2008, **33**(14), 3915.
67. Y. Nakamori, G. Kitahara, K. Miwa, N. Ohba, T. Noritake, S. Towata and S. Orimo, Hydrogen Storage Properties of Li-Mg-N-H Systems, *J. Alloys Compd.*, 2005, **404–406**, 396.
68. J. Lu, Z. Z. Fang, Y. J. Choi and H. Y. Sohn, Potential of Binary Lithium Magnesium Nitride for Hydrogen Storage Applications, *J. Phys. Chem. C*, 2007, **111**, 12129.

69. H. Leng, T. Ichikawa and H. Fujii, Hydrogen Storage Properties of Li-Mg-N-H Systems with Different Ratios of LiH/$Mg(NH_2)_2$, *J. Phys. Chem. B*, 2006, **110**(26), 12964.
70. T. Ichikawa, S. Isobe, N. Hanada and H. Fujii, Lithium Nitride for Reversible Hydrogen Storage, *J. Alloys Compd.*, 2004, **365**(1–2), 271.
71. S. Isobe, T. Ichikawa, Y. Kojima and H. Fujii, Characterization of Titanium Based Catalysts in the Li-N-H Hydrogen Storage System by X-Ray Absorption Spectroscopy, *J. Alloys Compd.*, 2007, **446–447**, 360.
72. W. I. F. David, M. O. Jones, D. H. Gregory, C. M. Jewell, S. R. Johnson, A. Walton and P. P. Edwards, A Mechanism for Non-Stoichiometry in the Lithium Amide/Lithium Imide Hydrogen Storage Reaction, *J. Am. Chem. Soc.*, 2007, **129**(6), 1594.
73. J. W. Makepeace, M. O. Jones, S. K. Callear, P. P. Edwards and W. I. F. David, In Situ X-Ray Powder Diffraction Studies of Hydrogen Storage and Release in the Li-N-H System, *Phys. Chem. Chem. Phys.*, 2014, **16**, 4061.
74. G. Lee, J. Shim, Y. Whan and K. S. Lee, Reversible Hydrogen Storage in $NaAlH_4$ Catalyzed with Lanthanide Oxides, *Int. J. Hydrogen Energy*, 2007, **32**, 1911.
75. K. Hoang, A. Janotti and C. G. Van De Walle, Decomposition Mechanism and the Effects of Metal Additives on the Kinetics of Lithium Alanate, *Phys. Chem. Chem. Phys.*, 2012, 2840.
76. S. Nayebossadri, K. F. Aguey-Zinsou and Z. X. Guo, Effect of Nitride Additives on Li-N-H Hydrogen Storage System, *Int. J. Hydrogen Energy*, 2011, **36**(13), 7920.
77. E. A. Nickels, M. O. Jones, W. I. F. David, S. R. Johnson, R. L. Lowton, M. Sommariva and P. P. Edwards, Tuning the Decomposition Temperature in Complex Hydrides: Synthesis of a Mixed Alkali Metal Borohydride, *Angew. Chem., Int. Ed.*, 2008, **47**(15), 2817.
78. D. Ravnsbœk, Y. Filinchuk, Y. Cerenius, H. J. Jakobsen, F. Besenbacher, J. Skibsted and T. R. Jensen, A Series of Mixed-Metal Borohydrides, *Angew. Chem., Int. Ed.*, 2009, **48**, 6659.
79. P. L. Bramwell, S. Lentink, P. Ngene and P. E. de Jongh, The Effect of Pore Confinement of $LiNH_2$ on Ammonia Decomposition Catalysis and the Storage of Hydrogen and Ammonia, *J. Phys. Chem. C*, 2016, **120**(48), 27212–27220.
80. P. Ngene, R. van den Berg, M. H. W. Verkuiklen, K. P. De Jong and P. E. de Jongh, Reversibility of the Hydrogen Desorption from NaBH4 by Confinement in Nanoporous Carbon, *Energy Environ. Sci.*, 2011, **4**, 4108.
81. Y. Yan, Y. S. Au, D. Rentsch, A. Remhof and P. E. de Jongh, Reversible Hydrogen Storage in $Mg(BH_4)_2$/Carbon Nanocomposites, *J. Mater. Chem. A*, 2013, **1**, 11177.
82. W. I. F. David, Effective Hydrogen Storage: A Strategic Chemistry Challenge, *Faraday Discuss.*, 2011, **151**, 399.
83. A. I. Bevan, A. Züttel, D. Book and I. R. Harris, Performance of a Metal Hydride Store on the “Ross Barlow” Hydrogen Powered Canal Boat, *Faraday Discuss.*, 2011, **151**, 353.

84. V. A. Yartys, M. V. Lototskyy, V. Linkov, S. Pasupathi, M. W. Davids, I. Tolj, G. Radica, R. V. Denys, J. Eriksen, K. Taube, J. Bellosta von Colbe, G. Capurso, M. Dornheim, F. Smith, D. Mathebula, D. Swanepoel and S. Suwarno, HYDRIDE4MOBILITY: An EU HORIZON 2020 Project on Hydrogen Powered Fuel Cell Utility Vehicles Using Metal Hydrides in Hydrogen Storage and Refuelling Systems, *Int. J. Hydrogen Energy*, 2021, **46**(72), 35896.
85. M. V. Lototskyy, I. Tolj, M. W. Davids, Y. V. Klochko, A. Parsons, D. Swanepoel, R. Ehlers, G. Louw, B. van der Westhuizen, F. Smith, B. G. Pollet, C. Sita and V. Linkov, Metal Hydride Hydrogen Storage and Supply Systems for Electric Forklift with Low-Temperature Proton Exchange Membrane Fuel Cell Power Module, *Int. J. Hydrogen Energy*, 2016, **41**(31), 13831.
86. M. Jehan, D. Fruchart, McPhy-Energy's proposal for solid state hydrogen storage materials and systems, *J. Alloys Compd.*, 2013, **580**(1), S343–S348.
87. J. Bellosta von Colbe, J. R. Ares, J. Barale, M. Baricco, C. Buckley, G. Capurso, N. Gallandat, D. M. Grant, M. N. Guzik, I. Jacob, E. H. Jensen, T. Jensen, J. Jepsen, T. Klassen, M. V. Lototskyy, K. Manickam, A. Montone, J. Puszkiel, S. Sartori, D. A. Sheppard, A. Stuart, G. Walker, C. J. Webb, H. Yang, V. Yartys, A. Züttel and M. Dornheim, Application of Hydrides in Hydrogen Storage and Compression: Achievements, Outlook and Perspectives, *Int. J. Hydrogen Energy*, 2019, **44**(15), 7780.
88. Toyota Motor Corporation, *Mirai Fuel Cell Vehicle*, 2017.
89. Hyundai, NEXO: Hyundai's hydrogen fuel-cell first, https://www.hyundai.com/au/en/news/vehicles/nexo-hyundai-hydrogen-fuel-cell-first#:~:text=Three high-pressure fuel tanks,a pressure of 700 bar, accessed on April 23, 2023.
90. Honda, Honda Clarity Fuel Cell Press Kit, 2017, https://hondanews.eu/eu/en/cars/media/pressreleases/106336/2017-honda-clarity-fuel-cell-press-kit, accessed on August 8, 2018.
91. M. V. Lototskyy, V. A. Yartys, B. G. Pollet and R. C. Bowman, Metal Hydride Hydrogen Compressors: A Review, *Int. J. Hydrogen Energy*, 2014, **39**, 5818.
92. M. V. Lototskyy, I. Tolj, L. Pickering, C. Sita, F. Barbir and V. Yartys, The Use of Metal Hydrides in Fuel Cell Applications, *Prog. Nat. Sci.:Mater. Int.*, 2017, **27**(1), 3.
93. L. K. Heung, Multistage Metal Hydride Compressor, in *Proceedings of the Solid Storage Science and Engineering Meeting*, Los Alamos, NM, 1985.
94. G. Friedlmeier, A. Manthey, M. Wanner and M. Groll, Cyclic Stability of Various Application-Relevant Metal Hydrides, *J. Alloys Compd.*, 1995, **231**(1–2), 880.
95. M. Wanner, G. Friedlmeier, G. Hoffmann and M. Groll, Thermodynamic and Structural Changes of Various Intermetallic Compounds during Extended Cycling in Closed Systems, *J. Alloys Compd.*, 1997, **253–254**, 692.

96. J. M. Ohi, N. Vanderborgh and Gerald Voecks Consultants, *Hydrogen Fuel Quality Specifications for Polymer Electrolyte Fuel Cells in Road Vehicles*, Report to the Safety, Codes and Standards Program, Fuel Cell Technologies Office, U.S. Department of Energy, 2016.
97. Z. Du, C. Liu, J. Zhai, X. Guo, Y. Xiong, W. Su and G. He, A Review of Hydrogen Purification Technologies for Fuel Cell Vehicles, *Catalysts*, 2021, **11**(3), 1.
98. S. Nayebossadri, J. D. Speight and D. Book, Hydrogen Separation from Blended Natural Gas and Hydrogen by Pd-Based Membranes, *Int. J. Hydrogen Energy*, 2019, **44**(55), 29092.
99. C. Zhao, B. Sun, J. Jiang and W. Xu, H_2 Purification Process with Double Layer Bcc-PdCu Alloy Membrane at Ambient Temperature, *Int. J. Hydrogen Energy*, 2020, **45**(35), 17540.
100. S. Uemiya, T. Matsuda and E. Kikuchi, Hydrogen Permeable Palladium-Silver Supported on Porous Ceramics Alloy Membrane, *J. Membr. Sci.*, 1991, **56**, 315.
101. M. D. Dolan, D. M. Viano, M. J. Langley and K. E. Lamb, Tubular Vanadium Membranes for Hydrogen Purification, *J. Membr. Sci.*, 2018, **549**, 306.
102. K. E. Lamb, D. M. Viano, M. J. Langley, S. S. Hla and M. D. Dolan, High-Purity H_2 Produced from NH_3 via a Ruthenium-Based Decomposition Catalyst and Vanadium-Based Membrane, *Ind. Eng. Chem. Res.*, 2018, 8.
103. S. Lasher, M. Stratonova and J. Thijssen, *Hydrogen Technical Analysis*, 2002.
104. X. L. Wang, K. Iwata and S. Suda, Hydrogen Purification Using Fluorinated $LaNi_{4.7}Al_{0.3}$ Alloy, *J. Alloys Compd.*, 1995, **231**(1–2), 860.
105. F. J. Liu and S. Suda, Properties and Characteristics of Fluorinated Hydriding Alloys, *J. Alloys Compd.*, 1995, **231**(1–2), 742.
106. Y. M. Sun, X. P. Gao, N. Araya, E. Higuchi and S. Suda, Duplicated Fluorination Technique for Hydrogen Storage Alloys, *J. Alloys Compd.*, 1999, **293**, 364.
107. C. K. Ho and A. Ambrosini, Thermal energy storage technologies, in *U.S. DOE Energy Storage Handbook*, United States Department of Energy, 2020, 1–21.
108. D. A. Sheppard, M. Paskevicius, T. D. Humphries, M. Felderhoff, G. Capurso, J. Bellosta von Colbe, M. Dornheim, T. Klassen, P. A. Ward, J. A. Teprovich, C. Corgnale, R. Zidan, D. M. Grant and C. E. Buckley, Metal Hydrides for Concentrating Solar Thermal Power Energy Storage, *Appl. Phys. A:Mater. Sci. Process.*, 2016, **122**(4), 1.
109. M. Felderhoff and B. Bogdanović, High Temperature Metal Hydrides as Heat Storage Materials for Solar and Related Applications, *Int. J. Mol. Sci.*, 2009, **10**(1), 335.
110. K. T. Møller, D. Sheppard, D. B. Ravnsbæk, C. E. Buckley, E. Akiba, H. Li and T. R. Jensen, Complex Metal Hydrides for Hydrogen, Thermal and Electrochemical Energy Storage, *Energies*, 2017, **10**, 1645.

111. The Royal Society, *Ammonia : Zero- Carbon Fertiliser, Fuel and Energy Store*, 2020.
112. C. Smith, A. K. Hill and L. Torrente-Murciano, Current and Future Role of Haber-Bosch Ammonia in a Carbon-Free Energy Landscape, *Energy Environ. Sci.*, 2020, **13**(2), 331–344.
113. R. Bañares-Alcántara, G. Dericks III, M. Fiaschetti, P. Grunewald, J. Masa Lopez, E. Tsang, A. Yang, L. Ye and S. Zhao, *Analysis of Islanded Ammonia-Based Energy Storage Systems*, 2015.
114. M. Ravi and J. W. Makepeace, Facilitating Green Ammonia Manufacture under Milder Conditions: What Do Heterogeneous Catalyst Formulations Have to Offer?, *Chem. Sci.*, 2022, **13**(4), 890.
115. K. Aika, M. Kumasaka, T. Oma, O. Kato, H. Matsuda, N. Watanabe, K. Yamazaki, A. Ozaki and T. Onishi, Support and Promoter Effect of Ruthenium Catalyst. III. Kinetics of Ammonia Synthesis over Various Ru Catalysts, *Appl. Catal.*, 1986, **28**, 57.
116. K. Aika, Role of Alkali Promoter in Ammonia Synthesis over Ruthenium Catalysts—Effect on Reaction Mechanism, *Catal. Today*, 2017, **286**, 14.
117. M. Kitano, Y. Inoue, Y. Yamazaki, F. Hayashi, S. Kanbara, S. Matsuishi, T. Yokoyama, S.-W. Kim, M. Hara and H. Hosono, Ammonia Synthesis Using a Stable Electride as an Electron Donor and Reversible Hydrogen Store, *Nat. Chem.*, 2012, **4**, 934.
118. M. Hara, M. Kitano and H. Hosono, Ru-Loaded C12A7:e^- Electride as a Catalyst for Ammonia Synthesis, *ACS Catal.*, 2017, **7**(4), 2313.
119. K. Lee, S. W. Kim, Y. Toda, S. Matsuishi and H. Hosono, Dicalcium Nitride as a Two-Dimensional Electride with an Anionic Electron Layer, *Nature*, 2013, **494**(7437), 336.
120. M. Kitano, Y. Inoue, H. Ishikawa, K. Yamagata, T. Nakao, T. Tada, S. Matsuishi, T. Yokoyama, M. Hara and H. Hosono, Essential Role of Hydride Ion in Ruthenium-Based Ammonia Synthesis Catalysts, *Chem. Sci.*, 2016, **7**(7), 4036.
121. M. Kitano, J. Kujirai, K. Ogasawara, S. Matsuishi, T. Tada, H. Abe, Y. Niwa and H. Hosono, Low-Temperature Synthesis of Perovskite Oxynitride-Hydrides as Ammonia Synthesis Catalysts, *J. Am. Chem. Soc.*, 2019, **141**(51), 20344.
122. K. Ooya, J. Li, K. Fukui, S. Iimura, T. Nakao, K. Ogasawara, M. Sasase, H. Abe, Y. Niwa, M. Kitano and H. Hosono, Ruthenium Catalysts Promoted by Lanthanide Oxyhydrides with High Hydride-Ion Mobility for Low-Temperature Ammonia Synthesis, *Adv. Energy Mater.*, 2021, **11**(4), 1.
123. Y. Jiang, R. Takashima, T. Nakao, M. Miyazaki, Y. Lu, M. Sasase, Y. Niwa, H. Abe, M. Kitano and H. Hosono, Boosted Activity of Cobalt Catalysts for Ammonia Synthesis with $BaAl_2O_{4-x}H_y$ Electrides, *J. Am. Chem. Soc.*, 2023, **145**(19), 10669.
124. A. J. Medford, A. Vojvodic, J. S. Hummelshøj, J. Voss, F. Abild-Pedersen, F. Studt, T. Bligaard, A. Nilsson and J. K. Nørskov, From the Sabatier

Principle to a Predictive Theory of Transition-Metal Heterogeneous Catalysis, *J. Catal.*, 2015, **328**, 36.
125. A. Vojvodic, A. J. Medford, F. Studt, F. Abild-Pedersen, T. S. Khan, T. Bligaard and J. K. Nørskov, Exploring the Limits: A Low-Pressure, Low-Temperature Haber-Bosch Process, *Chem. Phys. Lett.*, 2014, **598**, 108.
126. V. J. Bukas and J. K. Nørskov, A Molecular-Level Mechanism of the Biological N_2 Fixation A Molecular-Level Mechanism of the Biological N_2 Fixation, *ChemRxiv*, 2019, 1.
127. P. Wang, F. Chang, W. Gao, J. Guo, G. Wu, T. He and P. Chen, Breaking Scaling Relations to Achieve Low-Temperature Ammonia Synthesis through LiH-Mediated Nitrogen Transfer and Hydrogenation, *Nat. Chem.*, 2016, **9**, 64.
128. W. Gao, P. Wang, J. Guo, F. Chang, T. He, Q. Wang, G. Wu and P. Chen, Barium Hydride-Mediated Nitrogen Transfer and Hydrogenation for Ammonia Synthesis: A Case Study of Cobalt, *ACS Catal.*, 2017, 7, 3654.
129. W. Gao, J. Guo, P. Wang, Q. Wang, F. Chang, Q. Pei, W. Zhang, L. Liu and P. Chen, Production of Ammonia via a Chemical Looping Process Based on Metal Imides as Nitrogen Carriers, *Nat. Energy*, 2018, **3**, 1067.
130. M. Ravi and J. W. Makepeace, Lithium-Nitrogen-Hydrogen Systems for Ammonia Synthesis: Exploring a More Efficient Pathway Using Lithium Nitride-Hydride, *Chem. Commun.*, 2022, **58**(41), 6076.
131. H. Sugiyama, T. Nakao, M. Miyazaki, H. Abe, Y. Niwa, M. Kitano and H. Hosono, Low-Temperature Methanol Synthesis by a Cu-Loaded LaH_{2+x} Electride, *ACS Catal.*, 2022, **12**(20), 12572.
132. M. Miyazaki, K. Ogasawara, T. Nakao, M. Sasase, M. Kitano and H. Hosono, Hexagonal $BaTiO_{(3-x)}H_x$ Oxyhydride as a Water-Durable Catalyst Support for Chemoselective Hydrogenation, *J. Am. Chem. Soc.*, 2022, **144**(14), 6453.
133. Electro Energy, *Nickel-Metal Hydride Battery Development*, 1995.
134. V. Yartys, D. Noreus and M. Latroche, Metal Hydrides as Negative Electrode Materials for Ni–MH Batteries, *Appl. Phys. A:Mater. Sci. Process.*, 2016, **122**(1), 1.
135. Y. Oumellal, A. Rougier, G. A. Nazri, J. Tarascon and L. Aymard, Metal Hydrides for Lithium-Ion Batteries, *Nat. Mater.*, 2008, 7, 916.
136. L. Aymard, Y. Oumellal and J. Bonnet, Metal Hydrides : An Innovative and Challenging Conversion Reaction Anode for Lithium-Ion Batteries, *Beilstein J. Nanotechnol.*, 2015, **6**, 1821.
137. R. Mohtadi and S. I. Orimo, The Renaissance of Hydrides as Energy Materials, *Nat. Rev. Mater.*, 2016, **2**(3), 1.
138. Y. Oumellal, C. Zlotea, S. Bastide, C. Cachet-Vivier, E. Léonel, S. Sengmany, E. Leroy, L. Aymard, J.-P. Bonnet and M. Latroche, Bottom-up Preparation of MgH_2 Nanoparticles with Enhanced Cycle Life Stability during Electrochemical Conversion in Li-Ion Batteries, *Nanoscale*, 2014, **6**(23), 14459.

139. W. Zaïdi, Y. Oumellal, J. Bonnet, J. Zhang, F. Cuevas, M. Latroche, J. Bobet and L. Aymard, Carboxymethylcellulose and Carboxymethylcellulose-Formate as Binders in MgH_2 – Carbon Composites Negative Electrode for Lithium-Ion Batteries, *J. Power Sources*, 2011, **196**, 2854.
140. M. Matsuo, Y. Nakamori, S. Orimo, H. Maekawa and H. Takamura, Lithium Superionic Conduction in Lithium Borohydride Accompanied by Structural Transition, *Appl. Phys. Lett.*, 2007, **91**, 224103.
141. H. Maekawa, M. Matsuo, H. Takamura, M. Ando and Y. Noda, Halide-Stabilized $LiBH_4$, a Room-Temperature Lithium Fast-Ion Conductor, *J. Am. Chem. Soc.*, 2009, **131**, 894.
142. T. J. Udovic, M. Matsuo, W. S. Tang, H. Wu, V. Stavila, A. V. Soloninin, R. V. Skoryunov, O. A. Babanova, A. V. Skripov, J. J. Rush, A. Unemoto, H. Takamura and S. Orimo, Exceptional Superionic Conductivity in Disordered Sodium Decahydro- Closo -Decaborate, *Adv. Mater.*, 2014, 7622.
143. W. S. Tang, A. Unemoto, W. Zhou, V. Stavila, M. Matsuo, H. Wu, S. Orimo and T. J. Udovic, Unparalleled Lithium and Sodium Superionic Conduction in Solid Electrolytes with Large Monovalent Cage-like Anions, *Energy Environ. Sci.*, 2015, **8**, 3637.
144. S. Higashi, K. Miwa, M. Aoki and K. Takechi, A Novel Inorganic Solid State Ion Conductor for Rechargeable Mg Batteries, *Chem. Commun.*, 2014, **50**, 1320.
145. R. Le Ruyet, R. Berthelot, E. Salager, P. Florian, B. Fleutot and R. Janot, Investigation of $Mg(BH_4)(NH_2)$ - Based Composite Materials with Enhanced Mg^{2+} Ionic Conductivity, *J. Phys. Chem. C*, 2019, **123**, 10756.
146. M. C. Verbraeken, C. Cheung, E. Suard and J. T. S. Irvine, High H– Ionic Conductivity in Barium Hydride, *Nat. Mater.*, 2015, **14**, 95.
147. K. Fukui, S. Iimura, A. Iskandarov, T. Tada and H. Hosono, Room-Temperature Fast H − Conduction in Oxygen-Substituted Lanthanum Hydride, *J. Am. Chem. Soc.*, 2022, **144**, 1523.
148. G. J. Irvine, R. I. Smith, M. O. Jones and J. T. S. Irvine, Order – Disorder and Ionic Conductivity in Calcium Nitride-Hydride, *Nat. Commun.*, 2023, **14**, 4389.
149. R. de Bruyn Ouboter, Heike Kamerlingh Onnes's Discovery of Superconductivity, *Sci. Am.*, 1997, **276**(3), 98.
150. N. Jun, N. Norimasa, M. Takahiro, Z. Yuji and A. Jun, Superconductivity at 39 K in Magnesium Diboride, *Nature*, 2001, **410**, 63.
151. L. Gao, Y. Y. Xue, F. Chen, Q. Xiong, R. L. Meng, D. Ramirez, C. W. Chu, J. H. Eggert and H. K. Mao, Superconductivity up to 164 K in $HgBa_2Ca_{m-1}Cu_mO_{2m+2+\delta}$ (M=1, 2, and 3) under Quasihydrostatic Pressures, *Phys. Rev. B*, 1994, **50**(6), 4260.
152. N. W. Ashcroft, Metallic Hydrogen: A High-Temperature Superconductor?, *Phys. Rev. Lett.*, 1968, **21**(26), 1748.
153. D. Duan, Y. Liu, Y. Ma, Z. Shao, B. Liu and T. Cui, Structure and Superconductivity of Hydrides at High Pressures, *Natl. Sci. Rev.*, 2017, **4**(1), 121.
154. N. W. Ashcroft, Hydrogen Dominant Metallic Alloys: High Temperature Superconductors?, *Phys. Rev. Lett.*, 2004, **92**(18), 1.

155. B. Stritzker, High Superconducting Transition Temperatures in the Palladium-Noble Metal-Hydrogen System, *Z. Phys.*, 1974, **268**(2), 261.
156. A. Leiberich, W. Scholz and W. J. Standish, Superconductivity in H-charged Cu-implanted Pd, *Phys. Lett.*, 1981, **87**(1), 57.
157. M. I. Eremets, I. A. Trojan, S. A. Medvedev, J. S. Tse and Y. Yao, Superconductivity in Hydrogen Dominant Materials: Silane, *Science*, 2008, **319**(5869), 1506.
158. S. N. Neal, D. Stacchiola and S. A. Tenney, Spatially Resolved Multimodal Vibrational Spectroscopy under High Pressures, *Phys. Chem. Chem. Phys.*, 2023, **25**, 31578.
159. A. P. Drozdov, M. I. Eremets, I. A. Troyan, V. Ksenofontov and S. I. Shylin, Conventional Superconductivity at 203 Kelvin at High Pressures in the Sulfur Hydride System, *Nature*, 2015, **525**(7567), 73.
160. S. Zhang, M. Zhang and H. Liu, Superconductive Hydrogen-Rich Compounds under High Pressure, *Appl. Phys. A:Mater. Sci. Process.*, 2021, **127**(9), 1.
161. L. Ma, K. Wang, Y. Xie, X. Yang, Y. Wang, M. Zhou, H. Liu, X. Yu, Y. Zhao, H. Wang, G. Liu and Y. Ma, High-Temperature Superconducting Phase in Clathrate Calcium Hydride CaH_6 up to 215 K at a Pressure of 172 GPa, *Phys. Rev. Lett.*, 2022, **128**(16), 167001.
162. E. Snider, N. Dasenbrock-Gammon, R. McBride, M. Debessai, H. Vindana, K. Vencatasamy, K. V. Lawler, A. Salamat and R. P. Dias, Room-Temperature Superconductivity in a Carbonaceous Sulfur Hydride, *Nature*, 2020, **586**(7829), 373.
163. N. Dasenbrock-Gammon, E. Snider, R. McBride, H. Pasan, D. Durkee, N. Khalvashi-Sutter, S. Munasinghe, S. E. Dissanayake, K. V. Lawler, A. Salamat and R. P. Dias, Evidence of Near-Ambient Superconductivity in a N-Doped Lutetium Hydride, *Nature*, 2023, **615**(7951), 244.
164. P. Shan, N. Wang, X. Zheng, Q. Qiu, Y. Peng and J. Cheng, Pressure-Induced Color Change in the Lutetium Dihydride LuH_2, *Chin. Phys. Lett.*, 2023, 5.
165. K. P. Hilleke, X. Wang, D. Luo, N. Geng, B. Wang and E. Zurek, Structure, Stability and Superconductivity of N-Doped Lutetium Hydrides at Kbar Pressures, *Phys. Rev. B*, 2023, **108**, 014511.
166. Z. Li, X. He, C. Zhang, K. Lu, B. Bin, J. Zhang, S. Zhang, J. Zhao, L. Shi, Y. Peng, S. Feng, Z. Deng, J. Song, Q. Liu, X. Wang, L. Wang, Y. Li, J. Bass, V. Prakapenka, S. Chariton, H. Liu and C. Jin, Superconductivity above 70 K Experimentally Discovered in Lutetium Polyhydride, *Sci. China:-Phys., Mech. Astron.*, 2023, **66**(6), 1.
167. P. Kong, V. S. Minkov, M. A. Kuzovnikov, A. P. Drozdov, S. P. Besedin, S. Mozaffari, L. Balicas, F. F. Balakirev, V. B. Prakapenka, S. Chariton, D. A. Knyazev, E. Greenberg and M. I. Eremets, Superconductivity up to 243 K in the Yttrium-Hydrogen System under High Pressure, *Nat. Commun.*, 2021, **12**(1), 2.
168. E. Snider, N. Dasenbrock-Gammon, R. McBride, X. Wang, N. Meyers, K. V. Lawler, E. Zurek, A. Salamat and R. P. Dias, Synthesis of Yttrium Superhydride Superconductor with a Transition Temperature up to 262 K

by Catalytic Hydrogenation at High Pressures, *Phys. Rev. Lett.*, 2021, **126**(11), 117003.

169. W. Chen, D. V. Semenok, X. Huang, H. Shu, X. Li, D. Duan, T. Cui and A. R. Oganov, High-Temperature Superconducting Phases in Cerium Superhydride with a T_c up to 115 K below a Pressure of 1 Megabar, *Phys. Rev. Lett.*, 2021, **127**(11), 117001.
170. D. V. Semenok, A. G. Kvashnin, A. G. Ivanova, V. Svitlyk, V. Y. Fominski, A. V. Sadakov, O. A. Sobolevskiy, V. M. Pudalov, I. A. Troyan and A. R. Oganov, Superconductivity at 161 K in Thorium Hydride ThH_{10}: Synthesis and Properties, *Mater. Today*, 2020, **33**, 36.
171. A. P. Drozdov, P. P. Kong, V. S. Minkov, S. P. Besedin, M. A. Kuzovnikov, S. Mozaffari, L. Balicas, F. F. Balakirev, D. E. Graf, V. B. Prakapenka, E. Greenberg, D. A. Knyazev, M. Tkacz and M. I. Eremets, Superconductivity at 250 K in Lanthanum Hydride under High Pressures, *Nature*, 2019, **569**(7757), 528.
172. D. V. Semenok, I. A. Troyan, A. G. Ivanova, A. G. Kvashnin, I. A. Kruglov, M. Hanfland, A. V. Sadakov, O. A. Sobolevskiy, K. S. Pervakov, I. S. Lyubutin, K. V. Glazyrin, N. Giordano, D. N. Karimov, A. L. Vasiliev, R. Akashi, V. M. Pudalov and A. R. Oganov, Superconductivity at 253 K in Lanthanum–Yttrium Ternary Hydrides, *Mater. Today*, 2021, **48**, 18.
173. G. Kavlak, J. Mcnerney and J. E. Trancik, Evaluating the Causes of Cost Reduction in Photovoltaic Modules, *Energy Policy*, 2018, **123**, 700.
174. A. Kojima, K. Teshima, Y. Shirai and T. Miyasaka, Organometal Halide Perovskites as Visible-Light Sensitizers for Photovoltaic, *J. Am. Chem. Soc.*, 2009, **131**, 6050.
175. H. Min, D. Y. Lee, J. Kim, G. Kim, K. S. Lee, J. Kim, M. J. Paik, Y. K. Kim, K. S. Kim, M. G. Kim, T. J. Shin and S. Seok, Il. Perovskite Solar Cells with Atomically Coherent Interlayers on SnO_2 Electrodes, *Nature*, 2021, **598**, 444–450.
176. KAUST Solar Center, KAUST team sets world record for tandem solar cell efficiency, https://ksc.kaust.edu.sa/news/detail/2023/04/13/kaust-team-sets-world-record-for-tandem-solar-cell-efficiency, accessed on May 01, 2023.
177. National Renewable Energy Laboratory, Interactive Best Research-Cell Efficiency Chart, https://www.nrel.gov/pv/interactive-cell-efficiency.html, accessed on May 09, 2023.
178. H. Ohsato, Microwave Dielectrics with Perovskite-Type Structure, in *Perovskite Materials - Synthesis, Characterisation, Properties, and Applications*, ed. L. Pan, and G. Zhu, IntechOpen, 2016.
179. M. Li, M. J. Pietrowski, R. A. De Souza, H. Zhang, I. M. Reaney, S. N. Cook, J. A. Kilner and D. C. Sinclair, A Family of Oxide Ion Conductors Based on the Ferroelectric Perovskite $Na_{0.5}Bi_{0.5}TiO_3$, *Nat. Mater.*, 2014, **13**, 31.
180. M. Yashima, T. Tsujiguchi, Y. Sakuda, Y. Yasui, Y. Zhou, S. Torii, T. Kamiyama and S. J. Skinner, High Oxide-Ion Conductivity through the Interstitial Oxygen Site in $Ba_7Nb_4MoO_{20}$-Based Hexagonal Perovskite Related Oxides, *Nat. Commun.*, 2021, **12**(556), 1.

181. C. Y. R. Vera, H. Ding, D. Peterson, W. T. Gibbons and M. Zhou, A Mini-Review on Proton Conduction of $BaZrO_3$ - Based Perovskite Electrolytes, *J. Phys. Energy*, 2021, **3**, 032019.
182. S. Hikami and Y. Matsude, High T c Superconductors of the Perovskite Structure Oxides, *Jpn. J. Appl. Phys.*, 1987, **26**, 1027.
183. M.-J. Pan and C. A. Randall, A Brief Introduction to Ceramic Capacitors, in *IEEE Electrical Insulation Magazine*, IEEE, 2008, pp. 44–50.
184. O. Goldschmidt, Die Gesetze Der Krystallochemie, *Naturwissenschaften*, 1926, **14**, 477.
185. C. J. Bartel, C. Sutton, B. R. Goldsmith, R. Ouyang, C. B. Musgrave, L. M. Ghiringhelli and M. Scheffler, New Tolerance Factor to Predict the Stability of Perovskite Oxides and Halides, *Sci. Adv.*, 2019, **5**, 1.
186. D. A. Egger and L. Kronik, Role of Dispersive Interactions in Determining Structural Properties of Organic – Inorganic Halide Perovskites: Insights from First- Principles Calculations, *J. Phys. Chem. Lett.*, 2014, **5**, 2728.
187. J. Y. Kim, J. Lee, H. S. Jung, H. Shin and N. Park, High-Efficiency Perovskite Solar Cells, *Chem. Rev.*, 2020, **120**, 7867.
188. J. H. Noh, S. H. Im, J. H. Heo, T. N. Mandal and S. Seok, Il. Chemical Management for Colorful, Efficient, and Stable Inorganic-Organic Hybrid Nanostructured Solar Cells, *Nano Lett.*, 2013, **13**(4), 1764.
189. S. J. Zelewski, Y. Liu, A. J. Mirabelli, Y. Zhang, J. Huang, Y. Wang, K. Gong, M. C. Lai, L. Zhang, D. Yang, J. Lin, E. M. Tennyson, C. Ducati, S. D. Stranks, L. Cui and N. C. Greenham, Bright and Stable Perovskite Light-Emitting Diodes in the near-Infrared Range, *Nature*, 2023, **615**, 830.
190. K. Ji, M. Anaya, A. Abfalterer and S. D. Stranks, Halide Perovskite Light-Emitting Diode Technologies, *Adv. Opt. Mater.*, 2021, **9**(18), 2002128.
191. Z. Tan, J. Li, C. Zhang, Z. Li, Q. Hu, Z. Xiao, T. Kamiya, H. Hosono, G. Niu, E. Lifshitz, Y. Cheng and J. Tang, Highly Efficient Blue-Emitting Bi-Doped Cs_2SnCl_6 Perovskite Variant : Photoluminescence Induced by Impurity Doping, *Adv. Funct. Mater.*, 2018, **28**, 1801131.
192. D. J. Slotcavage, H. I. Karunadasa and M. D. Mcgehee, Light-Induced Phase Segregation in Halide- Perovskite Absorbers, *ACS Energy Lett.*, 2016, **1**, 1199.
193. J. Cao and F. Yan, Recent Progress in Tin-Based Perovskite Solar Cells, *Energy Environ. Sci.*, 2021, **14**, 1286.
194. X. Jiang, Z. Zang, Y. Zhou, H. Li, Q. Wei and Z. Ning, Tin Halide Perovskite Solar Cells : An Emerging Thin-Film Photovoltaic Technology, *Acc. Mater. Res.*, 2021, **2**, 210.
195. H. Zhu, K. Miyata, Y. Fu, J. Wang, P. P. Joshi, D. Niesner, K. W. Williams, S. Jin and X.-Y. Zhu, Screening in Crystalline Liquids Protects Energetic Carriers in Hybrid Perovskites, *Science*, 2016, **353**(6306), 1409.
196. T. Chen, W. Chen, B. J. Foley, J. Lee, J. P. C. Ruff, J. Y. P. Ko, C. M. Brown, L. W. Harriger, D. Zhang, C. Park, M. Yoon, Y. Chang, J. J. Choi and

S.-H. Lee, Origin of Long Lifetime of Band-Edge Charge Carriers in Organic – Inorganic Lead Iodide Perovskites, *Proc. Natl. Acad. Sci. U. S. A.*, 2017, **114**(29), 7519.

197. T. M. Koh, K. Fu, Y. Fang, S. Chen, T. C. Sum, N. Mathews, S. G. Mhaisalkar, P. P. Boix and T. Baikie, Formamidinium-Containing Metal-Halide: An Alternative Material for Near-IR Absorption Perovskite Solar Cells, *J. Phys. Chem. C*, 2014, **118**, 16458.
198. Q. Wang, X. Zheng, Y. Deng, J. Zhao, Q. Wang, X. Zheng, Y. Deng, J. Zhao, Z. Chen and J. Huang, Stabilizing the a -Phase of $CsPbI_3$ Perovskite by Sulfobetaine Zwitterions in One-Step Spin-Coating Films, *Joule*, 2017, **1**(2), 371.
199. J. Xue, J. Lee, Z. Dai, O. Lin, M. S. Goorsky, Y. Yang, J. Xue, J. Lee, Z. Dai, R. Wang, S. Nuryyeva, M. E. Liao, S. Chang, L. Meng, D. Meng, P. Sun, O. Lin and M. S. Goorsky, Surface Ligand Management for Stable $FAPbI_3$ Perovskite Quantum Dot Solar Cells, *Joule*, 2018, **2**(9), 1866.
200. D. N. Johnstone, A. N. Iqbal, D. Guo, K. Frohna, M. Danaie, E. M. Tennyson, S. Macpherson, A. Abfalterer, M. Anaya, Y. Chiang, P. Crout, F. S. Ruggeri, S. M. Collins, C. P. Grey, A. Walsh, P. A. Midgley and S. D. Stranks, Stabilized Tilted-Octahedra Halide Perovskites Inhibit Local Formation of Performance-Limiting Phases, *Science*, 2021, **374**(6675), 1598.
201. X. Zhao, Z. Zhao-karger, M. Fichtner and X. Shen, Halide-Based Materials and Chemistry for Rechargeable Batteries, *Angew. Chem., Int. Ed.*, 2020, **59**, 5902.
202. J. Zhao, L. Liao, F. Shi, T. Lei, G. Chen, A. Pei, J. Sun, K. Yan, G. Zhou, J. Xie, C. Liu, Y. Li, Z. Liang, Z. Bao and Y. Cui, Surface Fluorination of Reactive Battery Anode Materials for Enhanced Stability, *J. Am. Chem. Soc.*, 2017, **139**, 11550.
203. X. Hua, A. S. Eggeman, E. Castillo-martínez, R. Robert, H. S. Geddes, Z. Lu, C. J. Pickard, W. Meng, K. M. Wiaderek, N. Pereira, G. G. Amatucci, P. A. Midgley, K. W. Chapman, U. Steiner, A. L. Goodwin and C. P. Grey, Revisiting Metal Fluorides as Lithium-Ion Battery Cathodes, *Nat. Mater.*, 2021, **20**, 841.
204. X. Hua, R. Robert, L. Du, K. M. Wiaderek, M. Leskes, K. W. Chapman, P. J. Chupas and C. P. Grey, Comprehensive Study of the CuF_2 Conversion Reaction Mechanism in a Lithium Ion Battery, *J. Phys. Chem. C*, 2014, **118**, 15169.
205. C. Wang, J. Liang, J. T. Kim and X. Sun, Prospects of Halide-Based All-Solid-State Batteries : From Material Design to Practical Application, *Sci. Adv.*, 2022, **8**, 1.
206. Q. Liu, Y. Wang, X. Yang, D. Zhou, X. Wang, P. Jaumaux and F. Kang, Rechargeable Anion-Shuttle Batteries for Low-Cost Energy Storage, *Chem*, 2021, 7(8), 1993.
207. J. Chable, B. Dieudonné, M. Body and C. Legein, Fluoride Solid Electrolytes: Investigation of the Tysonite-Type Solid Solutions $La_{1-x}Ba_xF_{3-x}$ ($x < 0.15$), *Dalton Trans.*, 2015, **44**, 19625.

208. L. N. Patro and K. Hariharan, Fast Fluoride Ion Conducting Materials in Solid State Ionics : An Overview, *Solid State Ionics*, 2013, **239**, 41.
209. G. Denes, T. Birchall, M. Sayer and B. MF, $BaSnF_4$ - A new fluoride ionic conductor with the Alpha-$PbSnF_4$ structure, *Solid State Ionics*, 1984, **13**, 213.
210. I. Mohammad, R. Witter, M. Fichtner and M. A. Reddy, Room-Temperature, Rechargeable Solid-State Fluoride-Ion Batteries, *ACS Appl. Energy Mater.*, 2018, **1**, 4766.
211. T. Xia, Y. Li, L. Huang, W. Ji, M. Yang and X. Zhao, Room-Temperature Stable Inorganic Halide Perovskite as Potential Solid Electrolyte for Chloride Ion Batteries, *ACS Appl. Mater. Interfaces*, 2020, **12**, 18634.

CHAPTER 5

Heteroanionic Materials: The Structures and Applications of Mixed Anion Materials

GEOFFREY HYETT

School of Chemistry and Chemical Engineering, University of Southampton, UK
E-mail: g.hyett@soton.ac.uk

5.1 Introduction—What is a HAM?

This chapter will review the area of solid-state materials chemistry that focuses on the synthesis and applications of mixed anion materials. As the name implies, these are compounds that contain at least two different anions, sometimes called *h*etero*a*nionic *m*aterials or HAMs. Researchers are interested in these materials as the use of two or more anions with different chemical properties allows for the formation of novel structures and compositions, expanding the scope of materials space that can be explored and optimised when compared with monoanionic compounds. The success of this approach has been borne out with work on mixed anion materials leading to exciting and unexpected breakthroughs in superconductivity, photocatalysis and other functional properties that we will discuss in the later sections of this chapter.

In theory, a mixed anion material could contain any combination of anions, but in practice most HAM research is conducted on materials with only two different anions, and where one of them is the oxide anion, such as oxysulfides or oxynitrides. Although examples are known containing three anions, or

Inorganic Materials Series
Non-oxide and Heteroanionic Materials
Edited by Duncan H. Gregory
A catalogue record for this book is available from the British Library

materials where the anions do not include oxide, such as nitrofluorides, these remain in the minority, and we shall maintain our focus on the 'oxyanion' materials that predominate the literature. We shall also limit our discussion to materials where there is a stoichiometric amount of each anion present in the lattice, excluding systems where the second anion is present only as a fractional dopant. Finally, the definition of a mixed anion material normally excludes polyanions or oxoanions such as CO_3^{2-} or SO_4^{2-}, as although they contain multiple non-metals, only the oxygen in these molecular ions is co-ordinated to the cations. For example, titanium nitrate would not be considered a mixed anion material, whereas titanium oxynitride would be, as the latter contains metal-oxide and metal-nitride bonds while the former only has metal-oxide bonds. As highlighted by Rondinelli *et al.*,[1] it is useful to introduce some nomenclature from co-ordination chemistry to help describe mixed anion compounds. The addition of a second anion means it is now possible for a given cation in the structure to be co-ordinated to a single type of anion, common in some layered structures, or to a mixture of the anions. In co-ordination chemistry a complex with only one type of ligand is referred to as homoleptic, whereas one with a mixture of ligands is heteroleptic. We can borrow these terms and apply them to mixed anion materials, such that a cation co-ordinated to only one of the available anions can be said to be homoleptic, whereas one bound to a mixture of both would be heteroleptic.

The breadth of research that has been carried out over the last 30 years, means that it is not possible to provide an encyclopaedic coverage of all of the known HAMs, but instead we will focus on a more limited selection—oxynitrides, oxychalcogenides, and oxyarsenides. These have been most frequently investigated and exhibit a range of interesting properties. We intend for this to provide a useful introduction to the reader and offer an overview of the key structural features, synthetic methods and properties of mixed anion materials that will deliver necessary background and context for further exploration of this topic in the scientific literature. There are also a number of excellent reviews on mixed anion compounds that would act as further starting points.[1–7]

Before beginning our discussion of the known mixed anion materials, it is useful to subdivide them into two distinct classes. The first group are the oxynitrides and oxyfluorides, where the oxide ion is paired with another 'light' anion from the first row of the p-block of the Periodic Table, *i.e.*, nitrogen or fluorine. In the second group the oxide ion is paired with a 'heavy' anion from the 2nd, 3rd or 4th row of the p-block. This distinction is useful because the structures adopted and properties are to a first approximation controlled by how similar the heteroanion is to the oxide ion. The anions that are used in the first group (fluoride, nitride and oxide) are all isoelectronic, relatively 'hard' ions, similar in size and with high electronegativity values. This means that the anions can occupy similar crystallographic sites, or even share crystallographic sites with disordered distribution across the lattice. These oxynitrides and oxyfluorides tend to adopt structure types that have already been found from amongst the monoanionic oxide compounds, but with the difference in anion charge accounted for with a change in the charge or oxidation state of the metal ions in the structure. For example, titanium oxyfluoride, $TiOF_2$, adopts the ReO_3 structure which in the pure oxide requires a cation with a formal charge of +6. Substitution of two of the oxide ions for two lower charge fluoride ions allows this structure to be adopted by the titanium ion with a formal +4 charge. The oxide and fluoride ions are randomly distributed across the anion sites, which they can share due to their similar size and chemical properties. The second group of HAMs are those where the oxide anion is paired with a 'heavy' anion, for example sulfide or selenide, or other from the 2nd, 3rd or 4th row of the p-block. These heteroanions are larger in size, and more polarisable and covalent in their bonding than oxide ions, and so the two different anions tend to segregate onto different crystallographic sites, combined with long range ordering of the anions across distinct sites. Therefore, these HAMs often adopt structures not found in monoanionic oxide compositions and often have layered structures with separate oxide and heavy anion layers.

In the first half of the chapter, we will discuss the typical features found in the HAMs containing only light anions with a focus on the oxynitrides, and discuss how they can be synthesised, the structures they adopt, the challenges in their characterisation, and some of their potential applications as pigments and photocatalysts. To represent the group of HAMs containing the oxide ion paired with a heavy anion, in the second half the chapter we will discuss the chemistry of the oxychalcogenides and oxypnictides, as holotypic examples of this class. We will discuss their synthesis and characterisation, the structures adopted, and the properties they display—including transparent conductivity and superconductivity.

5.1.1 Nomenclature and Naming Conventions for Mixed Anion Materials

Before beginning our discussion of the HAMs in detail, it is worth considering some naming conventions. The starting point should always be the official IUPAC guide to the Nomenclature of Inorganic Chemistry (the Red Book) and as directed here we will notate common structure types using italics, such as the *perovskite* structure or CsCl type. The guidance also states that empirical formulae should be listed in order of increasing electronegativity, so that formally an oxysulfide such as lanthanum copper oxysulfide should be given a formula of LaCuSO and be referred to as a sulfide oxide. However, a convention has arisen in the literature for the use of term oxysulfide and similar for other mixed anion materials, and also for formulae that break the conventional ordering to highlight features of the structure. So, the lanthanum copper oxysulfide might be given a formula of LaCuOS to reflect the more common name of the class and highlight that it is a mixed anion material, or it might be formulated as LaOCuS, because, as we shall see later, this material forms layers of lanthanum oxide and copper sulfide, and this technically incorrect formula conveys this, and is thus preferred. For the purposes of this chapter, we will adhere to the latter convention on formulae, although the reader should be aware that both might be found in the literature.

5.2 Oxynitrides: Structures, Synthesis and Properties

As nitrogen and oxygen are adjacent in the Periodic Table their anions are sufficiently similar in size that they can occupy the same crystallographic sites, so that nitride ions (r_N = 1.46 Å) can substitute directly for oxide ions (r_O = 1.36 Å) in many structure types.[8] As they are also isoelectronic the X-ray scattering contrast between the ions is poor and X-ray diffraction methods cannot readily distinguish between the ions. This means that it is actually a significant challenge to determine if a given anion site in a structure is occupied by an oxide or a nitride ion, and therefore impossible to determine if the two different ions are ordered or disordered across the anion sub-lattice. The neutron scattering lengths do provide better contrast, 9.36 fm for nitride and 5.81 fm for oxide, so neutron diffraction studies can be used to probe this question.[9] Despite this similarity in size and electron configuration, there are still significant differences which allow oxynitrides to have distinct properties from either monoanionic oxides or nitrides. Nitrogen is more electropositive than oxygen (3.0 *vs.* 3.4 on the Pauling scale), so that replacement of oxide ions by nitride often leads to an increase in the energy of the valence band maximum and thus a reduction in the band gap. Nitrogen is also more polarisable (1.1 Å^3 *vs.* 0.8 Å^3), which leads to increased covalent interaction in the average metal–anion bond giving larger crystal field splitting in transition metal compounds.[10] The presence of nitride ions also leads to increased local degrees of freedom, and localised asymmetry, even in compounds where there is no long range order. For example, in *perovskite* structured $LaZrO_2N$ the average zirconium environment is a pseudo-octahedral ZrO_4N_2 which locally must be either *cis* or *trans* geometry, even if there is no long-range order. This will reduce the zirconium polyhedra symmetry from O_h to either C_{2v} or D_{4h} in the oxynitride, distorting the local environment due to small differences in bond length, and in the case of the *cis*-C_{2v} configuration introducing a local dipole. The final difference between the oxide ion and nitride ion is, of course, the formal charge. Substitution of nitride into the oxide sublattice requires charge compensation, either though vacancies, change in the oxidation state of the metal, or co-substitution of a higher charge cation. In principle, population of metal interstitial sites could also act as a charge compensation mechanism. All of these will lead to changes in the band structure or position of the fermi level, and hence the properties of the material as we will discuss later.

5.2.1 Synthesis

Some oxynitrides can be made by the direct ceramic route—that is the combination and reaction of oxide and nitride precursors in stoichiometric ratios—for example Nd_2AlO_3N has been reported from the reaction of Nd_2O_3 and AlN at 1350 °C under a nitrogen atmosphere.[11] However, the majority of oxynitrides are not made using this method, and are instead synthesised by nitridation of an oxide precursor.[3] This requires a nitriding agent, the most common of which is ammonia, although other nitrogen sources have been used, such as urea.[12] Elemental nitrogen would be an extremely convenient nitriding agent due to its low cost and ready availability. Unfortunately, N_2 proves to be unsuitable, in the majority of cases due to the strength of the nitrogen–nitrogen triple bond. As a result, the gas is highly inert except at extremely high temperature (which is challenging to reach using conventional furnaces) or in select cases when combined with highly reactive metals (such as those in the s-block to form metal nitrides).

For the most common ammonolysis route, an oxide precursor powder is placed in a crucible inside a tube furnace, and a flow of ammonia is maintained over the sample during its heating, reaction and cooling cycle. The oxide precursor can be synthesised by solid state routes, with a flux mineraliser or *via* sol–gel synthesis and isolated prior to nitridation or generated *in situ* and nitrided in a single step reaction. For example, in the synthesis of $LaTiO_2N$ (eqn (5.1)) crystalline $La_2Ti_2O_7$ is used as a precursor, which is first synthesised by direct 2:1 reaction of TiO_2 and La_2O_3 at 1300 °C and the purity of which can be confirmed by powder X-ray diffraction. The $La_2Ti_2O_7$ can then be reacted with ammonia at 950 °C to yield $LaTiO_2N$.[13] However, attempting to adapt this procedure for the zirconium analogue to synthesise $LaZrO_2N$ from ammonolysis of $La_2Zr_2O_7$ is not successful as the more electropositive zirconium is resistant to nitridation. Instead, a citrate sol–gel route must be used, sometimes called the Pechini or polymerisable complex method, as shown in eqn (5.2). In this method, soluble sources of the metal ions, such as chlorides, nitrides or alkoxides are co-ordinated by citric acid in solution. A glycol source is then added, which can form a condensation polymer gel with the citrate in which the metal ions are randomly distributed.[14] For the $LaZrO_2N$ example La_2O_3 and $ZrO(NO_3)_2$ are dissolved in nitric acid, co-ordinated by citrate, and polymerised with the addition of ethylene glycol. The gel can then be dried and fired at a relatively low temperature in air (600 °C) to burn off the organic components and leave an *amorphous* solid with nominal composition '$La_2Zr_2O_7$'. The amorphous oxide precursor can then be more readily nitrided with a high flow rate of ammonia, due to its higher surface area, and relative thermodynamic instability of the amorphous precursor relative to the crystalline oxide.[13] This potentially allows the Pechini method to utilise lower temperatures and shorter reaction times.

$$La_2O_3 + 2TiO_2 \xrightarrow{1300°C} La_2Ti_2O_7 + 2NH_3 \xrightarrow{950°C} 2LaTiO_2N + 3H_2O \quad (5.1)$$

$$La_2O_3 + 2ZrO(NO_3)_2 \xrightarrow{\text{citrategel}, 600°C, -NO_2, -H_2O, -CO_2} 'La_2Zr_2O_7'_{(am,s)} \xrightarrow{+NH_3, 950°C} 2LaZrO_2N \quad (5.2)$$

$$2BaCO_3 + Nb_2O_5 + 2NH_3 \xrightarrow{1000°C} 2BaNbO_2N + 3H_2O \quad (5.3)$$

$$La_2Ta_2O_8 + 4NH_3 \xrightarrow{CaCl_2 Flux, 1000°C} 2LaTaON_2 + 6H_2O \quad (5.4)$$

Further examples of oxynitride synthesis are given in eqn (5.3) and (5.4). In eqn (5.3) the synthesis of $BaNbO_2N$ is shown where the oxide precursor is produced *in situ* from the simultaneous reaction of $BaCO_3$ and Nb_2O_5 in the presence of ammonia.[15] This is similar to the first example except the oxide intermediate is not isolated and characterised prior to ammonolysis, but is instead formed and reacted directly *in situ*. In the final example the oxide precursor ($La_2Ta_2O_8$) is nitrided with the aid of a flux—a low melting point alkali halide which acts as a mineraliser and allows for an increased reaction rate between the precursor and ammonia gas.[16]

Although there can be challenges in identifying the optimum conditions and precursors, ammonolysis still remains the most common and successful approach for the synthesis of oxynitrides. It is therefore worth considering the mechanism. At elevated temperatures ammonia is in equilibrium with hydrogen and nitrogen, as shown in Figure 5.1, with the dissociation increasing with temperature and almost complete by 350 °C. The mole fraction of ammonia at the temperatures used for ammonolysis (750–1000 °C) is less than 0.02%. Given that molecular nitrogen is an extremely poor nitriding agent due to the strength of its bond, this leads to a conundrum—how can ammonia be used as a nitriding agent at high temperatures, after it is almost completely decomposed to nitrogen? The answer is that the ammonolysis reactions are not carried out at equilibrium, but under dynamic conditions where there is a constant flow of relatively cool gas mixture (NH_3, H_2, N_2) encountering the much hotter surface of the precursor oxide particles.[3] This allows a mix of molecular ammonia and hydrogen to reach the precursor

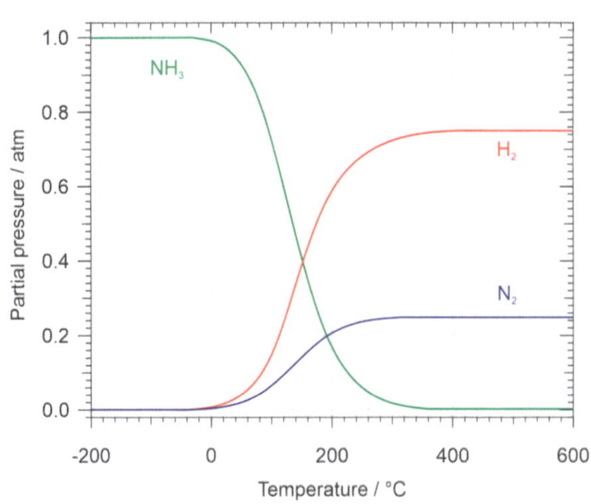

Figure 5.1 Plot showing the decomposition of ammonia into nitrogen and hydrogen as function of temperature at standard pressure, through partial pressure of each gas. Notice that under equilibrium conditions the mole fraction of ammonia begins to rapidly drop above 100 °C, and decomposition into nitrogen and hydrogen is almost entirely complete by 350 °C.

oxide particles, where surface reactions lead to formation of radical species $N^{\cdot}$, $NH^{\cdot}$, and $NH_2^{\cdot}$.[17] It is these radical species that are the active nitriding agents, while surface H_2 extracts oxide ions in the solid to form H_2O, which is also the thermodynamic driving force of the reaction.[18] The high gas flow ensures contact between molecular ammonia and the oxide precursor, and also minimises the partial pressure of H_2O in the reactor gas mix favouring the forward reaction to the solid oxynitride product. Overall, the ammonolysis process occurs at non-equilibrium, and therefore optimum conditions require a balance of both gas flow and reaction temperature, which must often be determined experimentally, as can be highlighted in the challenges of synthesizing tantalum oxynitride. Like many oxynitrides, tantalum oxynitride is widely reported as being synthesised from the direct ammonolysis of the metal oxide, in this case Ta_2O_5.[19] However, the oxynitride is a kinetic product and reaction for longer, or at higher temperatures, leads to full nitridation to the more thermodynamically stable Ta_3N_5.[20,21] In order to selectively form only the partially nitrided product, TaON, it is necessary to introduce a small partial pressure of water into the ammonia stream, by passing the ammonia vapour through a water bubbler. This produces a 'wet' ammonia flow favouring the formation of the oxynitride.[22]

Having considered some of the approaches to the synthesis of oxynitrides, we will now discuss the compositions and structures that oxynitrides are found to adopt. As we shall see, predominantly these phases contain early transition metals. We shall start with the simplest possible oxynitrides, the ternary phases where one metal ion is paired with the anions, before moving onto more complex four and five component compositions which for the oxynitrides are dominated by the *perovskite* structure. In the following discussion a number of element placeholders will be used. M for a general metal ion, A for a group 1, 2 or lanthanide ion, Ln for a lanthanide specifically, B for a transitional metal ion, and X for an unspecified mix of oxide and nitride ions.

5.2.2 Ternary Oxynitride Structures

Arguably the simplest of the ternary oxynitrides are those that adopt the *rock salt* structure.[23–25] These are formed by many of the early transition metals, with examples known for titanium, zirconium, vanadium, niobium and tungsten. As these transition metals all can have variable oxidation states this allows their oxynitrides to form a solid solution between the end member oxide, $M^{II}O$, and nitride, $M^{III}N$, rather than having a single stable composition. Additionally, the *rock salt* oxynitrides, like their end members, are highly defective, with both the cation and anion sites supporting up to ~25% vacancies, and disorder of the oxide and nitride ions across the anion sites. They therefore have a general formula of $M_{(1-\delta)}O_xN_y$ where δ can be as high as 0.25, and the sum of x and y ranges from ¾ to 1. The *rock salt* oxynitrides can be made through ammonia nitridation of oxide precursors, and the resulting oxynitrides are typically metallic, with partially filled d-bands.

In addition to the defect rich *rock salt* oxynitrides, a number of the early transition metals can also form compositionally stable oxynitrides with the metal in its maximum oxidation state. For example, hafnium and zirconium, as metals with stable +4 oxidation states, can form an M_2ON_2 oxynitride, adopting the *bixbyite* structure. The *bixbyite* structure is related to the *calcite* structure, with a distorted cubic close packed array of metal ions, with ¾ of the tetrahedral sites occupied by anions. The tetrahedral site *vacancies* are ordered across a 2 × 2 × 2 supercell, but neutron diffraction studies of Zr_2ON_2 indicate that the oxide and nitride ions are disordered across the occupied anion sites. The Group 5 metals niobium and tantalum can form a stable stoichiometric MON oxynitride, which adopts the monoclinic ZrO_2 structure, or *baddeleyite* structure.[26,27] In contrast to many other oxynitrides, in this structure the oxide and nitride ions are found to be ordered across discrete crystallographic sites. This structure is also based on a distorted *calcite* structure, with nitride ions found in ½ the tetrahedral holes, in alternating planes perpendicular to the [001] direction, with the oxide ions occupying the other planes but in unconventional three co-ordinate trigonal planar sites. Overall, the metal environment is a seven co-ordinate MN_4O_3 polyhedron. The first-row transition metals titanium and vanadium have been found to form defect *pseudobrookite* structures, in addition to their *rock salt* structures.[28,29] In both cases a somewhat unconventional synthesis was used, and the samples are metastable. The approximate compositions are $V_3O_{4.61}N_{0.21}$ and $Ti_{2.9}O_4N$. These structures contain edge sharing MX_6 octahedra, with two anion sites being four co-ordinate, and one anion site three co-ordinate to the metal. Neutron diffraction studies of $Ti_{2.9}O_4N$ indicate that the nitride ions are preferentially found in the four co-ordinate site, as in the *baddeleyite* structure, consistent with Pauling's 2nd rule.

Finally, the p-block metals gallium and aluminium are also found to form ternary oxynitrides.[30,31] These are defect *spinels* with nominal composition of M_3O_3N, which can, unusually, be made from a direct melt of respective oxide and nitride precursors. In these structures the oxide and nitride ions form a disordered cubic close packed lattice, as determined by neutron diffraction, with the M^{3+} ions occupying ½ of the octahedral and ⅛ of all the tetrahedral holes, however as there as twice as many tetrahedral holes as octahedral holes, the aluminium or gallium ions are found to occupy them in a ratio of 1 : 2, *e.g.*, $M^{(tet)}M\text{-}^{(oct)}{}_2O_3N$. This description represents the ideal structure; the majority of real samples contain further disordered vacancies on the tetrahedral sites, requiring an increase in the oxygen to nitrogen ratio to provide charge compensation, such that the composition is closer to $M_{2.8}O_{3.5}N_{0.5}$ in most samples. Aluminium oxynitride *spinel* is a transparent ceramic with isotropic transparency across the UV and near IR and also has good mechanical properties, and has been considered as a material for bullet-proof or blast resistant window applications.[32]

5.2.3 Quaternary Oxynitrides

5.2.3.1 *Oxynitrides Adopting the Perovskite Structure*

The reported quaternary oxynitrides are dominated by materials that adopt the ABX_3 *perovskite* structure, composed of vertex sharing BX_3 octahedra with the A ion in a 12 co-ordinate site located between the BX_3 octahedra. There are at least 30 unique oxynitride *perovskite* compositions that have been identified, with an early transition metal on the B site and with the A site occupied by a Group 2 metal or lanthanide ion.[3] The early transition metals found on the B site include all of the Group 4, 5 and 6 elements, with the exception of chromium and hafnium. The most widely investigated are the oxynitrides containing tantalum, with ten compositions known, and which we will use as a case study to highlight the structural features of the *perovskite* oxynitrides.[16,33–37] Tantalum is found in its maximum oxidation state on the B site in these *perovskites*, with the oxygen to nitrogen ratio determined by the choice of A counter ion. For example, in $ATaO_2N$, where $A^{2+} = Ca^{2+}$, Sr^{2+}, Ba^{2+}, Eu^{2+}, the dication dictates the 2 : 1 ratio of oxygen to nitrogen for charge neutrality, while replacement of the A^{2+} ion with a lanthanide ion (La^{3+}, Nd^{3+}, Gd^{3+}, Ce^{3+}, Pr^{3+}, Sm^{3+} examples are known), leads to the more nitrogen rich $LnTaON_2$ composition. Similar arguments apply to the Nb^{5+} examples, while samples containing a formal +4 ion on the B site, such as Ti^{4+} or Zr^{4+}, are only found with a lanthanide on the A site, $LnBO_2N$, with a more oxygen rich composition for charge balance.

The ideal *perovskite* structure has a cubic $Pm\bar{3}m$ symmetry, which additionally for an oxynitride would also require a random distribution of the oxide and nitride across the anion sites, and this has indeed been identified and confirmed by neutron diffraction for a number of examples, including $BaTaO_2N$, $EuNbO_2N$ and $BaNbO_2N$.[15,33] However, both anion ordering and structural distortions can lower this symmetry. The structural distortions are tilting of the octahedra to help accommodate an ion on the *A* site, which is smaller than the optimum size and were first observed in the oxide *perovskites*. There are a number of possible tilting combinations, leading to different space group symmetries, and these were first characterised and notated by Glazer, who identified 23 possible combinations,[38] which were later reduced to 15 possible combinations that can be observed in real crystals.[39] The glazer notation assigns possible tilts using a three letter code with superscripts. An example would be the $a^0a^0c^-$ tilt. The three letters correspond to the *x*, *y* and *z* axis of the underlying ideal $Pm\bar{3}m$ cubic *perovskite*, and describe the tilt along each of these axes. The superscript symbol indicates the type of tilt: a '+' symbol indicates an in-phase tilt, a tilt in the same direction and the same magnitude in adjacent polyhedra along that axis, while a '−' symbol indicates an anti-tilt, that is successive tilts are equal in magnitude but in opposite directions; finally, a zero ('0') indicates that there is no tilt along that axis. Figure 5.2 shows a schematic diagram of some of the possible tilts. The letters used indicate the respective magnitude of the tilt,

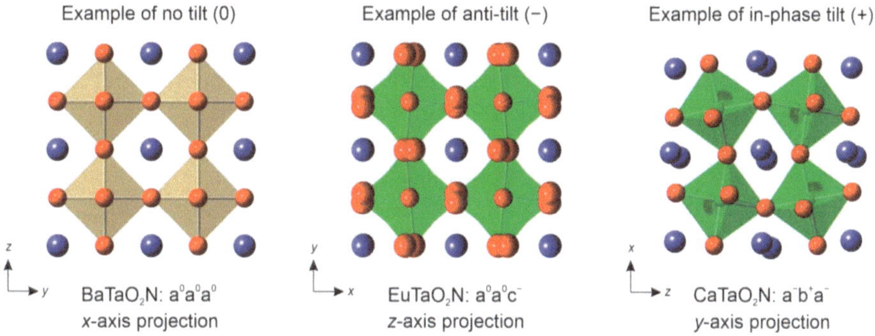

Figure 5.2 Tilting of octahedra in *perovskite* oxynitrides: The Glazer notation can be used to describe the tilting along each of the underlying perovskite axes, including an in-phase tilt (+), out of phase tilt (−) or no tilt (0). Examples of each are shown using the structures of $BaTaO_2N$, $EuTaO_2N$ and $CaTaO_2N$ structures, which are displayed with projections down appropriate axes to highlight the tilts.

with repeated letters indicating a tilt of equal magnitude. For example, in the tilting system $a^+a^+a^+$ there are in-phase rotations around the x, y and z axis, and this rotation is of the same magnitude along each axis. In contrast the $a^-b^+a^-$ would have an anti-tilt of the same magnitude on the x and z axis, with an in-phase tilt of different size on the y axis.

Despite there being 15 possible combinations, only four are commonly observed among the *perovskite* oxynitrides, which are $a^0a^0a^0$, $a^0a^0c^-$, $a^-b^0a^-$ and $a^-b^+a^-$, in order of increasing tilting. The first represents the ideal cubic system, and in the remaining three ($a^0a^0c^-$, $a^-b^0a^-$ and $a^-b^+a^-$) each has a 'unique axis'. For example, the in-phase tilt in $a^-b^+a^-$, is unique and aligned along the y axis, whereas an out of phase tilt is observed parallel to x and z. It should be noted that the tilt axes are directed along pseudo-Cartesian axes aligned along the connectivity of the octahedra, which may differ from the actual crystallographic axes of the unit cell due to reduction in symmetry. In systems with no anion ordering, or where the structure is determined from XRD data such that the oxide and nitride anions cannot be distinguished, these tilts correspond to space group symmetries of $Pm\bar{3}m$, $I4/mcm$, $Imma$ and $Pnma$ in order of increasing tilting.

If we consider the known tantalum containing *perovskites* that have been investigated by X-ray diffraction, where anion ordering cannot be ascertained and the oxide and nitride ions must be assumed to be disordered, $BaTaO_2N$ is found to adopt the ideal cubic structure, whereas $SrTaO_2N$ and $EuTaO_2N$ adopt distorted structures. This is because there is a good match for the barium ion radius of 1.61 Å on the larger *A* site with the much smaller tantalum ion (r = 0.64 Å) on the B site. Replacement of barium with the smaller Eu^{2+} or Sr^{2+} ions (both approximately 1.35 Å radius) leads to the structures adopting $I4/mcm$ symmetry with one tilt axis in order to reduce the size of the A site to better fit these ions, while the even smaller calcium ion in $CaTaO_2N$

forces the structure to adopt *Pnma* with three tilt axes.[40,41] A similar trend is observed in the $LnTaON_2$ compositions containing Ln^{3+} lanthanide ions. The largest of these, La^{3+} adopts the *Imma* double tilt, whereas the increasingly smaller lanthanides, Ce–Gd all adopt the three tilt *Pnma* structure.[34] So far we have assumed that long range anion ordering is absent, but it may well be present within a material, and it can be determined by neutron diffraction. If it is present in an $ATaO_2N$ sample, then the local octahedral geometry of each tantalum ion must be either *cis*- or *trans*-TaO_4N_2. (In the other common composition, $LnTaON_2$, the local geometry is either *cis*- or *trans*-TaO_2N_4 which is symmetrically equivalent, so will be ignored for now.)

Using $ATaO_2N$ as an example in the absence of tilting, ordering of *trans*-TaO_4N_2 octahedra leads to *P4/mmm* symmetry, with the linear N–Ta–N connectivity of the octahedra creating the unique tetragonal axis. For the local *cis* geometry long range order can only be created by arranging the *cis* nitrogen pairs into zig-zig chains aligned along the 110 plane of the original cell, which reduces the symmetry to orthorhombic *Pmna*. When one of the observed tilt systems is present, each with a unique *tilt axis*, there are two possible symmetries, either the unique tilt axis is aligned parallel to the direction of the long range *anion ordering* or perpendicular to it.[36] For example, with the $a^-b^+a^-$ system the unique tilt axis is the axis with the in-phase tilt, and with *trans*-anion ordering the ordering axis can be either parallel to the in-phase tilt axis, giving *Pnma* symmetry, or perpendicular to the tilt axis, giving $P2_1/m$ symmetry. This is shown Figure 5.3, along with a summary of all the other possible cell symmetries for combinations of tilting and anion ordering.

This analysis of the possible tilt and anion ordering combinations is useful in providing the theoretical framework to the experimentalist for understanding the observed diffraction data and providing a series of starting models for Rietveld refinement against X-ray or neutron diffraction data. However, the differences in the diffraction patterns resulting from these models can be challenging to identify in real data. For example, one of the first reports on $LaTaO_2N$ found that it had $a^-b^0a^-$ tilting, combined with a *trans* arrangement of the TaO_4N_2 octahedra and the long range anion order axis aligned perpendicular to the unique zero tilt axis, giving a *C2/m* symmetry.[16] However, a more recent study confirmed the tilting order, but comparing the long range anion ordered and disordered structural models identified that a better fit to the data could be found assuming the randomised *Imma* model. The difference was small, a fit parameter of R_{wp} = 2.54% for the disordered model and 3.15% for the ordered.[36] These are small margins on which to determine the correct structure, but in this case was confirmed by the presence of a few weak reflections in the neutron diffraction pattern that were only accounted for by the disordered model.

In the most recent literature a consensus is emerging that oxynitride *perovskites* tend to long range disorder and that *trans* ordering of the anions is particular unfavourable.[42,43] Theoretical arguments based on covalent bonding interactions show that the preferred local configuration of the anions around the tantalum ion is to have the stronger Ta–N bonds *cis* to each other,

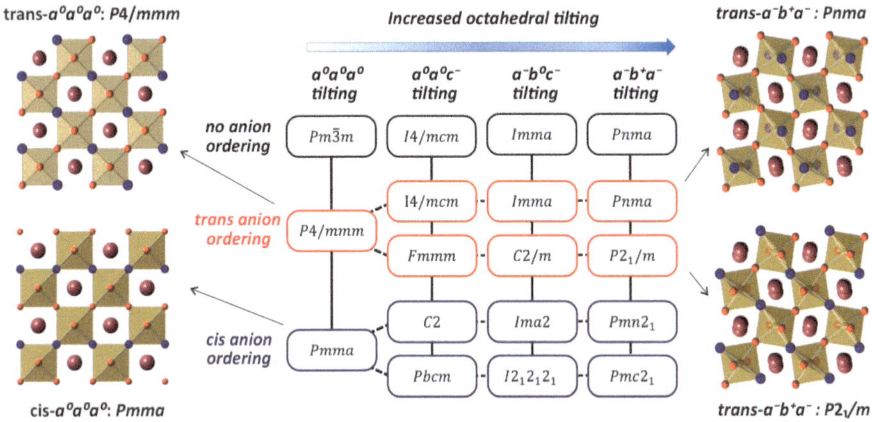

Figure 5.3 Relationship between Glazer tilting and anion ordering and the effect on cell symmetry. Cell diagrams show some possible arrangments for ABO_2N structures with blue nitride ions shown slightly larger than the red oxide ions to demonstrate the *cis* or *trans* relationship within each octhahedra, and how these can be aligned for long range order. Strucutre diagrams on the left show the possible *trans* and *cis* long-range order without tilting. The diagrams on the right show the $a^-b^+a^-$ tilting system with the two possible *trans* long-range ordering alignments that result in *Pnma* and $P2_1/m$ symmetry, respectively. The central diagram shows the space group symmetry for each of the observed tilting systems for *perovskite* oxynitrides, combined with each of the possible *cis* or *trans* long-range orders. Adapted from ref. 36 with permission from American Chemical Society, Copyright 2014.

giving *cis*-TaO_4N_2 octahedra, with *trans* N–Ta–N linkages between octahedra. This combination of localised order restrictions leads to zig-zig chains of the tantalum to nitride bonds throughout the lattice, but which tend to disorder on a longer scale as there are multiple choices for the 90° turn at each tantalum ion in the zig-zag chain, only one of which leads to long range order. It is possible to statistically model these chains using a self-avoiding walk and is similar to the models used to align the chains of linear polymers. The result is either fully random or only partial order of the anions at the scale length investigated by neutron diffraction experiments. However, this shorter-range order can be observed and confirmed in electron diffraction experiments.

In addition to the *perovskite* oxynitrides there are a limited number of structures that adopt the related *Ruddlesden–Popper* (RP) structures. These are layered structures, with *rock salt* layers separating *perovskite* blocks, with general formula $A_{n+1}M_nO_{3n+1}$ where n represents the thickness of the *perovskite* layer. Structures are known for Sr_2NbO_3N, Sr_2TaO_3N, and Ba_2TaO_3N which have the $n = 1$ RP structure,[44,45] and $Eu_3Ta_2N_4O_3$, which is an $n = 2$ structure.[46] The RP phases are much less common than the *perovskites* as the layered structures cannot adopt the tilt distortions necessary to accommodate the smaller A ions, as the *perovskite* layers need to align to the symmetry of the *rock salt* layer. Therefore, examples are only known for the larger barium, strontium and europium ions which are best matched in size to the typical B ions.

5.2.3.2 Other Quaternary Oxynitride Structure Types: Pyrochlore and Wurtzite

Although the *perovskite* and *perovskite*-related structures are the most common quaternary oxynitrides, a small number adopt the *pyrochlore or wurtzite* structures, which will be summarised here. The *pyrochlore* structure is adopted by some molybdenum and tantalum containing compounds: $Sm_2Ta_2O_5N_2$, $Sm_2Mo_2O_{3.8}N_{3.2}$, $Y_2Mo_2O_{4.5}N_{2.5}$ and $Eu_2Ta_2O_6N$.[47–50] The *pyrochlore* structure has a large cubic unit cell and can be seen as vertex sharing MX_6 octahedra, where M = Mo or Ta, with an interpenetrating lattice of vertex sharing XLn_4 tetrahedra. There is a preference for oxide ions to occupy the anion sites of the MX_6 octahedra, and the nitride ions to be four coordinate to the lanthanide, but there is some site mixing and disorder in most of the observed compositions. The samples are synthesised using the ammonolysis of an oxide precursor route. The hexagonal *wurtzite* structure is a relatively simple structure composed of a hexagonal close packed lattice of anions with cations in half the tetrahedral holes. Oxynitrides examples include $Zn_{1.2}Ge_{0.7}O_{0.8}N_{1.2}$, $Li_{0.8}Ga_{1.2}N_{0.4}O_{1.6}$ and $Ga_{0.34}Zn_{0.66}O_{0.66}N_{0.33}$. These are all synthesised *via* ammonolysis of oxide precursors, with these oxides made using solid state or citrate sol–gel routes. Where conducted, neutron powder diffraction data suggests that both cation and anion sites are fully disordered.[51–53]

5.2.4 Quintenary Oxynitrides

The majority of published quintenary oxynitrides adopt the *perovskite* structure, with the additional element incorporated by partial substitution of an extra metal ion onto the A site, for example $Ca_{0.5}Sr_{0.5}WO_2N$, or B site as in $CaNb_{0.3}Ta_{0.7}O_2N$. These phases can be considered as solid solutions of the oxynitride *perovskites* discussed earlier, and there is no evidence that the additional metals are ordered throughout the lattice. However, at the time of writing there no examples to our knowledge of what might be considered 'true' five component oxynitrides, with three distinct metal environments—of the kind that are known for the oxychalcogenides discussed in the second half of the chapter. It is inevitable that future research will identify these, and it will be interesting to see what structure types these are found to adopt.

From this structural overview it is clear that the majority of oxynitrides adopt a variant of a known oxide structure, *rock salt*, *bixbyite*, *baddeleyite*, *spinel*, *pyrochlore*, *wurtzite*, and in particular *perovskite,* with the oxide and nitride ions typically disordered across the anion sites. This was what would be expected and can be rationalised based on the similarity of the size, chemistry and other properties of the two anions. In the next section we will consider some of the functional features of the oxynitride materials which arise from presence of both oxide and nitride ions in their structures, and how the differences between them, which are relatively small in structural terms, can actually have a large effect on the electronic and bonding properties which are then reflected in the unique features and applications of oxynitrides.

5.2.5 Properties and Applications of Oxynitrides

5.2.5.1 Pigments

As colourants, inorganic pigments have several advantages over conventional organic dyes, for example, they are more likely to be resistant to fading when exposed to light, and they can also be used in applications such as glasses and ceramic glazes, where there is a need for stability at high temperatures. An important type of inorganic pigment are the historic cadmium pigments, such as cadmium red and cadmium yellow, which are based on the solid solution of CdS and CdSe, and are thus sometimes referred to as cadmium selenosulfides.[54] The end members have band gaps of 2.42 eV and 1.75 eV, respectively, with the solid solution allowing for precise control of the band gap across this range, corresponding to colours of yellow to orange to red.[55] However there has been concern about the potential toxicity due to the cadmium found in these pigments.[56]

Many oxynitrides also have band gaps of between 1.8 eV and 3.0 eV, meaning they can absorb photons falling in the visible range of the electromagnetic spectrum, which contrasts with the much larger or smaller band gaps observed in equivalent pure oxides and nitrides. For example, tantalum oxide has a band gap of 3.9 eV, tantalum nitride a band gap of 2.1 eV, with tantalum oxynitride falling between these at 2.4 eV.[57] The reason for this is that in pure oxides the highly electronegative oxygen forms a deeply buried valence band, but substitution of the more electropositive nitrogen raises the energy of the valence band reducing the band gap and allowing visible light absorption. Due to this, oxynitrides have been proposed as non-toxic replacement pigments, with Jansen and Letschert identifying that a solid solution of tantalum containing *perovskite* oxynitrides can be used to cover the entire spectral range of the cadmium selenosulfates.[58] The end members of this solution, $CaTaO_2N$ and $LaTaON_2$ are yellow and red in colour, with band gaps of 1.75 eV and 2.00 eV, covering almost the same range as the cadmium pigments, and with a similar spectral profile (see Figure 5.4 for a comparison of cadmium red and $LaTaON_2$). The oxynitrides $CaTaO_2N$ and $LaTaON_2$ can form a fully miscible solid solution, $Ca_{1-x}La_xTaO_{2-x}N_{1+x}$, with compositional variation leading to a linear change in the band gap, allowing for materials with equivalent colour to all the most common cadmium selenosulfide pigments, and with equivalent or better thermal stability. The similar size of the La^{3+} (1.35 Å) and Ca^{2+} ions (1.34 Å) explains the stability of the solid solution

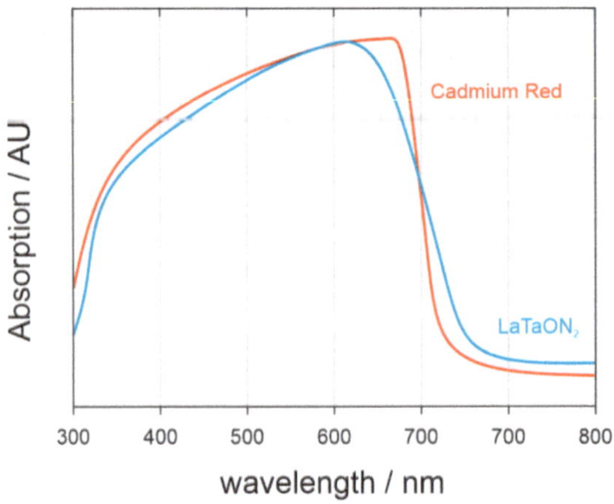

Figure 5.4 Comparison of the visible absorption spectra of cadmium red and the non-toxic oxynitride alternative, $LaTaON_2$. For contrast in the diagram the spectrum of 'Cadmium Red' is shown in red, while that of $LaTaON_2$ is shown in blue, however do not be confused! Both display an almost identical red colour as pigments, and as can be seen the oxynitride very closely mimics the spectral features of the older cadmium red pigment. Adapted from ref. 58 with permission from Springer Nature, Copyright 2000.

between the two end members, with both adopting distorted *perovskite* structures and only a 7% expansion across the series, principally due to increasing fraction of the larger nitride ion. The band gaps of the members in the solid solution follow the expected trend of decreasing, *e.g.*, with increased mixing of the more electropositive nitrogen 2s/2p states into the conduction band.

Similar colour trends covering the 'earthy' red-orange-brown tones were observed in the tantalum and niobium *perovskites* ABO_2N (A = Ca^{2+}, Sr^{2+}, Ba^{2+}; and B = Nb^{5+}, Ta^{5+}) that have also been considered as pigments,[59] although empirically the reflectivity of the tantalum compounds was found to be higher than the niobium containing compounds, making them brighter and more suitable for use. In this series of compounds, the oxide ion to nitride ion ratio remains fixed, and instead the compounds were found to have band gaps and colours dependant on the size of the alkali metal ion. For the tantalum compounds the colours were red with barium (1.8 eV), orange with strontium (2.1 eV) and yellow with calcium (2.4 eV), while for niobium the colours were black (1.7 eV), brown (1.9 eV) and orange (2.0 eV) with barium, strontium and calcium respectively, as found in Table 5.1. In both series a clear increase in band gap is observed with decreasing size of the A ion, related to increased structural titling. In $BaTaO_2N$ the size of the A ion is well matched, and the compound adopts the ideal perovskite with a primitive cubic cell. Replacement of barium with the smaller alkaline earth metals distorts the structure to tetragonal *I4/mcm* with strontium and then orthorhombic *Pbnm* with calcium. The local tantalum oxynitride octahedra remain

Table 5.1 Colours and band gap of the perovskite oxynitride compounds, incldu-ing quantificaiton of the colour using the Lab notation system.

Material	Colour	Band gap/eV	*L* a* b**	Swatch
$CaTaO_2N$	Yellow	2.4	78, −7, 84	
$Ca_{0.5}La_{0.5}O_{1.5}N_{1.5}$	Orange	2.2	52, 29, 74	
$LaTaON_2$	Red	2.0	26, 37, 30	
$SrTaO_2N$	Orange	2.1	68, 44, 74	
$BaTaO_2N$	Red	1.8	54, 80, 69	
$SrNbO_2N$	Brown	1.9	35, 36, 19	
$BaNbO_2N$	Black	1.7	0, 0, 0	

relatively unchanged with similar tantalum oxide and tantalum nitride bond lengths, but with the tilting distortions changing the relative alignment of the octahedra to each other. In the cubic $BaTaO_2N$ structure the Ta–O–Ta bond angle between octahedra is of course 180°, but with tilting this reduces to 169.9° and then 153.3° for $SrTaO_2N$ and $CaTaO_2N$, respectively. This decrease in inter-octahedra bond angle leads to poorer overlap of the Ta 5d states that make up the conduction band, leading to a narrower band width and hence a larger band gap. A similar trend is observed in the niobium containing series, but with a smaller band gap in each case due to the presence of the more electronegative niobium creating a lower average energy for the position of the conduction band minimum.

5.2.5.2 *Visible Light Photocatalysis*

The observation of strong visible light absorption in oxynitrides, which prompted their consideration as pigments, has subsequently led to their investigation in another class of functional material where the ability to absorb light is key—as photocatalysts. A photocatalyst is a semiconductor that can be used to catalyse surface reactions through the absorption of ultra-band gap photons, resulting in excited electron–hole pairs that then migrate to the particle surface where they can initialise or catalyse redox reactions (Figure 5.5). The classic example is ‘self-cleaning’ titanium dioxide,[60] which when exposed to light will degrade any surface organic pollutant such as grease. The *anatase* polymorph of titanium dioxide has a band gap of 3.2 eV and so can absorb ultra-violet photons, generating highly energetic electrons and holes in its bands. The majority of these electrons and holes recombine, but a minority will migrate to the particle surface. Here the holes can directly oxidise surface organic pollutants, while the electrons can activate atmospheric oxygen by reducing it to the highly reactive superoxide radical, $O_2^{\cdot -}$, which can further oxidise surface organics. The evidence for

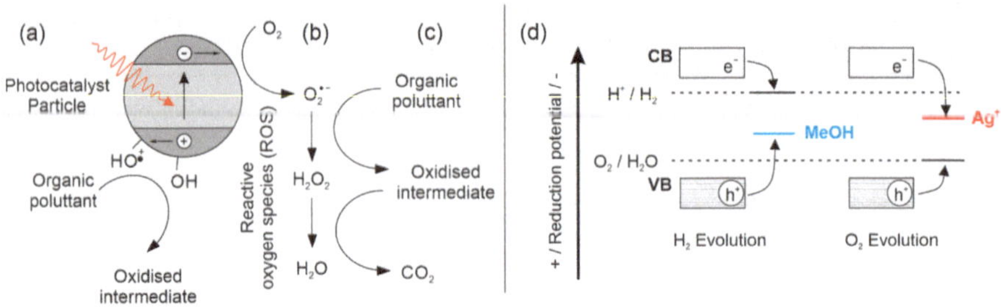

Figure 5.5 On the left is shown a schematic diagram of the process of photo-oxidation of organic pollutant on the surface of a photocatalyst particle, with CO_2 and H_2O as the ultimate products. The key steps to photo-oxidation are: (a) absorption of a photon to generate electron–hole pairs, and migration of holes to form surface hydroxyl radicals which can begin oxidation of a pollutant to radical intermediates; (b) migration of electrons to the surface to reduce atmospheric oxygen to reactive oxygen species, which can then; (c) oxidise the target pollutant. On the right-hand side of the figure: (d) an energy level diagram of the reduction potential of hydrogen and oxygen for water splitting (dotted lines), showing position of the band edges of the photocatalyst, which must straddle the two reduction potentials to have sufficient overpotential to carry out the half-reactions. Also shown are the reduction potentials of the sacrificial reagents, methanol and silver ions, showing that respective oxidation and reduction of these are easier to achieve for a given photocatalyst. Adapted from ref. 64 with permission from American Chemical Society, Copyright 2007.

this is provided by quenching experiments which show that photocatalytic activity is reduced in the presence of hole and superoxide scavengers.[61,62] The overall reaction is oxidation of surface organic contaminants to water and carbon dioxide with any heteroatoms converting into water soluble mineral acids. These atmospheric oxidations are thermodynamically favourable reactions, spontaneous, but which do not occur at room temperature due to the large kinetic activation barrier of non-labile triplet oxygen. The photocatalyst mediates this by using surface redox reactions as lower energy pathways, and providing additional energy from the photons to overcome the energy barrier. Titanium dioxide has also been shown to be able to destroy bacteria through this mechanism, where the photocatalytic reaction decomposes the lipids in the cell membrane leading to cell death, so that self-cleaning photocatalysts are also anti-microbial.[63]

There is also significant interest in the ability of titanium dioxide and other photocatalysts to drive the 'water splitting' reaction—the simultaneous oxidation and reduction of water to form oxygen and hydrogen.[65] Unlike the self-cleaning reactions this is a non-spontaneous process with a large positive Gibbs free energy ($\Delta G_r^\circ = +238$ kJ mol^{-1}) that requires energy input from a light source, mediated *via* the photocatalyst, not only to overcome the kinetic barrier but also to provide the energy to drive the reaction. In the case of titanium dioxide, the reaction also requires the presence of a platinum co-catalyst to promote the formation of hydrogen. Overall, water splitting photocatalysis allows for the energy available in a light source to be converted into stored chemical energy in the form of hydrogen gas. When water splitting was first identified it was hoped that it would allow sunlight to generate hydrogen as a sustainable source of 'solar fuel' or artificial photosynthesis. However, as titanium dioxide must be activated by UV light, and this is only a small fraction of solar energy available, even with ideal efficiency titania and other UV activated photocatalysts are not practical catalysts for solar fuel generation.

It is clear that a photocatalyst with a smaller band gap that could be activated by visible light—a so-called visible light photocatalyst—would have a significant advantage for water splitting, being able to use more of the energy available in sunlight. A visible light photocatalyst is also essential for self-cleaning applications if the intended application is in enclosed or indoor environments where there is little or no UV light. There is however a trade-off: smaller band gap materials are indeed able to use more of the photons in the visible spectrum, but this is balanced against a loss of overpotential to drive the desired reactions.[66] In other words, the photocatalyst needs to retain a sufficiently large band gap to have a conduction band with enough negative potential to generate superoxide radicals or reduce water to hydrogen, and a sufficiently positive conduction band to oxidise organics or generate oxygen from water. In practice this is equivalent to a band gap of at least 1.9 eV. As the band gap range of oxynitride materials is from approximately 1.8 eV to 3.0 eV this makes them suitable candidates for visible light photocatalysis.

A common approach in the literature to carry out initial assessment of possible water splitting photocatalysts is to make use of sacrificial reagents, which can be used to apply a chemical bias and test effectiveness of the catalyst at each half reaction (either oxygen formation or hydrogen formation) with a lower overpotential than required for complete water splitting. These tests are useful for gatekeeping—they are easier to conduct and materials which cannot evolve hydrogen or oxygen with sacrificial reagents will not be able to achieve full water splitting. Typical reagents include silver ions which can be used as sacrificial electron acceptor to assess oxygen evolution, while methanol can be used as a sacrificial electron donor to assess hydrogen production, as shown in eqn (5.5) and (5.6).

$$2Ag^{+} + H_2O \xrightarrow{h\nu \text{ photocatalyst}} 2Ag + 2H^{+} + \tfrac{1}{2}O_2 \tag{5.5}$$

$$CH_3OH \xrightarrow{h\nu \text{ photocatalyst}} CH_2O + H_2 \tag{5.6}$$

One of the first oxynitrides assessed as a visible light photocatalyst was tantalum oxynitride, with a band gap of 2.4 eV. Initial experiments focused on the possibility of using the catalyst for the water splitting reaction. Tantalum oxynitride, TaON, was found to be effective at evolving oxygen and hydrogen under visible light irradiation with sacrificial reagents. Oxygen evolution occurred with a quantum yield of 10%,[67] but although hydrogen evolution was observed in the presence of methanol this also required surface platinum loading as a co-catalyst, and the rate was lower. Tantalum oxynitride was also investigated for its potential as a self-cleaning coating, and it was found to destroy greasy organic test contaminants when activated by visible light. The rates were found to be reasonable under bright sunlight conditions, with a layer of contaminant equivalent to a fingerprint being completely removed in 24 h, but too low to be effective under indoor lighting alone, which is orders of magnitude less intense than sunlight.[68] Visible light activated photocatalytic activity has also been observed in *spinel* structured Ga_3O_4N,[69] and *wurtzite* structured gallium zinc oxynitride,[70] however the best performing oxynitride materials have been found from amongst those adopting the *perovskite* structure.

The structural and compositional diversity present in the *perovskite* oxynitrides has provided a greater range of materials to select from for water splitting photocatalysis, and also the opportunity to tune properties through doping and the formation of solid solutions. Much of the initial work was carried out by the Domen research group, using a standard set of conditions with a filtered Xe lamp as the light source. Sacrificial reagents were used to assess hydrogen and oxygen evolution separately, with a fixed quantity of catalyst and the use of a platinum co-catalyst for the hydrogen tests. This approach was used to screen the majority of the tantalum, niobium and titanium containing quaternary oxynitrides disccussed above,[64] and found that $CaTaO_2N$, $SrTaO_2N$, $BaTaO_2N$, and $LaTaON_2$, were all able to evolve hydrogen, but not oxygen.[71] $LaNbON_2$ and $CaNbO_2N$ were able to evolve oxygen but only very small amounts of hydrogen,[72,73] while $LaTiO_2N$ was able to generate moderates amount of both hydrogen and oxygen.[74] In all cases small amounts of N_2 were also detected, indicating partial photo-oxidation of the oxynitride.

Following on from the tests using sacrificial reagents, it was shown that overall water splitting without these reagents can be achieved with an oxynitride photocatalyst, as long as appropriate co-catalysts are used. $CaTaO_2N$ loaded with Pt as a hydrogen catalyst and $RhCrO_x$ as an oxygen catalyst was able to generate of both oxygen and hydrogen from pure water, although nitrogen evolution was still observed, and the overall quantum efficiency for hydrogen formation was extremely small, around 0.0003%. Improved performance was found by making use of quintenary *perovskite* solid solutions as they have been found to be effective for enhancing the photocatalytic activity of the tantalum containing samples through band engineering. The solid-solution of $BaTa_{1-x}Zr_xO_{2+x}N_{1-x}$, formed by partial substitution of $BaZrO_3$ into $BaTaO_2N$, allows for an opening up of the band gap by simultaneously shifting the position of both valence and conduction band to increase the overpotential for the redox reactions, although with a penalty of reduced spectral absorption range. An optimum was found at 5% doping, which led to a six-fold enhancement in hydrogen evolution.[76] Changes to band structure were also observed with the solid solution of $LaTaO_2N–SrTiO_3$. In this case, substitution of $SrTiO_3$ into $LaTaO_2N$ maintains the band gap of the oxynitride, by simultaneously lowering the relative energy of both the conduction band (with exchange of tantalum for titanium) and the valence band (with exchange of nitrogen for oxygen), as seen in Figure 5.6. This increases the overpotential for oxygen evolution, but reduces the overpotential for hydrogen formation.[75] This was seen as an acceptable compromise, as undoped $LaTaO_2N$ is active for hydrogen evolution only, while with $SrTiO_3$ doping the solid solution can evolve oxygen with increasing rates as the strontium and titanium fractions increase, although the hydrogen evolution rate decreases. In the solid solution, substitution of between 40% and 50% provides a comprise that allows for reasonable rates of both hydrogen and oxygen evolution.

The most successful water splitting photocatalytic system identified so far is based on $LaTaO_2N$ with partial substitution of tantalum for magnesium. Replacement of Ta^{5+} with the much lower charge Mg^{2+} in $LaMg_xTa_{1-x}O_{1+3x}N_{2-3x}$ leads to a significant increase in the oxygen to nitrogen ratio, shifting the position of the valence band to increase the overpotential for oxygen evolution, without sacrificing hydrogen evolution activity.[77] To achieve overall water splitting this system also made use of a $RhCrO_x$ hydrogen co-catalyst, and an amorphous titanium oxide coating layer. The titanium oxide coating layer suppresses the loss of nitrogen from the photocatalyst surface, and so stabilises the catalyst from corrosive photooxidation, while still allowing the transport of H_2O, O_2 and H_2 to and from the catalyst and co-catalyst surface.[78] The optimum composition was found to be $x = 0.33$, $LaMg_{0.33}Ta_{0.67}O_2N$, where overall water splitting from pure water with a quantum efficiency of 0.3% was

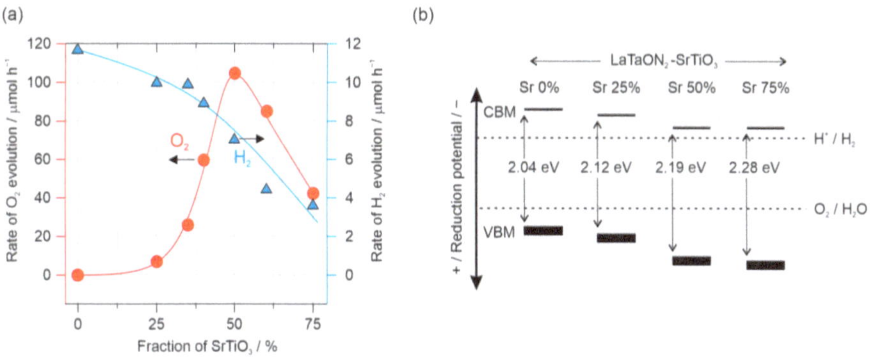

Figure 5.6 The solid solution $LaTaON_2$–$SrTiO_3$ was assessed for water splitting, using sacrificial reagents. As shown in (a) $LaTaON_2$ alone (0% fraction of $SrTiO_3$) is capable of generating hydrogen (shown by blue triangles), but does not allow for oxygen evolution (red circles). Increased mixing of $SrTiO_3$ into the solution decreases the hydrogen evolution rate, but allows for oxygen evolution up to a peak at 50% doping. Analysis of the effect of substitution on the band edges in (b) shows that both CBM and VBM become more positive with increasing $SrTiO_3$, giving improved overpotential for oxygen evolution while retaining a relatively stable bad gap size, at the cost of the loss of some hydrogen evolution overpotential. This effect peaks at 50% substitution. Adapted from ref. 75 with permission from the Royal Society of Chemistry.

observed. Although a low overall efficiency, it is still a significant improvement on previous results, and without the need for sacrificial reagents. Water splitting photocatalysis is still very much an active research field, with the most recent papers showing the value of more complex optimised composition from solid solutions, but also the importance of co-catalysts and protective layers in enhancing overall efficiency.

5.3 Oxychalcogenides and Oxypnictides: The Heavy Layer HAMs

In the previous section we reviewed the structures and application of the oxynitrides, to represent the class of HAMs where the oxide ion is paired with another 'light' anion from the first row of the p-block. In this section we will look at the second key type of HAM, those where the oxide ion is paired with what we might term a *heavy anion*, one found in the 2nd or 3rd row of the p block, that is any anion except for nitride, fluoride or carbide. These materials include oxychalcogenides (containing oxide alongside sulfide or selenide), oxypnictides (paired with phosphorus or arsenic) and oxyhalides (paired with chloride or bromide). Mixed anion materials where the oxide ion is found alongside a 4th row anion, such as telluride or iodide, are possible, but less common. To highlight this type of HAM we shall focus our discussion on the oxychalcogenide and oxypnictide materials as these are the most investigated of the heavy anion mixed anion materials.

When oxygen is paired with a heavy anion there is a large discrepancy in size in contrast to what we observed in the oxynitrides. For example, the six coordinate ionic radius of the oxide ion is 1.4 Å, identical to the nitride ion, whereas the heavy anions in the same coordination vary in radius from 1.8 Å to 2.2 Å. This much larger difference in size means that the two anions almost never occupy the same crystallographic sites, and there is often also a segregation of the cations due to differences in the bonding characteristics of the anions. For example, differences in polarisability cause harder cations to preferentially bond to the oxide anion, while the softer cations preferentially bond to the more polarisable heavier anion. As we shall see when we discuss the structures adopted by these compounds, this often results in layered structures, with homoleptic metal ion environments, and with structure types not adopted by homoanionic oxides. Instead, we shall see new archetypes unique to HAMs. In the final section, we will discuss how these new structure types and the mixing of anions with more distinct chemistries leads to the properties of p-type transparency and superconductivity observed in the heavy anion containing HAMs.

5.3.1 Synthesis

The most common route for the synthesis of oxysulfides and oxypnictides is a modification of the standard ceramic synthesis used for oxides. In this established approach for the synthesis of complex metal oxides, precursor binary oxides, carbonates or elements are combined in appropriate proportions, and ground together using a pestle and mortar to provide an intimate mixture of the particles. These are then heated in an alumina crucible in air to facilitate a solid–solid reaction where ion diffusion between the particles leads to formation of the desired phase—a temperature of approximately two-thirds of the melting point of the material is required in order to have sufficient ion diffusion. An example for the formation of Sr_2RuO_4, is given in eqn (5.7).[79] To achieve a complete reaction intermittent regrinding and heating may be necessary, and after the first step the mixture might also be pressed into a pellet to provide improved contact between the particles after any evolution of gas has occurred.

$$2SrCO_3 + Ru + O_2 \xrightarrow[1200°C]{} Sr_2RuO_4 + 2CO_2 \tag{5.7}$$

The example oxide synthesis of Sr_2RuO_4 relies on exposure of the sample to atmosphere, both to provide oxygen as a reagent, and also to allow decomposition of strontium carbonate with the loss of carbon dioxide. For HAMs this approach must be modified to exclude air, as although the compounds are typically air stable at room temperature, at the elevated temperatures required for the solid-state reaction the samples are liable to oxidation—for example, materials containing the sulfide ion would oxidise to sulfate. To avoid this in HAM synthesis the ground and pelletised samples are placed inside an alumina crucible, which is then melt sealed in a silica ampoule under vacuum. Once sealed the sample is protected from atmosphere, and the ampoule can be placed in a furnace to react the pelletised reagents at temperatures between 700 °C and 1200 °C. Additional precautions can include drying the silica tube prior to sealing by heating under vacuum,[80] and due to the possible air sensitivity of compounds, carrying out all manipulation of precursors in an inert gas filled glove box. The need to seal the material in silica prevents the use of any reagent which is intended to decompose with evolution of gas, such as carbonates and nitrates used in oxide synthesis, as these could cause the ampoules to crack or explode, a potential hazard, and also exposing the samples to air. This is unfortunate, as carbonates and nitrates are often available at low cost in very high purity. Examples of how elements or binary compounds can be combined to form more complex quaternary or quintenary HAM compounds using this adapted ceramic synthesis are given in eqn (5.8)–(5.10).[81–83]

$$2BaO + Co + 2Ag + 2Se \xrightarrow[825°C]{} Ba_2CoO_2Ag_2Se_2 \tag{5.8}$$

$$2SrO + Ga_2O_3 + 2SrS + Cu_2O \rightarrow 2Sr_2GaO_3CuS \tag{5.9}$$

$$CeAs + NiAs + Ni + CeO_2 \xrightarrow[\text{1450 K for 40 hours}]{} 2CeONiAs \tag{5.10}$$

As is it extremely challenging to purify non-soluble solid state mixed anion materials it is important to ensure that the reagents used are as high purity as can be acquired (99.9% or higher), and that great care is taken when weighing to ensure that the correct stoichiometric ratio of precursors is achieved. Ideally the researcher should use a four decimal place balance, and take care that material is not lost to spillage or in weighing boats. If high purity samples of the binary precursors cannot be provided by suppliers, these are sometimes synthesised by the researchers prior to use. For example, barium sulfide can be synthesised by reaction of barium carbonate with carbon disulfide gas, while copper sulfide can be formed by direct reaction of copper and sulfur. Alkaline earth oxides can be made from vacuum decomposition of the appropriate carbonates. When using elemental reagents, particularly metals, there is a concern about surface oxidation that will introduce additional oxygen to the reaction, and so alter the desired stoichiometric ratio. To counter this, it is often best to avoid the use of elemental precursors, even if they are nominally stable and high purity. If oxidisable elemental precursors have to be used, it is preferable to use pellet rather than powder reagent as this has lower surface area, although it can be harder to grind and combine using a pestle and mortar.

It is also possible to attempt to grow single crystals of target materials, to access the more definitive structural characterisation that single crystal X-ray diffraction offers over powder diffraction techniques. A typical flux growth method makes use of an excess of NaCl/KCl flux added to the standard precursors, which is then heated in a sealed ampoule to a temperature above the melting point of the flux. Slow cooling of the flux mixture may then lead to single crystals of a sufficiently large size for single crystal diffraction to grow.[84]

5.3.2 Structural Review

In this section we will review some of the more common structures adopted by oxychalcogenides and oxypnictides. As described earlier, many of these are structure types not adopted by an equivalent oxide, but which can be considered as containing layers or fragments of common oxide structures such *perovskite*, *rock salt* and *litharge*, or their anti-types.

We will start our review of the structures of these HAMs with the simple ternary compositions, before moving onto the more complex quaternary and quintenary compositions. In our descriptions of the structures, we will use the same conventions for labelling of placeholder elements, A, Ln and B as found in the oxynitride discussion, but where B is specifically a hard transition metal or post transition metal, binding principally to oxygen. In addition, we will introduce M as a softer late transition metal ion, often bound homoleptically to the heavy ion, Q, representing the pnictide, halide or chalcogenide.

Each structure will be defined by its composition using this convention, and also listed with its *Inorganic Crystal structure Database (ICSD)* structure type in italics, if one has been assigned. Compositions will be listed to emphasis structural features such as layering, rather than following strict IUPAC convention.

5.3.2.1 Ternary Compounds

5.3.2.1.1 A_2O_2Q: La_2O_3 Type. The majority of the ternary lanthanide oxy-chalcogenides, Ln_2O_2Q (Ln = La–Lu, Q = S or Se), adopt a trigonal structure related to the structure of La_2O_3 as shown in Figure 5.7. In the La_2O_3 structure the lanthanum ions form a hexagonal close packed lattice, with layers of oxide ions in alternating tetrahedral and octahedral holes, along the ABAB... packing sequence of the hexagonal close packed lattice. In the Ln_2O_2Q structure, sulfur or selenium replaces the octahedrally coordinated oxide ion, to create layers of oxide and chalcogenide within the structure. The lanthanide ion is in an heteroleptic seven coordinate LnO_4Q_3 environment.

The extended structure can also usefully be considered as $[Ln_2O_2]^{2+}$ sheets of edge sharing Ln_4O tetrahedra, separated by layers of a hexagonal array of chalcogenide ions. Each Ln_4O tetrahedra shares three of its edges with adjacent tetrahedra, creating the layer shown in the [001] projection in Figure 5.7(c), which can be described as a bilayer of interpenetrating hexagonal nets of tetrahedra of alternating orientation. A related version of this structure is adopted by Sc_2O_2S, with the same oxide layers, but displaced relative to each other to place the sulfur in a six coordinate trigonal prism environment, rather than the octahedral co-ordination found in the standard La_2O_3 structure.

5.3.2.1.2 A_2O_2Q: Th_2TeN_2 Type. Bi_2O_2Se and the lanthanide oxytellurides Ln_2O_2Te, (Ln = La–Ho) adopt the tetragonal layered Th_2TeN_2 structure type,

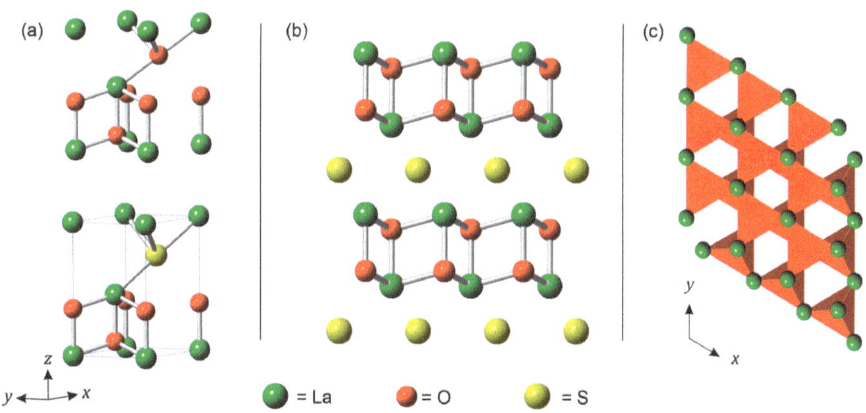

Figure 5.7 Diagrams of the structure of La_2O_2S, highlighting different features. (a) Trigonal unit cell of La_2O_2S (bottom) and comparison with parent compound La_2O_3 (top), showing hexagonal packed lattice of lanthanum ions, with oxide and sulfide ions in tetrahedral and octahedral holes in alternating layers. (b) Diagram showing expanded lattice of La_2O_2S, highlighting interpretation as a $[La_2O_2]^{2+}$ layer separated by S^{2-} ions, and (c) [001] view of the oxide layer showing tetrahedral connectivity. This structure is also adopted by Ln_2O_2Q, Ln = La–Lu, Q = S or Se.

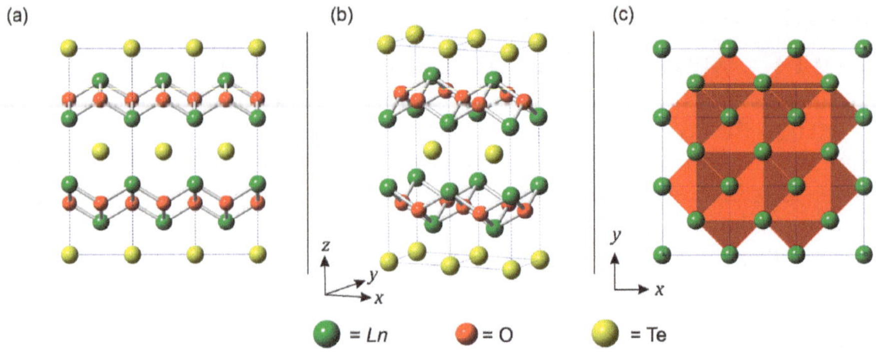

Figure 5.8 Diagrams showing the tetragonal structure adopted by Ln_2O_2Te and Bi_2O_2Se, comprising a $[Ln_2O_2]^{2+}$ litharge layer, a key feature of many HAM structures, separated by telluride ions. Figures showing (a) [100] view of cells, (b) offset [100] view of cells and (c) [001] view of the layers showing connectivity of the lanthanum oxide tetrahedra.

depicted in Figure 5.8. This is similar to the La_2O_2S structure, as it also contains a layer of Ln_4O tetrahedra, but in this case each A_4O unit is edge-sharing with four rather than three adjacent tetrahedra to create a denser extended layer, as seen in Figure 5.8(c). The structure of the oxide layer can be described as adopting the *litharge* structured fragment of $[Ln_2O_2]^{2+}$, which is then separated by layers of chalcogenide ions. The lanthanide ion is in a square prismatic co-ordination LnO_4Q_4. As we will see in the next section, this *litharge* fragment structural motif is an important one found in a number of more complex layered HAMs.

5.3.2.1.3 AOQ: PbClF Type. The heavier actinide elements (Np, U, Th) with stable A^{4+} ions can form oxychalcogenides with the PbFCl structure, which like the La_2O_2Te structure contains *litharge* structured metal oxide layers as discussed earlier, but now separated by a double layer of chalcogenide ions with a van der Waals interaction between the chalcogenides.[85] The composition can be thought of as being AOQ = $A_2O_2Q_2$, to compare to the single layer of chalcogenide in the A_2O_2Q structure adopted by the lanthanide oxytellurides. The PbClF structure is depicted in Figure 5.9(a) and provides a co-ordination environment for the metal ion that can be viewed as a square prism, AO_4S_4. Unexpectedly, this is the structure adopted by ZrOS—a rare example of a transition metal containing ternary oxychalcogenide structure,[86] as few other transition metals have a stable +4 oxidation state with an ionic radius large enough to fill the A site in this structure.

5.3.2.1.4 A_2OQ_2: Yb_2S_2O Type. The Yb_2S_2O type is a less common structure type, which is chalcogenide rich compared to the compositions adopted by the majority of the lanthanide oxysulfides, with a 2 : 1 sulfide to oxide ratio, rather than the more common 1 : 2 ratio. Examples of Ln_2S_2O are known for

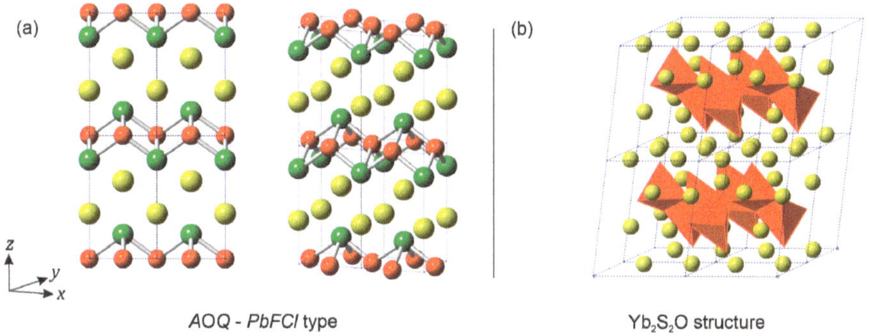

Figure 5.9 Structures of two of the less common ternary oxychalcogenides: (a) The PbFCl structure adopted by the actinide oxysulfides and oxyselenides and a key structure type related to many of the quintenary oxychalcogenide structures; (b) the YbS_2O structure adopted by some of the heavy lanthanide oxychalcogenides.

the heavy lanthanide elements where Ln = Gd–Yb, if they are synthesised at high pressure in a belt type apparatus. Under ambient pressure synthesis conditions they adopt the La_2O_2S structure type.[87] As with the previous ternary oxychalcogenides, the Yb_2S_2O structure contains oxide layers composed of A_4O tetrahedral, forming in edge sharing pairs which connect *via* vertices to form the layers, with sulfide ions between the layers. Unlike the *litharge* layers, this oxide layer type is not observed again in the more complex structures. The metal environment is seven co-ordinate, LnS_4O_3.

5.3.2.2 Quaternary Phases

5.3.2.2.1 AOBQ: ZrCuSiAs Type. The most common of the quaternary structures for compounds with a simple 1 : 1 : 1 : 1 or AOBQ stoichiometry, is the ZrCuSiAs type. The structure contains *litharge* $[A_2O_2]^{2+}$ layers, separated by $[B_2Q_2]^{2-}$ *anti-litharge* layers, giving a pleasing symmetry to the structure as seen in Figure 5.10(a). This maintains the tetrahedral coordination of the oxide ion to the *A* ion, seen in the majority of the ternary oxychalcogenides, while in the *anti-litharge* this is reversed, with the second *cation* in homoleptic tetrahedral coordination to the heavy Q anion. The A ion links the two layers in a square anti-prismatic heteroleptic AO_4Q_4 environment. The structure can be derived from the PbClF structure by insertion of the B ion into the tetrahedral hole inside the chalcogenide double layer. The *anti-litharge* $[B_2Q_2]^{2-}$ layer seen in this structure is an extremely common feature of many of the more complex layered HAMs.

The ZrCuSiAs structure is very widely adopted, with at least 80 known examples typically with a lanthanide ion on the A site, and different pairs of ions in what can be termed the 'heavy' $[B_2Q_2]^{2-}$, *anti-litharge* layer.[88] Copper chalcogenides *anti-litharge* layers are very common, with 20 examples such

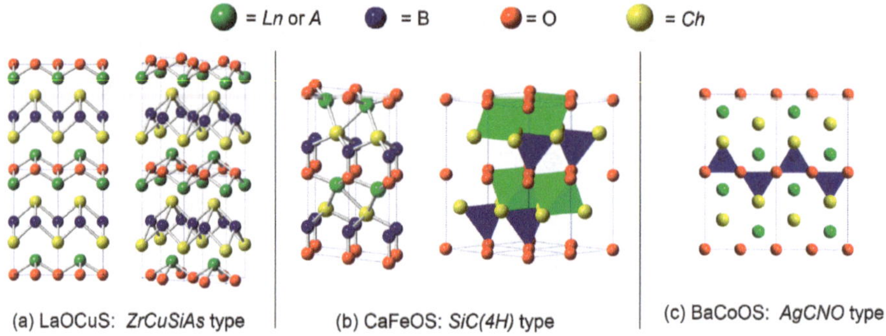

Figure 5.10 Unit cell diagrams of the structures of the ABOQ mixed anion compounds. (a) Two viewpoints of by far the most common ABOQ structure, the ZrCuSiAs type with alternating litharge and anti-litharge layers. (b) The CaFeOS structure, based on close packed oxide and sulfide layers, with iron in tetrahedral holes and calcium in octahedral holes. (c) The more unusual BaCoOS structure, found for a small number of compounds, containing a more distorted layering than is common amongst HAM structures.

as LaOCuS or YOCuSe, as are iron pnictide layers, with 11 examples such as LaOFeP or GdOFeAs.[88] The overall structural diversity is high, with at least 25 different heavy layers recorded. However, in all these examples the 'light' oxide layer, $[A_2O_2]^{2+}$, retains its +2 charge through the inclusion of a lanthanide ion on the *A* site, while the $[B_2Q_2]^{2-}$ layer has a balancing −2 charge, through appropriate pairing of either singly charged B^+ ions with a chalcogenide or doubly charged B^{2+} ions with a pnictide. There are a small number of cases where the light layer contains either the nitride or fluoride ion, for example ThNFeAs and BaFCuS, allowing for inclusion of an A^{4+} and A^{2+} ion in the *A* site respectively, but the overall layer charge remains constant. The segregation of the metal ions and the reduced dimensionally present in the ZrCuSiAs compositions affects the electronic structure, and has important implications in the applications of the layered HAMs, which will be discussed in the final sections of this chapter.

There are also some variations possible on the ZrCuSiAs structure, for example $La_2O_2FeSe_2$ where only half of the B sites are occupied in the heavy anion layer. This reduces the dimensionality of the layer, so that there are edge sharing chains of FeS_4 instead of a fully connected two-dimensional layer. There are also two alternative types which share the AOBQ stoichiometry but with different structures, which we will discuss further.

5.3.2.2.2 AOBQ: SiC(4H) Type. This is a rare structure type that has been found to be adopted by materials with Ca^{2+} ions on the A site and a late transition metal dication on the B site, in the compounds CaFeOS, CaCoOS and CaZnOS.[89] The structure is based on close packing of hexagonal layers of sulfide and oxide ions adopting a four layer repeat ABCB… stacking order as found in the eponymous SiC(4H) structure shown in Figure 5.10(b). The 4H in the structure name is the Ramsdell notation for SiC polytypes and highlights hexagonal (H) packing, with the four-layer repeat. The interstitial holes are filled in alternating layers, with the transition metal ions in tetrahedral holes, and the calcium ions in octahedral holes. Both metals are in heteroleptic coordination, BOS_3 and CaO_3S_3, respectively, so unlike the ZrCuSiAs structure, there are no homoleptic cation with exclusive co-ordination to either oxide or sulfide ions.

5.3.2.2.3 AOBQ: AgCNO Type. The second rare alternative AOBQ type is the AgCNO structure type adopted by cobalt and zinc containing oxysulfides when the A site is occupied by a barium ion, in for example BaCoOS and BaZnOS, shown in Figure 5.10(c). This structure is preferentially adopted over the SiC(4H) type, for barium containing compositions, because the barium ion is too large for the octahedral site found in the SiC(4H) structure. The AgCNO structure contains sheets of vertex sharing zinc or cobalt BO_2S_2 tetrahedra, with links between the tetrahedra through both the oxide and sulfide vertex ions to form the layers. The barium ions sit between the layers, in an eight coordinate BaO_4S_4 environment.[90]

5.3.2.2.4 $A_2O_3B_2Q_2 = A_2O_2B_2OQ_2$: $Fe_2La_2O_3Se_2$ Type. The $Fe_2La_2O_3Se_2$ structure is adopted by a range of compounds with lanthanide ions on the A site, with iron or cobalt on the B site, and with either sulfur or selenide on the Q site, as shown in Figure 5.11(a). Like many of the previous examples of HAM structures this also contains the lanthanide oxide *litharge* layers, separated by a single layer fragment of the *anti-perovskite* structure, and can be formulated

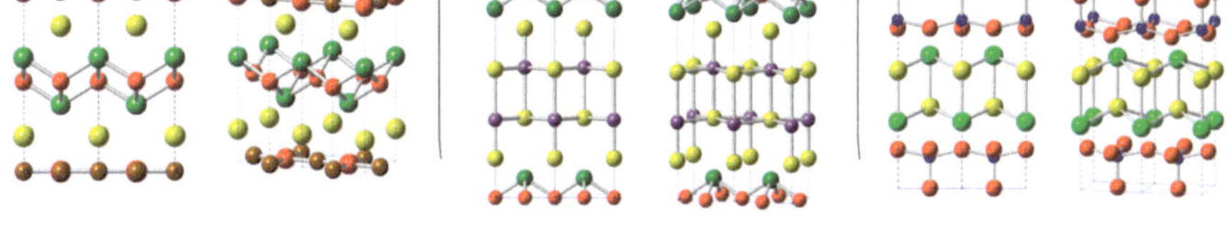

(a) $Fe_2La_2O_3Se_3 = La_2O_2Fe_2O_3Se_2$ (b) $CeBiOS_2 = Ce_2O_2Bi_2S_4$ (c) $Sm_2Ti_2O_5S_2$

Figure 5.11 Diagrams showing the structure of further quaternary oxychalcogenide structure types, from [100] and isometric projection: (a) $Fe_2La_2O_3Se_3$ type, (b) $CeBiOS_2$ type and (c) $Sm_2Ti_2O_5S_2$ type. The first two contain the common litharge layer, while $Sm_2Ti_2O_5S_2$ can be derived from the *Ruddlesden–Popper* structure.

as $[Ln_2O_2][B_2OCh_2]$ to highlight this.[91] The B site in the *anti-perovskite* layer is in heteroleptic octahedral, BO_2S_4 co-ordination. The octahedra are edge sharing to form the layer. The chalcogenide ions are in a square prismatic co-ordination to both metal ions, while the oxide ions in the anti-perovskite layer are in square planar geometry to the B ions.

5.3.2.2.5 $AOBQ_2 = A_2O_2B_2Q_4$: $CeBiOS_2$ Type. This structure type is adopted by the bismuth oxychalcogenides, $AOBiCh_2$, with the A site occupied by either lanthanum or cerium. The structure type shown in Figure 5.11(b) has a *litharge* layer $[A_2O_2]^{2+}$ separated by bismuth chalcogenide layers, $[Bi_2Ch_4]^{2+}$, adopting the *rock salt* structure. This can be considered as a modification of the PbClF structure, but with a double layer of *rock salt* inserted into the chalcogenide double layer. The tin sulfide, $LaOSnS_3 = La_2O_2Sn_2S_3$, adopts a similar but distorted structure with only a single layer of *rock salt* SnS.

5.3.2.2.6 $A_2B_2O_5Q_2$: $Sm_2Ti_2O_5S_2$ Type. The last of the quaternary phases we will consider is the $Sm_2Ti_2O_5S_2$ type, adopted by a number of titanium compounds with a lanthanide on the A site and sulfur on the Q site, as shown in Figure 5.11(c).[92] The structure can be considered a modification of a *Ruddlesden–Popper* phase. The *Ruddlesden–Popper* (RP) phases adopt the *perovskite* structure but are separated into layers by sections of *rock salt* structure. They have a general formula of $A_{n+1}M_nO_{3n+1}$, with $n = 1$ to 3 being the most common. The $Sm_2Ti_2O_5S_2$ structure is an $n = 2$ RP phase, but with an A site vacancy in the centre of the *perovskite* block, and the apical oxides in the *rock salt* layer replaced with sulfide ions, such that the RP composition $A_3M_2O_7$ is converted to $A_2M_2O_5S_2$. The geometry of the lanthanide in the structure is a nine coordinate capped twisted square prism, AO_4S_5, with this unusual geometry arising from the lanthanide spanning the *perovskite* and *rock salt* layer. The titanium ion is in a square based pyramid co-ordination to oxygen, or octahedral geometry if the interaction to the apical sulfur is considered short enough to be a bonding interaction.

5.3.2.3 *Quintenary Phases*

5.3.2.3.1 $A_2BO_2M_2Q_2$: $Sr_2Mn_3Sb_2O_2$ Structure. The $Sr_2Mn_3Sb_2O_2$ structure type in Figure 5.12(a) is one of the most frequently investigated quintenary layered mixed anion phases, with at least 20 unique compositions reported, and can also be considered as an archetype for many of the other quintenary structures discussed in this section. It consists of a 'heavy anion' layer, $[M_2Q_2]^{2-}$, adopting the *anti-litharge* structure containing a late transition metal and the heavy anion previously seen in the *ZrCuSiAs* structure, separated by layers of $[A_2BO_2]^{2+}$, which can be considered as a single layer slice of the *perovskite* structure, where A is an alkaline earth ion and B is an early transition metal. This can also be considered as the anti-type of the structure adopted by $La_2O_3Fe_2S_2$ that had alternating *litharge* and *anti-perovskite* layers, whereas $Sr_2Mn_3Sb_2O_2$ is composed of alternating *anti-litharge* and

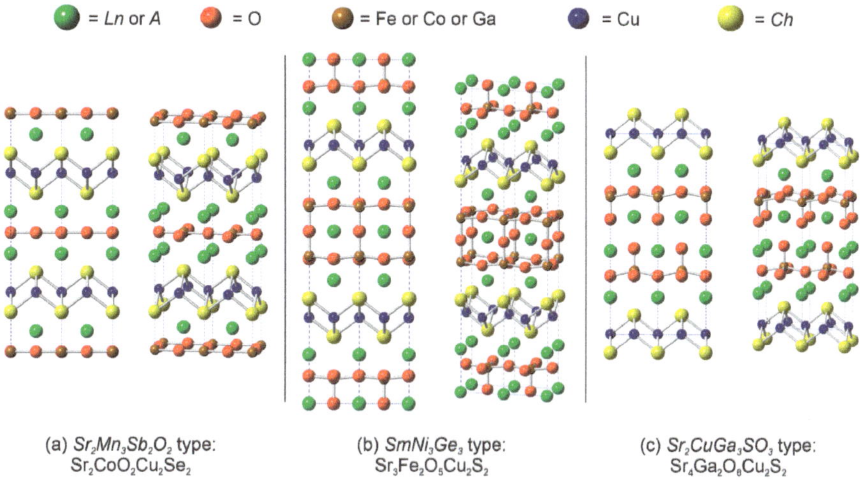

Figure 5.12 Diagrams showing the structure of three key quintenary oxychalcogenide structure types, all containing the *anti-litharge* layer, from [100] and isometric projection: (a) $Sr_2Mn_3Sb_2O_2$ type, adopted by $Sr_2CoO_2Cu_2Se_2$; (b) $SmNi_3Ge_3$ type, adopted by $Sr_3Fe_2O_5Cu_2S_2$; and (c) $Sr_2CuGa_3SO_3$ type, which can reformulated as $Sr_4Ga_2O_6Cu_2S_2$ to highlight the structure and make it easier to compare with the other layered compounds.

perovskite layers. As previously seen, the *anti-litharge* layers contain the late transition metal in homoleptic tetrahedral coordination to the heavy anion. The transition metal in the oxide layer on the B site is also homoleptic, with square planar coordination to the oxide ions. The two transition metals in the heavy and light layers have strong covalent interactions to the oxide and chalcogenide anions, respectively, while the A site alkaline earth ion bridges the two layers with a square anti-prism AO_4Q_4 coordination, with a predominantly ionic interaction.

In all currently reported samples of HAMs adopting the $Sr_2Mn_3Sb_2O_2$ structure the A site is occupied by an alkaline earth ion, while for charge balance the B site is occupied by an early transition metal ion in the +2 oxidation state, such as in $Sr_2MnO_2Cu_2S_2$.[93] The heavy anion layers that are most common in this structure type are silver selenides and copper chalcogenide layers, although others have been reported such as the zinc arsenide layer in $Ba_2MnO_2Zn_2As_2$.[94]

5.3.2.3.2 $A_3B_2O_5M_2Q_2$: $SmNi_3Ge_3$ and Other Related Structures. The $SmNi_3Ge_3$ structure can be derived from the $A_2BO_2M_2Q_2$ composition, as it also contains $[M_2Q_2]^{2-}$ *anti-litharge* layers separated by a *perovskite* fragment, but in this case increased to a double layer thickness with an $[A_3B_2O_5]^{2+}$ composition, as shown in Figure 5.12(b). A typical example would be $Sr_3Fe_2O_5Cu_2S_2$.[95] The increased thickness of the layer changes the B site coordination to square based pyramidal, and also introduces a second A site in addition to the inter layer A site. This intra layer site is located at the centre of the *perovskite* block with the 12 coordination to oxygen expected for the *perovskite* structure. Most examples have either Ba^{2+} or Sr^{2+} on the two A sites, and then either Sc^{3+}, Fe^{3+} or a Group 13 metal ion on the B site. Different heavy layers reported for the structure include iron arsenide, cobalt arsenide, and copper and silver chalcogenide layers.

There are also examples of further materials related to the $A_2BO_2M_2Q_2$ structure where the perovskite block is extended to a triple layer, for example $Sr_4Mn_3O_{7.5}Cu_2Ch_2$,[96] which has a formal oxygen vacancy to charge balance the Mn^{3+} ion, but which has been found to have further vacancy and disorder on the manganese and oxygen sites in the central layer of the perovskite block.[97] Fully occupied, the structure would require a mixed $M^{3+/4+}$ site with composition $A_4B_3O_8M_2Q_2$. A general formula for the materials with varying thickness perovskite blocks is $A_{n+1}B_nO_{3n-1}M_2Q_2$, where the n = 1 and n = 2 members are the $Sr_2Mn_3Sb_2O_2$ and $SmNi_3Ge_3$ types, respectively, with numerous examples. Only a few examples of the n = 3 structure have been reported, and at the time of writing only single reports of n = 4, and n = 5 examples. These higher order members of the series, $Ca_5(Sc_{0.5}Ti_{0.5})_4O_{11}Fe_2As_2$ (n = 4) and $Ca_6(Sc_{0.4}Ti_{0.6})_5O_{14}Fe_2As_2$ (n = 5), could not be made phase pure and required mixed ion occupation of the B site to target the formal +3.5 and +3.6 oxidation state required for charge balance.[98] One reason for the instability of the higher order examples is that it becomes increasingly difficult to stabilise the larger perovskite blocks as the *B* site charge tends to that in the 'infinite' layer *perovskite* itself. Therefore, segregation into the pure *perovskite* and a lower order layered compound can become more favourable, as shown for a theoretical case in eqn (5.11), where a hypothetical attempt at a n = 4 material can phase separate into *perovskite* and an n = 2 phase.

$$Ca_5(Sc_2Ti_2)O_{11}Fe_2As_2 \rightarrow 2CaTiO_3 + Ca_3Sc_2O_5Fe_2As_2 \qquad (5.11)$$

There are also examples of compounds derived from the $A_2BO_2M_2Q_2$ structure type where the single layer of perovskite is retained, but the thickness of the tetrahedrally co-ordinated heavy anion layer is increased. The examples published so far have copper sulfide layers and are $Sr_2MnO_2Cu_{4-\delta}S_3$ and $Sr_2MnO_2Cu_{6-\delta}S_4$, which also contain copper vacancies associated with increasing the thickness of the layer.[99] These can be considered as m = 2 and m = 3 of the general formula for increased thickness copper sulfide layers of $Sr_2MnO_2Cu_{2m-\delta}S_{m+1}$, with m = 1 being the archetype $Sr_2MnO_2Cu_2S_2$.

5.3.2.3.3 $A_2BO_3MQ = A_4B_2O_6M_2Q_2$: $Sr_2CuGa(SO_3)$ Type. The formula of the $Sr_2CuGa(SO_3)$ structure type can be rewritten as $Sr_4Ga_2O_6Cu_2S_2$, and in this form the relationship to the previous compounds can be seen, with the common $[Cu_2S_2]^{2-}$ *anti-litharge* layers separated by an oxide block with composition $[Sr_4Ga_2O_6]^{2+}$, as shown in Figure 5.12(c). The structure of this oxide layer is more complex then we have covered in prior examples, but can be seen as two single layers of strontium gallium oxide *perovskite* sandwiched around a SrO *rock salt* layer. The structure has a square based pyramid environment for the gallium ion, coordinated to oxide, while the strontium embedded in the oxide layers has a capped anti-square prismatic, nine coordinate geometry, resulting from spanning the *perovskite* and *rock-salt* layers. $Sr_4Ga_2O_6Cu_2S_2$ was the first reported example,[82] but to date over 20 compositions have been identified, with a number of different pnictide and chalcogenide heavy layers. The composition of the oxide block is typically an alkaline earth on the A site, with the B site occupied by a +3 ion, either an early transition metal such as scandium, iron or manganese, or a post transition metal from Group 13. It should be noted that these charge requirements are identical to those found for the $SmNi_3Ge_3$ structure type, and for some ion combinations it is possible to synthesise both structure types, for example, $Sr_4Sc_2O_6Cu_2S_2$ and $Sr_3Sc_2O_5Cu_2S_2$ are both stable. However, this is not always the case, as there is no $SmNi_3Ge_3$ structure equivalent to the stable $Sr_4Ga_2O_6Cu_2S_2$. There is no clear understanding yet to predict if either or both of the structures can be formed for a given set of ions,[100] but there does seem to be an overall trend that the $Sr_2CuGa(SO_3)$ structure is favoured by smaller ions.

5.3.2.3.4 Structural Summary. The layered compounds of the oxychalcogenides and oxypnictides discussed above show a great deal of diversity, but also a number of common features, including the *litharge* and *anti-litharge* layers, and *perovskites* or related oxide layers. The connections between this range of structures are summarised in Figure 5.13, showing how the addition and removal of ions or layers can transform each structure into the next, and hopefully provide the reader with a sense of the relationship between them.

5.3.3 Applications of HAMs with Heavy Anions

In this final section we will consider how the presence of the heavy anion in this type of HAM, and the unique layering that this induces in the structures, generates unusual optical and electrical properties in the materials. We shall illustrate this with two examples, the p-type conductivity seen in some oxychalcogenides, and the superconductivity observed in iron containing oxypnictides. In both cases the unique transport properties are found in layered compounds and associated with the heavy anion containing layer.

5.3.3.1 p-Type Conductivity in Layered Copper Chalcogenides

One of the more unusual properties that has been identified in heavy anion HAMs, and specifically those with copper sulfide and copper selenide *anti-litharge* layers, is transparent conductivity. This is significant because transparent conducting films are now indispensable to modern technology, being widely used in display devices and touch screens, as anti-frost coatings on aircraft windows, and are being considered for use as charge collection layers in photovoltaic devices.[101,102] The most widely used material for these coatings

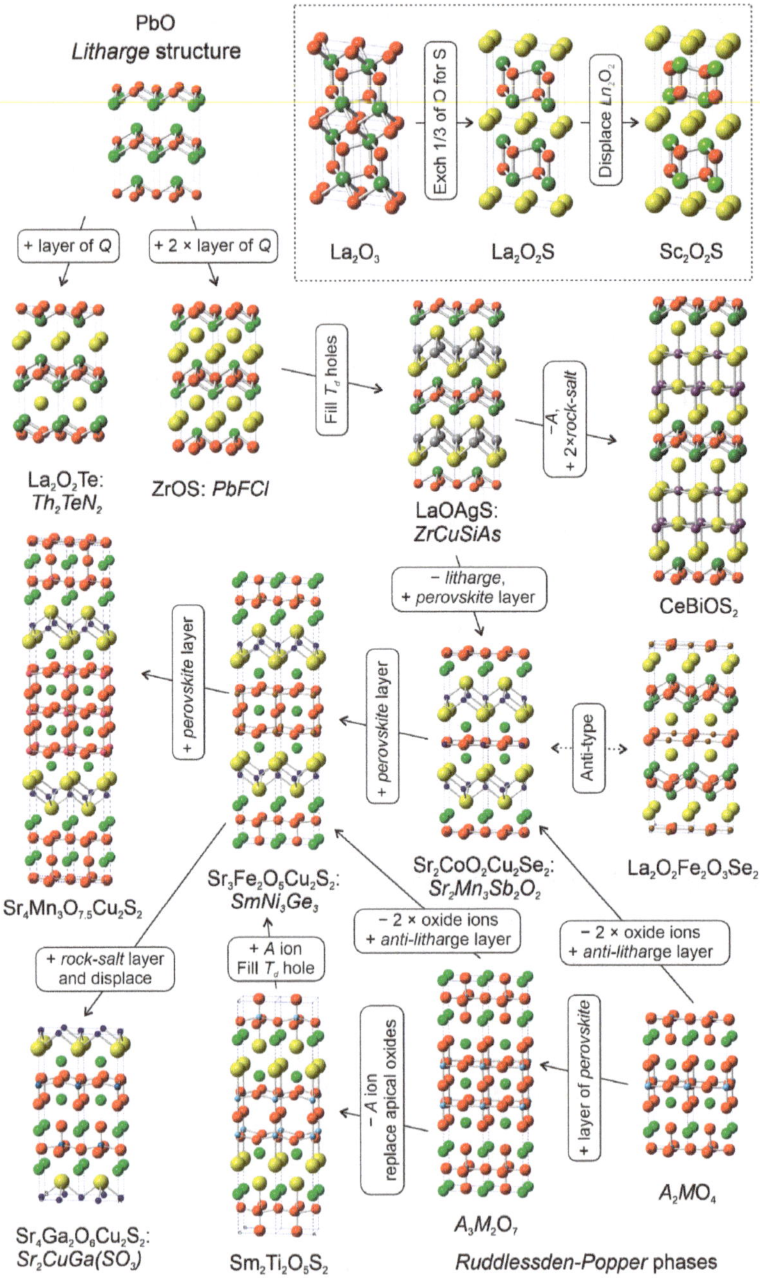

Figure 5.13 Summary of the key structures commonly adopted by oxychalcogenides and oxypnictides, and the structural relationships between them. Key starting points are the litharge structure (top left) and the *Ruddlesden–Popper* phases (bottom right). The majority adopt a primitive or body centred tetragonal space group. The structures in the inset on the top right are the trigonal phases, based on the La_2O_3 structure. Adapted from ref. 4 with permission from American Chemical Society, Copyright 2008.

is tin doped indium oxide, known as indium tin oxide (ITO), which combines visible light transparency with high electrical conductivity. Pristine indium oxide is a transparent wide band gap (3.75 eV) insulator, which can be converted into a transparent conductor by doping with tin to introduce shallow donor states and charge carriers. At sufficient concentration the tin states allow for degenerate doping of the conduction band, while leaving the band gap intact and retaining the transparency of the material. The high mobility of the electrons in the indium oxide conduction band (between 40–60 $cm^2 V^{-1} s^{-1}$) combined with the charge carriers provided by donor tin states provides an overall conductivity (10^4 S cm^{-1}) that is equivalent to a 'poor' metal such as steel or bismuth, and an order of magnitude higher than a typical semiconductor.[103] As the tin dopant is an electron donor, ITO is an n-type conductor with electrons as the majority carriers, as are all of the so called 'transparent conducting oxides' (or TCOs) which also include aluminium doped zinc oxide (AZO) and fluorine doped tin oxide (FTO). Attempts to produce a p-type transparent conductor from the TCO materials through doping with electron acceptors to generate holes have been thwarted because although they have excellent electron mobility in their conduction bands, from overlap of more dispersed metal sp orbitals, they have much lower hole mobility. Hole transport occurs in the valence band which in the TCOs is composed of orbitals originating from the highly electronegative oxide ions, with poor overlap and hence poor hole mobility.

A key exception to the trend that oxides have low hole mobility is copper(I) oxide, which has a hole mobility of ~100 $cm^2 V^{-1} s^{-1}$, but too small a band gap (2.2 eV) to make it transparent. The origin of the unexpectedly large hole mobility in copper(I) oxide can be explained by the high effective nuclear charge of copper as a result of its position in the Periodic Table. As a late transition metal, the high effective nuclear charge of copper lowers the energy of the Cu 3d orbitals so that they can hybridise with O 2p orbitals and form part of the valence band. The Cu 3d orbitals are more delocalised, allowing for improved hole mobility. However, the high effective nuclear charge of copper which lowers the 3d orbitals also lowers the position of the 4s and 4p states which make up the conduction band, leading to a band gap too small to allow visible light transmission. Investigation of alternative copper containing materials found that the layered *delafossites* $CuAlO_2$, $CuScO_2$ and $CuGaO_2$ had sufficiently large band gaps (>3.1 eV) to be transparent and with reasonable hole conductivities of between 10^{-1} and 10^0 $cm^2 V^{-1} s^{-1}$. This allowed them to be declared p-type transparent conductors, with appropriate doping,[104,105] with the best case being $CuScO_2$ with a conductivity of 30 S cm^{-1}.[106] The *delafossite* structure is layered, with linear CuO_2 units separated by MO (M = Al, Sc, or Ga) sheets, which reduces the dimensionality compared to copper(I) oxide and so through a quantum confinement effect increases the band gap while retaining some of the conductivity—although far less than 10^4 S cm^{-1} required for commercial transparent films.

The next step towards viable p-type transparent conductors was the realisation that the layered copper oxychalcogenide HAMs would be even more suitable, as they have reduced dimensionality to increase the band gap, but with more covalent chalcogenide ions which can hybridise with the copper d-orbitals for improved dispersion and increased hole mobility. The first materials considered were the ZrCuSiAs structured lanthanum oxychalcogenides, LaOCuCh (Ch = S, Se and Te).[107,108] These were found to have increasing hole mobility with increasing weight of chalcogen, rising from 0.5 $cm^2 V^{-1} s^{-1}$ for the oxysulfide, to 8 $cm^2 V^{-1} s^{-1}$ for the oxyselenide, and finally 80 $cm^2 V^{-1} s^{-1}$ for the oxytelluride. The trend in band gaps is the reverse, with the band gap decreasing from 3.1 eV to 2.8 eV to 2.3 eV from the oxysulfide to the oxytelluride. This means that despite its excellent mobility, LaOCuTe is not viable as a transparent conductor due to its small band gap, while LaOCuS has good optical properties but poor mobility. The oxyselenide can be considered as a compromise material, with reasonable conductivity, and transparency across most of the visible spectrum, although some absorption of the blue.

Modelling of the band structure reveals that the LaOCuCh materials have direct band gaps, and that the conduction band minimum is composed of the La 5d and Cu 4s orbitals,[107] while the valence band maximum is a hybridisation of the Cu 3d and Ch p states, shown schematically in Figure 5.14(a). As the electronegativity of the chalcogenide decreases down the group this leads to the energy of the chalcogenide orbitals rising, and the observed decrease in band gap. There will also be increases in covalency, and greater orbital overlap for the heavier chalcogenides, explaining the increasing hole mobility in the sequence of oxysulfide to oxytelluride. The mobility itself

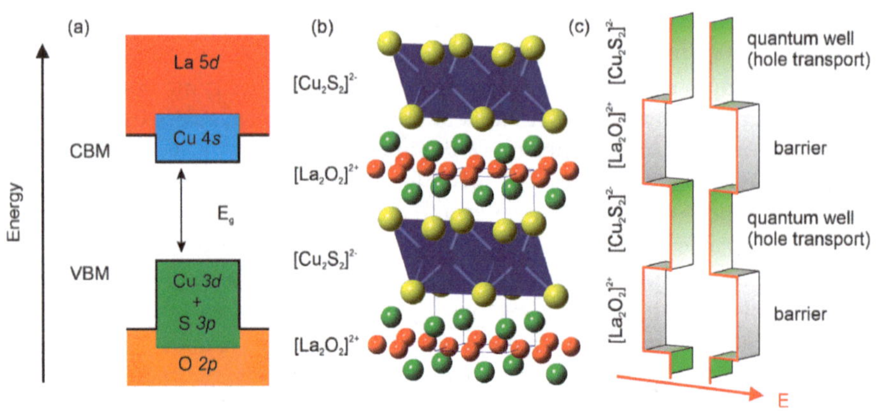

Figure 5.14 Possible interpretations of the band structure of LaOCuS to help explain the p-type conductivity observed. (a) Schematic of the band structure showing composition of the band edges. The valence band maximum (VBM) is made up of the hybridisation of the copper and sulfur orbitals. The high dispersion of these allows for good hole mobility. (b) The covalent copper sulfide layers are separated by the more ionic lanthanum oxide layers. (c) The lanthanum oxide layers can be considered as potential energy barriers, leading to the quantum well effect. Diagrams adapted from ref. 107 with permission from Elsevier, Copyright 2006.

is found to be confined to the $[Cu_2Ch_2]^{2-}$ layers, with the lanthanum oxide block, $[La_2O_2]^{2+}$, acting as a barrier layer. There is strong covalency within the copper chalcogenide layer, but the interaction to the oxide layers is predominantly ionic—as can be observed in the large La–Ch distance, which in each case is considerably larger than the sum of the ionic radii, Figure 5.14(b). The LaOCuCh electronic structure has also been described as a multiple quantum well, with easy charge carrier mobility within the layers, (the well), but the lanthanum oxide layers acting as a potential energy barrier—this is shown in Figure 5.14(c). This quantum confinement can also be used to explain the increase in the band gap for the layered LaOCuCh copper chalcogenides compared to the simple Cu_2Ch compounds with an extended network of copper chalcogenide bonds. It is well established that for nanoparticles smaller than the exciton radius of the material—quantum dots—there is an increase in the size of the band inversely proportional to the size of the particle. For example, bulk copper(I) sulfide has a band gap of 1.1 eV, whereas copper sulfide quantum dots with a radius of 5 nm have band gaps of 1.6 eV. The copper sulfide layer in LaOCuS can be considered as an extreme example of this as the layer is only two copper–sulfide bonds thick (approximately 0.5 nm) such that the quantum confinement leads to an increase of the band gap up to 3.1 eV.

For a transparent conductor high carrier mobility is vital, but it is also necessary to increase the carrier concentration through doping to introduce acceptor levels near the valence band. At sufficiently high concentration the acceptor levels will be able to overlap with each other, and the valence band state, forming a donor band and leading to partial vacancy at the top of valence band and classic degenerate semiconductor behaviour, where the transport properties begin to resemble those of a metal. This includes much higher carrier concentration, but also a loss of dependence on temperature for the conductivity. This has been achieved in LaOCuS through partial replacement of the lanthanum with strontium in $La_{1-x}Sr_xOCuS$. This is shown in Figure 5.15, where a strontium doping of $x = 0.03$ gives a relatively stable conductivity of 18 S cm^{-1} across the temperature range measured, indicative of degenerate behaviour.[109] In contrast, the undoped sample has a much lower conductivity, and one which is heavily temperature dependent. This would be expected for a non-degenerate semiconductor, where there is a dependence of conductivity on thermal promotion of charge carriers. Doping studies have also been carried out on the higher mobility oxyselenide, LaOCuSe, where magnesium was found to be the most effective dopant to generate shallow acceptor levels. Magnesium doping of 20% to give $Mg_{0.2}La_{0.8}CuOSe$, yielded a sample with a conductivity of 140 S cm^{-1}, an improvement on the oxysulfide, but this is still at least an order of magnitude below what is required to match the n-type TCOs.[110]

The LaOCuCh compounds demonstrated that layered copper oxychalcogenide compounds can produce materials with hole charge carriers and large band gaps, and work has continued with the investigation of the more 'complex' layered quintenary oxychalcogenides discussed in the structural

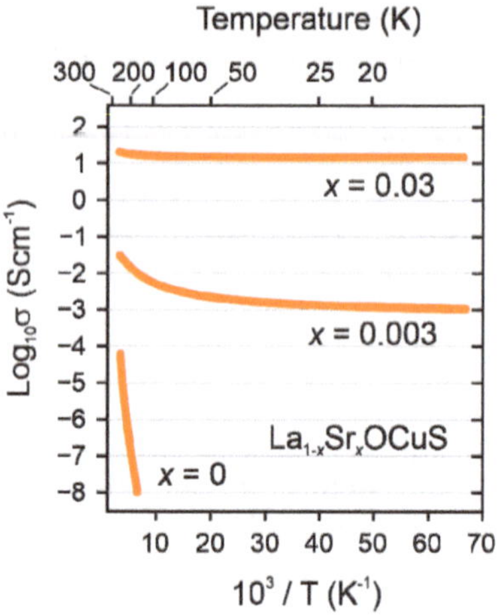

Figure 5.15 Plot of the conductivity (log 10 scale) of strontium doped LaOCuS as a function of reciprocal temperature. From this it can be seen that doping significantly increases the room temperature conductivity, but also decreases the temperature dependence as degenerate doping is achieved. Adapted from ref. 109 with the permission of AIP Publishing.

review, where the lanthanum oxide layer is replaced with other oxide blocks, but the copper chalcogenide layer retains the *anti-litharge* structure. In order to maintain optical transparency, the oxide layer must contain closed-shell, hard ions, but despite this limitation there are still many possible compositions. The more complex oxide layers also provide three different sites for doping in order to optimise the optoelectronic properties. This strategy has identified further wide band gap materials which also have p-type conductivity in their copper sulfide layers including $Sr_2ZnO_2Cu_2S_2$ and $Sr_4Ga_2O_6Cu_2S_2$. Sodium doping to replace 5% of the strontium in both materials gave respective conductivities of 0.1 S cm^{-1} and 2.4×10^{-2} S cm^{-1}, too low for commercial application, but evidence of the viability of the approach.[111] Better success was found with $Sr_3Sc_2O_5Cu_2S_2$, which *undoped* displayed a conductivity of 2.8 S cm^{-1}, with a hole mobility of 150 cm^2 V^{-1} s^{-1}, which exceeds the mobility even of the best n-type TCOs.[112] The investigation of the layered oxychalcogenides as transparent conductors remains an area of research interest,[113,114] and also highlights the importance of investigating mixed anion materials to identify properties which cannot be found in homoanionic oxide compounds. However, another area has gathered even greater research interest.

5.3.3.2 *Superconductivity*

Superconductivity was first identified in a HAM in 2006 by the Hosono group in the oxyphosphide compound LaOFeP, which adopts the ZrCuSiAs structure.[115] The compound was found to have a critical temperature (T_c) of 4 K, which to the uninitiated may seem at first unimpressive but superconductivity was entirely unprecedented for this class of material. The Hosono group quickly followed up this discovery with other related ZrCuSiAs compositions also displaying superconductivity, culminating in 2008 with the identification of a fluorine doped iron arsenide compound, $LaO_{0.89}F_{0.11}FeAs$, with a T_c of 26 K.[116] The publication of these reports unleashed a flurry of research activity within the solid state community, which resulted in the identification of new materials with even higher critical temperatures culminating after only a few months in the discovery of $SmO_{1-x}F_xFeAs$ with a T_c of 55 K.[117] Subsequent research has since only identified one HAM material with a higher critical temperature, the similar but hydride doped $SmO_{1-x}H_xFeAs$ with a T_c of 56 K.[118] The key feature shared by all of the compounds in this novel class of superconductor is that they contain iron in a pnictide layer adopting the *anti-litharge* structure fragment, and hence are known as the iron based superconductors (IBSC). A number of instructive reviews have been published in recent years on the IBSCs, which can be recommended.[6,119,120] To fully understand the significance of these discoveries and why they generated such excitement and research focus, we need to consider the history of superconductivity (Figure 5.16).

This history begins in 1911 when select elemental metals were found at very low temperature, typically below 10 K, to have electrical resistances of zero—that is they become conductors with no loss of energy to resistance.[121] Subsequent research on alloys, in particular those of niobium, identified materials with critical temperatures as high as 23 K, and also found that as well as zero resistance a superconductor can completely exclude a magnetic field and are thus are also perfect diamagnets, leading to discovery of the Meissner effect.[122] In the late 1950s Bardeen, Cooper and Schrieffer developed a model to describe superconductivity in these low temperature superconductors, now known as BCS theory, which is based on the formation of mobile, charge carrying electrons into 'Cooper pairs', which are bound to each other through coupling with lattice phonons, such that they are not scattered by the phonons—the normal mechanism of resistance in a metal. Within the Cooper pairs the spins are aligned anti-parallel such that their total spin angular momentum is zero ($S = 0$) and their orbital angular momentum is also zero, or *s*-symmetric.[123,124] The ability of the niobium alloy superconductors to hold large currents and generate electromagnetic fields without loss of energy, when cooled with liquid helium, allowed the development of the large superconducting magnets used in magnetic resonance instruments from the 1960s onwards. The next major discovery in superconductivity occurred in 1986 when a series of layered copper containing oxides were found to be superconducting at much higher temperatures, often above 30 K, and including $YBa_2Cu_3O_{7-x}$ with a T_c of 93 K.[125] This was significant for practical reasons because temperatures above 78 K can be achieved with the much lower cost liquid nitrogen cryogenic, rather than liquid helium. The copper-based superconductors were also unusual as they were found to be semi-conducting above their critical temperatures whereas the original superconductors were all metallic above their critical temperatures. These so

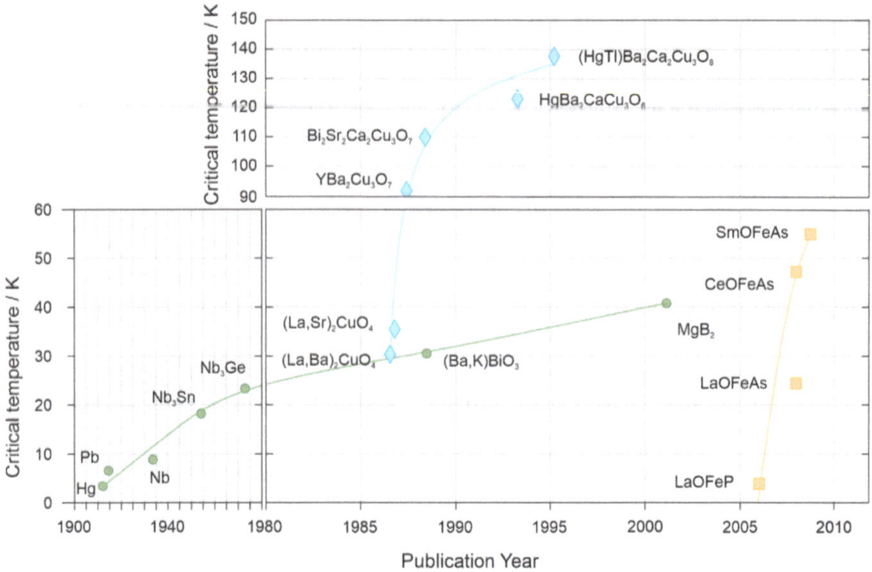

Figure 5.16 The history of superconductor discovery. A plot of the critical temperatures of a range of superconductors, against their date of discovery and publication. For simplicity only the 'parent' composition is given, ignoring the dopant or oxygen vacancy, which is essential for high temperature superconductivity. Temperature scale and time line shown on split scales, and types classified by colour. The green circles represent 'conventional' superconductors that can be explained using BCS theory. The blue diamonds are the high temperature cuprates, while orange squares are the iron-based superconductors. The diagram also highlights the changes in pace of research discovery, with the possible limits of the iron-based superconductors being reached in less than five years, compared to close on a century for the conventional superconductors.

called high temperature superconductors could not be fully explained using the classical BCS theory, but they still contain Cooper pairs, with d-wave symmetry.[126] Further examples of high temperature superconductors have been found, all containing copper, for example $Hg_{0.8}Tl_{0.2}Ba_2Ca_2Cu_3O_{8.5}$ with a T_c value of approximately 130 K.[127] At this point there were two types of superconductor: the 'conventional' superconductors, typically simple alloys with transition temperatures below 30 K which can be explained by BCS theory; and the 'unconventional' high temperature copper oxide containing superconductors.

It is within this context that we can now place the discovery of superconductivity in the iron pnictide compounds. The iron-based superconductors are unconventional superconductors, which cannot be fully explained using BCS theory, like the copper containing superconductors. However, they differ from the cuprate superconductors as their Cooper pairs are found to have an a isotropic sign reversing $s^{\pm}$ symmetry, rather than d wave symmetry, implying that they have a different mechanism and are in fact a third type of superconductor.[128] Up until the discovery of the IBSC it seemed that 'high-temperature' superconductivity, where the T_c is above 40 K, was exclusive to materials containing cuprate layers. The discovery of the IBSC was therefore so exciting because they disprove this hypothesis, and opened up another direction for the investigation of superconductive materials. There is still no theory that can fully explain conventional superconductivity, the high temperature cuprates, and the iron-based superconductors. However, there are a number of features that are common amongst the IBSCs that seem to be key to their superconductivity and may be useful in ultimately understanding the phenomena. They are certainly useful in the search for further novel superconductors, and we will discuss these below.

One of the most important 'parent' compounds to the IBSCs is the undoped oxyarsenide, LaOFeAs. At room temperature LaOFeAs is metallic, although with a conductivity that places it at the poorer end of the metals—a feature that is also common amongst the conventional superconductors. What is more unusual is that investigation of its band structure indicates it to be a 'semi-metal' as it contains an indirect gap between the conduction band minimum and valence band maximum. The respective bands overlap in energy but not in *k*-space so that the band structure contains both electron and hole pockets as shown in Figure 5.17(a). Below 150 K the structure undergoes a tetragonal to orthorhombic distortion, which is also closely related with an anti-ferromagnetic ordering at 134 K. The magnetic ordering originates from localised spins on the iron(II) ions, aligned within the iron arsenide sheets parallel to the *b* direction to generate chains of aligned spins that are then aligned anti-parallel and alternating in the *a* direction. There is also an anti-parallel alignment to nearest neighbours between the iron arsenide layers, across the lanthanum oxide barrier layer, in the *c* direction. Superconductivity can be induced into LaOFeAs by doping to introduce either holes or electrons, due to the pockets in the band structure, and can be achieved through direct doping within the layer or indirect doping in the oxide layer. For example, indirect doping through partial replacement of lanthanum with strontium introduces holes in $Sr_xLa_{1-x}OFeAs$, while replacement of oxygen with fluorine indirectly introduces electrons in $LaO_{1-x}F_xFeAs$. In both cases the materials have superconductive transition temperatures.[116,129] The doping can also occur with replacement of the iron, for example with replacement of cobalt onto the iron sites in $LaOCo_xFe_{1-x}As$, which also induces superconductivity.[130] This is in direct contrast to the cuprate superconductivity, where any significant substitution of copper will lead to a loss of superconductivity. In all cases it is found that the optimum doping with either electrons or holes is between 0.1 to 0.2 charge carriers per iron atom.

The appearance of superconductivity in the IBSCs through doping is always associated with the suppression or elimination of the orthorhombic structural distortion and loss of antiferromagnetic ordering. As the dopant is introduced the transition temperatures for the orthorhombic and antiferromagnetic transition temperatures are reduced, at the point when

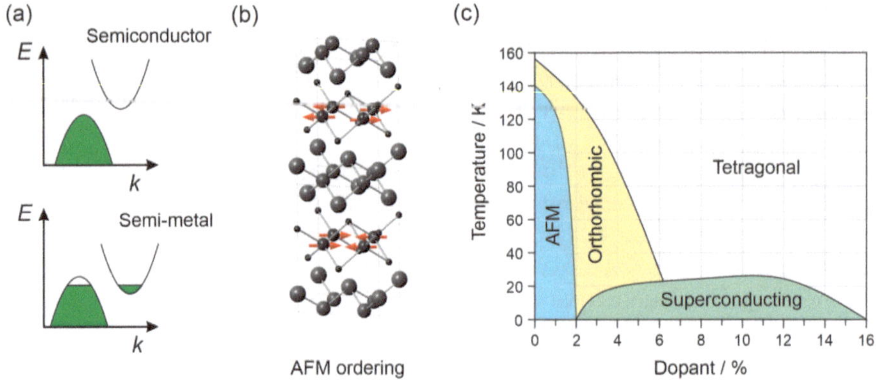

Figure 5.17 Understanding the key electronic and magnetic features of an iron-based superconductor. (a) Schematic *E–k* diagrams for an indirect semiconductor, with offset band edges and a small gap (top) and the related semi-metal (bottom). The iron-based superconductors (IBSCs) are semi metals. They have offset band edges in *k*-space, but which overlap in energy across the fermi level, leading to electron and hole 'pockets'. (b) The undoped IBSC parent compounds also tend to display anti-ferromagnetic ordering of the spins on their iron ions, and an orthorhombic distortion at low temperature. In the diagram a unit cell of the typical LaOFeAs is shown, with the direction of the ordering of spins highlighted with red arrows. (c) The appearance of superconductivity requires doping of the samples to introduce electrons and holes. This is normally associated with a suppression of the antiferromagnetic (AFM) and tetragonal to orthorhombic transition temperatures, with a simultaneous increase in the superconducting transition temperature, up to an optimum dopant concentration. (a) and (b) Adapted from ref. 120 with permission from the Royal Society of Chemistry, and (c) adapted from ref. 116 with permission from American Chemical Society, Copyright 2008.

antiferromagnetic ordering is lost superconductivity will appear, with further doping increasing the T_c, until some optimum concentration after which it will begin to drop. This is known as the superconducting dome, and can be seen schematically in Figure 5.17(c). In the case of $LaO_{1-x}F_xFeAs$, a superconductive state is observed close to absolute zero at 2% doping, with the T_c rising rapidly to around 22 K with doping up to 5%, and then reaches a maximum at 11% doping with a T_c of 26 K before declining at higher doping.

Pressure has also been found to affect the supercomputing transition temperature. For $LaO_{1-x}F_xFeAs$ with doping optimised to give a T_c of 26 K, application of pressure was found to increase this further to 43 K. When lanthanum was replaced by the smaller samarium ion and indirectly hole doped with hydride ions in $SmFeAsO_{1-x}H_x$, at optimum doping a T_c of 56.3 K was found, but application of pressure to this composition, in contrast to the doped lanthanum composition, *decreased* the critical temperature.[118] The explanation for this seems to be the effect that pressure and substitution have on the iron pnictide geometry. A review of a large number of iron arsenides superconductors found that, all other factors being equal, the superconducting transition temperature increases when either pressure *or* site substitution cause changes to the lattice that leads to the $FeAs_4$ tetrahedra in the anti-litharge layer transforming to a more *regular* shape—that is the geometry approaches the tetrahedral ideal where all bond angles are 109.48°.[131] Due to the symmetrical constraints within the ZrCuSiAs structure, for a non-regular tetrahedral iron arsenide layer there are two unique bond angles. These are defined in Figure 5.18(a) and (b): the α angle is between the iron and *cis* arsenic atoms with respect to the long *c*-axis of the cell, while the β As–Fe–As angle is *trans* with respect to this axis. In the ideal tetrahedral geometry $\alpha = \beta$, but if the cell is expanded in the basal *a* lattice direction this will lead to an increase in α and decrease in β, and *vice versa* if the cell is compressed. This was demonstrated in the series $LnFeAsO_{1-x}$, with differing lanthanide ions and a slight oxygen vacancy to act as electron dopant. The differing sizes of the lanthanide ion can be used to control the geometry of the iron arsenide layer, and increasing T_c is observed as the α and β angles converge towards the tetrahedral ideal, shown in Figure 5.18(c). This is sometimes called 'chemical' pressure, and explains the initial the observations in the $LaO_{1-x}F_xFeAs$ and $SmFeAsO_{1-x}H_x$ systems. In the fluorine doped lanthanum iron oxyarsenide at ambient pressure the α angle is larger than the ideal, and when pressure is applied this compresses the cell and the iron environment, bringing it closer to ideal geometry

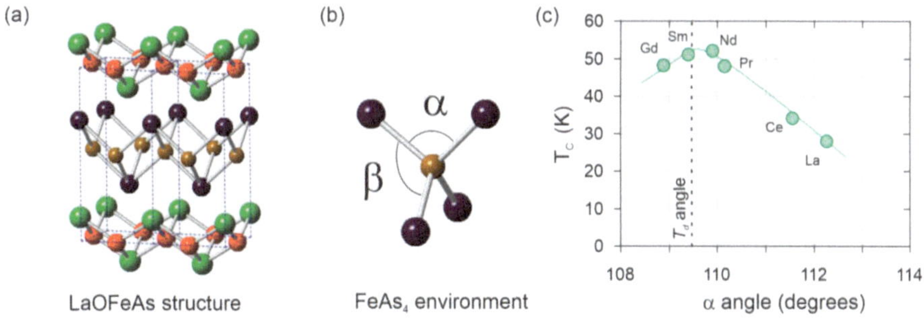

Figure 5.18 Possible dependence of the superconducting transition temperature on bond angle in the iron pnictide layer. (a) Left hand figure shows unit cell and structure of a typical LnOFeAs superconductor. (b) Changing the size of the lanthanide can affect the geometry of the iron arsenide layer, which can be measured using the α angle, or β angle for the As–Fe–As bonds. It has been hypothesised that higher transition temperatures can be achieved if the iron geometry is brought more closely to the ideal tetrahedral angle, *i.e.*, when $\alpha = \beta = 109.48°$. (c) Data supporting this are shown where the α angle (in samples at 10 K) is plotted against T_c for a number of LnOFeAs materials. Dotted line shows the ideal $\alpha = 109.48°$ value, with a clear decline in T_c values for samples further from this point.

and raising the critical temperature. In contrast the samarium containing material already has close to ideal geometry, so application of pressures distorts away from this, and lowers the transition temperature.

Superconductivity has also been identified in more complex structures with an expanded or larger blocking oxide layer. For example, a large number of compounds have been identified that adopt the $Sr_2CuGaSO_3$ structure. These have the advantage that the additional complexity of the blocking layer allows for greater opportunity for tuning of the structure. These include $Sr_4Sc_2O_6Fe_2P_2$, and $Sr_4V_2O_6Fe_2As_2$ with critical temperatures of 16 K and 26 K.[132] It was observed in the latter that increasing the pressure on the sample raised the critical temperature which, as with the ZrCuSiAs materials, was associated with an increased regularity of iron arsenide tetrahedra. This effect was utilised by Clarke *et al.*, who conducted doping studies on $Sr_4V_2O_6Fe_2As_2$ to partially substitute the Sr^{2+} ion with the smaller Ca^{2+} to apply 'chemical pressure' to reduce the size of the unit cell and change the iron arsenide geometry. At a dopant concentration of 5% they observed an increase in the critical temperature to 30 K with a closer approach to the ideal tetrahedra geometry. Over-doping from this point led to a distortion away from the ideal geometry, and a decrease in the critical temperature.[133] Initially these $Sr_2CuGaSO_3$ phases were thought to be able to display superconductivity without doping, however later studies indicated that they contain a small oxygen deficiency, for example $Sr_4V_2O_{6-\delta}Fe_2As_2$, which acts as an n-type dopant. The effect of the amount of oxide vacancy was studied, and the maximum T_c identified for $0.06 < \delta < 0.29$.[134] Superconductivity was also observed in $Sr_4MgTiO_6Fe_2As_2$, where the nominal B^{3+} site in the structure is occupied by equal amounts of Mg^{2+} and Ti^{4+} ions. The ratio of titanium can then be increased to provide n-type doping, while also modulating the lattice size, due to the difference in ionic radii between the magnesium (0.72 Å) and the titanium ions (0.61 Å) with a maximum T_c onset identified for $Sr_4Mg_{0.4}Ti_{1.6}O_6Fe_2As_2$ of 37 K.[135]

The investigation of iron pnictides with the $Sr_2CuGaSO_3$ structure shows that the superconductivity occurring within the layers is not suppressed by increasing the thickness of the blocking layers. Further investigations were conducted to push this to its limit by considering the homologous series of layered HAMs with an increasing thickness of *perovskite* blocks, with general formula $A_{n+1}B_nO_{3n-1}Fe_2As_2$. For the n = 3, 4, and 5 samples considered in the study, the nominal charge on the B site for electroneutrality needs to be +3.33, +3.5 and +3.6. This was achieved by using an appropriate mix of Sc^{3+} and Ti^{4+} ions on the B site, a ratio of 2 : 1, 1 : 1 and 2 : 3, respectively. These were all found to be superconducting with T_c values of between 33 K and 42 K, with the highest value for the material with the largest blocking layer, $Ca_6(Sc_{0.4}Ti_{0.6})_5O_{14}Fe_2As_2$, which has an interlayer gap between the iron arsenide layers of 24.6 Å. The observation of a high critical temperature in a compound with such a large spacing between the layers confirms that the superconductivity is controlled by the iron arsenide layer alone.

5.4 Summary

In this chapter, we have highlighted the importance of mixed anion compounds to modern materials science, and reviewed some of the more recent discoveries that have resulted from the study of heteroanionic materials, and how the advantages of using multiple anions have led to breakthroughs and an impetus in a number of functional areas, including photocatalysis and superconductivity. We have shown how it can be useful to divide the HAMs into two classes, those containing only light anions and those that also contain a heavy anion. We have reviewed their structures and highlighted how they differ from each other, and crucially how the presence of multiple anions can control their properties.

References

1. J. K. Harada, N. Charles, K. R. Poeppelmeier and J. M. Rondinelli, *Adv. Mater.*, 2019, **31**(19), 1805295.
2. S. Tippireddy, P. K. D S, S. Das and R. C. Mallik, *ACS Appl. Energy Mater.*, 2021, **4**(3), 2022.
3. S. G. Ebbinghaus, H.-P. Abicht, R. Dronskowski, T. Mueller, A. Reller and A. Weidenkaff, *Prog. Solid State Chem.*, 2009, **37**(2–3), 173.
4. S. J. Clarke, P. Adamson, S. J. C. Herkelrath, O. J. Rutt, D. R. Parker, M. J. Pitcher and C. F. Smura, *Inorg. Chem.*, 2008, **47**(19), 8473.
5. S. D. N. Luu and P. Vaqueiro, *J. Materiomics*, 2016, **2**(2), 131.
6. D. C. Johnston, *Adv. Phys.*, 2010, **59**(6), 803.
7. H. Kageyama, K. Hayashi, K. Maeda, J. P. Attfield, Z. Hiroi, J. M. Rondinelli and K. R. Poeppelmeier, *Nat. Commun.*, 2018, **9**(1), 772.
8. R. D. Shannon, *Acta Crystallogr., Sect. A*, 1976, **32**(SEP1), 751.
9. V. F. Sears, *Neutron News*, 1992, **3**(3), 26.
10. A. Fuertes, *Dalton Trans.*, 2010, **39**(26), 5942.
11. R. Marchand, R. Pastuszak, Y. Laurent and G. Roult, *Rev. Chim. Miner.*, 1982, **19**(6), 684.
12. A. Gomathi, S. Reshma and C. N. R. Rao, *J. Solid State Chem.*, 2009, **182**(1), 72.
13. S. J. Clarke, B. P. Guinot, C. W. Michie, M. J. C. Calmont and M. J. Rosseinsky, *Chem. Mater.*, 2001, **14**(1), 288.
14. M. Kakihana, *J. Ceram. Soc. Jpn.*, 2009, **117**(1368), 857.
15. F. Pors, R. Marchand, Y. Laurent, P. Bacher and G. Roult, *Mater. Res. Bull.*, 1988, **23**(10), 1447.
16. E. Gunther, R. Hagenmayer and M. Jansen, *Z. Anorg. Allg. Chem.*, 2000, **626**(7), 1519.
17. A. Hellwig and A. Hendry, *J. Mater. Sci.*, 1994, **29**(18), 4686.
18. S. H. Elder, F. J. DiSalvo, L. Topor and A. Navrotsky, *Chem. Mater.*, 2002, **5**(10), 1545.
19. M. de Respinis, M. Fravventura, F. F. Abdi, H. Schreuders, T. J. Savenije, W. A. Smith, B. Dam and R. van de Krol, *Chem. Mater.*, 2015, **27**, 7091.

20. M. Harb, P. Sautet, E. Nurlaela, P. Raybaud, L. Cavallo, K. Domen, J.-M. Basset and K. Takanabe, *Phys. Chem. Chem. Phys.*, 2014, **16**(38), 20548.
21. M. Hara, E. Chiba, A. Ishikawa, T. Takata, J. N. Kondo and K. Domen, *J. Phys. Chem. B*, 2003, **107**(48), 13441.
22. M. Hara, T. Takata, J. N. Kondo and K. Domen, *Catal. Today*, 2004, **90**(3–4), 313–317.
23. B. Wang, B. C. Chakoumakos, B. C. Sales and J. B. Bates, *J. Solid State Chem.*, 1996, **122**(2), 376.
24. R. Kiessling and L. Peterson, *Acta Metall.*, 1954, **2**(5), 675.
25. N. Schonberg, *Acta Chem. Scand.*, 1954, **8**(2), 208.
26. D. Armytage and B. E. F. Fender, *Acta Crystallogr., Sect. B: Struct. Sci.*, 1974, **30**(MAR15), 809.
27. X.-J. Wang, F. Krumeich, M. Wörle, R. Nesper, L. Jantsky and H. Fjellvåg, *Chem. – Eur. J.*, 2012, **18**(19), 5970.
28. S. Nakhal, W. Hermes, T. Ressler, R. Pöttgen and M. Lerch, *Z. Anorg. Allg. Chem.*, 2009, **635**(12), 2016.
29. O. Y. Wu, I. P. Parkin and G. Hyett, *J. Solid State Chem.*, 2012, **190**, 169.
30. H. X. Willems, G. de With, R. Metselaar, R. B. Helmholdt and K. K. Petersen, *J. Mater. Sci. Lett.*, 1993, **12**(18), 1470.
31. E. Soignard, D. Machon, P. F. McMillan, J. Dong, B. Xu and K. Leinenweber, *Chem. Mater.*, 2005, **17**(22), 5465.
32. I. G. Crouch, G. V. Franks, C. Tallon, S. Thomas and M. Naebe, *Glasses and Ceramics*, Butterworth-Heinemann, Oxford, 2017, pp. 331–393.
33. T. Motohashi, Y. Hamade, Y. Masubuchi, T. Takeda, K.-i. Murai, A. Yoshiasa and S. Kikkawa, Structural phase transition in the perovskite-type tantalum oxynitrides, $Ca_{1-x}Eu_xTa(O,N)_3$, *Mater. Res. Bull.*, 2009, **44**(9), 1899.
34. N. Cordes and W. Schnick, *Chem. – Eur. J.*, 2017, **23**(47), 11410.
35. J. Jacobs, F. Oehler, J. Schettlock and S. G. Ebbinghaus, *Z. Anorg. Allg. Chem.*, 2018, **644**(24), 1832.
36. S. H. Porter, Z. Huang and P. M. Woodward, *Cryst. Growth Des.*, 2014, **14**(1), 117.
37. K. Hibino, M. Yashima, T. Oshima, K. Fujii and K. Maeda, *Dalton Trans.*, 2017, **46**(43), 14947.
38. A. M. Glazer, *Acta Crystallogr., Sect. B*, 1972, **28**(11), 3384.
39. C. J. Howard and H. T. Stokes, *Acta Crystallogr., Sect. B: Struct. Sci.*, 1998, **54**, 782.
40. N. Cordes, T. Bräuniger and W. Schnick, *Eur. J. Inorg. Chem.*, 2018, **2018**(46), 5019.
41. Y.-I. Kim, P. M. Woodward, K. Z. Baba-Kishi and C. W. Tai, *Chem. Mater.*, 2004, **16**(7), 1267.
42. P. J. Camp, A. Fuertes and J. P. Attfield, *J. Am. Chem. Soc.*, 2012, **134**(15), 6762.
43. M. Yang, J. Oró-Solé, J. A. Rodgers, A. B. Jorge, A. Fuertes and J. P. Attfield, *Nat. Chem.*, 2011, **3**(1), 47.

44. S. J. Clarke, K. A. Hardstone, C. W. Michie and M. J. Rosseinsky, *Chem. Mater.*, 2002, **14**(6), 2664.
45. G. Tobías, D. Beltrán-Porter, O. I. Lebedev, G. Van Tendeloo, J. Rodríguez-Carvajal and A. Fuertes, *Inorg. Chem.*, 2004, **43**(25), 8010.
46. N. Cordes, M. Nentwig, L. Eisenburger, O. Oeckler and W. Schnick, Ammonothermal Synthesis of the Mixed-Valence Nitrogen-Rich Europium Tantalum Ruddlesden-Popper Phase $Eu^{II}Eu^{III}{}_2Ta_2N_4O_3$, *Eur. J. Inorg. Chem.*, 2019, **2019**(17), 2304.
47. F. Pors, R. Marchand and Y. Laurent, *J. Solid State Chem.*, 1993, **107**(1), 39.
48. M. J. Martínez-Lope, M. T. Casais and J. A. Alonso, *Z. Naturforsch., B: J. Chem. Sci.*, 2006, **61**(2), 164.
49. B. Anke, S. Hund, C. Lorent, O. Janka, T. Block, R. Pöttgen and M. Lerch, *Z. Anorg. Allg. Chem.*, 2017, **643**(23), 1824.
50. G. M. Veith, M. Greenblatt, M. Croft and J. B. Goodenough, *Mater. Res. Bull.*, 2001, **36**(7), 1521.
51. P. Bacher, G. Roult, M. Ghers, O. Merdrignac, J. Guyader and Y. Laurent, *Mater. Chem. Phys.*, 1989, **21**(3), 223.
52. S. Kikkawa, K. Nagasaka, T. Takeda, M. Bailey, T. Sakurai and Y. Miyamoto, *J. Solid State Chem.*, 2007, **180**(7), 1984.
53. D. P. Chen, J. C. Neuefeind, K. M. Koczkur, D. L. Bish and S. E. Skrabalak, *Chem. Mater.*, 2017, **29**(15), 6525.
54. H. Müller, W. Müller, M. Wehner and H. Liewald, Artists' Colors, in *Ullmann's Encyclopedia of Industrial Chemistry*, Wiley-VCH, Weinheim, 2003.
55. A. Pan, R. Liu, F. Wang, S. Xie, B. Zou, M. Zacharias and Z. L. Wang, *J. Phys. Chem. B*, 2006, **110**(45), 22313.
56. G. Genchi, M. S. Sinicropi, G. Lauria, A. Carocci and A. Catalano, *Int. J. Environ. Res. Public Health*, 2020, **17**(11), 3782.
57. W.-J. Chun, A. Ishikawa, H. Fujisawa, T. Takata, J. N. Kondo, M. Hara, M. Kawai, Y. Matsumoto and K. Domen, *J. Phys. Chem. B*, 2003, **107**(8), 1798.
58. M. Jansen and H. P. Letschert, Inorganic yellow-red pigments without toxic metals, *Nature*, 2000, **404**(6781), 980.
59. R. Aguiar, D. Logvinovich, A. Weidenkaff, A. Rachel, A. Reller and S. G. Ebbinghaus, *Dyes Pigm.*, 2008, **76**(1), 70.
60. A. Mills, A. Lepre, N. Elliott, S. Bhopal, I. P. Parkin and S. A. O'Neill, *J. Photochem. Photobiol., A*, 2003, **160**(3), 213.
61. D. Wu, B. Wang, W. Wang, T. An, G. Li, T. W. Ng, H. Y. Yip, C. Xiong, H. K. Lee and P. K. Wong, *J. Mater. Chem. A*, 2015, **3**(29), 15148.
62. L. Ye, Y. Su, X. Jin, H. Xie and C. Zhang, *Environ. Sci.: Nano*, 2014, **1**(2), 90.
63. K. Sunada, T. Watanabe and K. Hashimoto, *J. Photochem. Photobiol., A*, 2003, **156**(1–3), 227.
64. K. Maeda and K. Domen, *J. Phys. Chem. C*, 2007, **111**(22), 7851.
65. A. Fujishima and K. Honda, *Nature*, 1972, **238**(5358), 37.

66. J. R. Bolton, S. J. Strickler and J. S. Connolly, *Nature*, 1985, **316**(6028), 495.
67. M. Hara, G. Hitoki, T. Takata, J. N. Kondo, H. Kobayashi and K. Domen, TaON and Ta_3N_5 as new visible light driven photocatalysts, *Catal. Today*, 2003, **78**(1), 555.
68. S. D. Cosham, V. Celorrio, A. N. Kulak and G. Hyett, *Dalton Trans.*, 2019, **48**(28), 10619.
69. H. A. N. Dharmagunawardhane, W. R. Woerner, Q. Wu, H. Huang, X. Chen, A. Orlov, P. G. Khalifah and J. B. Parise, *J. Mater. Chem. A*, 2014, **2**(45), 19247.
70. T. Hisatomi, K. Hasegawa, K. Teramura, T. Takata, M. Hara and K. Domen, *Chem. Lett.*, 2007, **36**(4), 558.
71. D. Yamasita, T. Takata, M. Hara, J. N. Kondo and K. Domen, *Solid State Ionics*, 2004, **172**(1–4), 591.
72. G. Hitoki, T. Takata, J. N. Kondo, M. Hara, H. Kobayashi and K. Domen, *Electrochemistry*, 2002, **70**(6), 463.
73. L. P. Wan, F. Q. Xiong, B. X. Zhang, R. X. Che, Y. Li and M. H. Yang, *J. Energy Chem.*, 2018, **27**(2), 367.
74. A. Kasahara, K. Nukumizu, G. Hitoki, T. Takata, J. N. Kondo, M. Hara, H. Kobayashi and K. Domen, *J. Phys. Chem. A*, 2002, **106**(29), 6750.
75. H. Kato, K. Ueda, M. Kobayashi and M. Kakihana, *J. Mater. Chem. A*, 2015, (3), 11824.
76. T. Matoba, K. Maeda and K. Domen, *Chem. – Eur. J.*, 2011, **17**(52), 14731.
77. C. Pan, T. Takata, M. Nakabayashi, T. Matsumoto, N. Shibata, Y. Ikuhara and K. Domen, *Angew. Chem., Int. Ed.*, 2015, **54**(10), 2955.
78. C. Pan, T. Takata and K. Domen, *Chem. – Eur. J.*, 2016, **22**(5), 1854.
79. J. S. Gardner, G. Balakrishnan and D. M. Paul, *Phys. C*, 1995, **252**(3), 303.
80. G. Hyett, Z. A. Gal, C. F. Smura and S. J. Clarke, *Chem. Mater.*, 2008, **20**(2), 559.
81. S. J. C. Herkelrath, J. N. Blandy and S. J. Clarke, *J. Solid State Chem.*, 2018, **264**, 119.
82. W. J. Zhu and P. H. Hor, *Inorg. Chem.*, 1997, **36**(17), 3576.
83. Y. Luo, H. Han, H. Tan, X. Lin, Y. Li, S. Jiang, C. Feng, J. Dai, G. Cao, Z. a. Xu and S. Li, *J. Phys.: Condens. Matter*, 2011, **23**(17), 175701.
84. H. Lincke, R. Glaum, V. Dittrich and M. H. Moller, *Z. Anorg. Allg. Chem.*, 2009, **635**(6–7), 936.
85. G. B. Jin, A. D. Raw, S. Skanthakumar, R. G. Haire, L. Soderholm and J. A. Ibers, *J. Solid State Chem.*, 2010, **183**(3), 547.
86. F. Jellinek, *Acta Chem. Scand.*, 1962, **16**, 791.
87. K.-J. Range, K. G. Lange and A. Gietl, *J. Less-Common Met.*, 1990, **158**(1), 137.
88. R. Pöttgen and D. Johrendt, *Z. Naturforsch., B: J. Chem. Sci.*, 2008, **63**(10), 1135.
89. T. Sambrook, C. F. Smura, S. J. Clarke, K. M. Ok and P. S. Halasyamani, *Inorg. Chem.*, 2007, **46**(7), 2571.
90. E. J. T. Salter, J. N. Blandy and S. J. Clarke, *Inorg. Chem.*, 2016, **55**(4), 1697.

91. N. Ni, S. Jia, Q. Huang, E. Climent-Pascual and R. J. Cava, *Phys. Rev. B: Condens. Matter Mater. Phys.*, 2011, **83**(22), 224403.
92. M. Goga, R. Seshadri, V. Ksenofontov, P. Gütlich and W. Tremel, *Chem. Commun.*, 1999, (11), 979.
93. W. J. Zhu and P. H. Hor, *J. Solid State Chem.*, 1997, **130**(2), 319.
94. T. Ozawa, M. M. Olmstead, S. L. Brock, S. M. Kauzlarich and D. M. Young, *Chem. Mater.*, 1998, **10**(1), 392.
95. W. J. Zhu and P. H. Hor, *J. Solid State Chem.*, 1997, **134**(1), 128.
96. W. J. Zhu and P. H. Hor, *J. Solid State Chem.*, 2000, **153**(1), 26.
97. G. Hyett, N. Barrier, S. J. Clarke and J. Hadermann, *J. Am. Chem. Soc.*, 2007, **129**(36), 11192.
98. H. Ogino, S. Sato, K. Kishio, J. Shimoyama, T. Tohei and Y. Ikuhara, *Appl. Phys. Lett.*, 2010, **97**(7), 3.
99. Z. A. Gál, O. J. Rutt, C. F. Smura, T. P. Overton, N. Barrier, S. J. Clarke and J. Hadermann, *J. Am. Chem. Soc.*, 2006, **128**(26), 8530.
100. D. O. Charkin, A. V. Sadakov, O. E. Omel'yanovskii and S. M. Kazakov, *Mater. Res. Bull.*, 2010, **45**(12), 2012.
101. K. Ellmer, *Nat. Photonics*, 2012, **6**, 809.
102. C. G. Granqvist, *Int. J. Nanotechnol.*, 2009, **6**(9), 785.
103. P. P. Edwards, A. Porch, M. O. Jones, D. V. Morgan and R. M. Perks, *Dalton Trans.*, 2004, (19), 2995.
104. H. Kawazoe, M. Yasukawa, H. Hyodo, M. Kurita, H. Yanagi and H. Hosono, *Nature*, 1997, **389**(6654), 939.
105. K. Ueda, T. Hase, H. Yanagi, H. Kawazoe, H. Hosono, H. Ohta, M. Orita and M. Hirano, *J. Appl. Phys.*, 2001, **89**(3), 1790.
106. M. K. Jayaraj, A. D. Draeseke, J. Tate and A. W. Sleight, *Thin Solid Films*, 2001, **397**(1), 244.
107. K. Ueda, H. Hiramatsu, M. Hirano, T. Kamiya and H. Hosono, *Thin Solid Films*, 2006, **496**(1), 8.
108. M. L. Liu, L. B. Wu, F. Q. Huang, L. D. Chen and J. A. Ibers, *J. Solid State Chem.*, 2007, **180**(1), 62.
109. H. Hiramatsu, M. Orita, M. Hirano, K. Ueda and H. Hosono, *J. Appl. Phys.*, 2002, **91**(11), 9177.
110. H. Hiramatsu, K. Ueda, H. Ohta, M. Hirano, T. Kamiya and H. Hosono, *Appl. Phys. Lett.*, 2003, **82**(7), 1048.
111. K. Ueda, S. Hirose, H. Kawazoe and H. Hosono, *Chem. Mater.*, 2001, **13**(5), 1880.
112. M.-L. Liu, L.-B. Wu, F.-Q. Huang, L.-D. Chen and I.-W. Chen, *J. Appl. Phys.*, 2007, **102**(11), 116108.
113. B. A. D. Williamson, G. J. Limburn, G. W. Watson, G. Hyett and D. O. Scanlon, *Matter*, 2020, **3**(3), 759.
114. D. O. Scanlon and G. W. Watson, *Chem. Mater.*, 2009, **21**(22), 5435.
115. Y. Kamihara, H. Hiramatsu, M. Hirano, R. Kawamura, H. Yanagi, T. Kamiya and H. Hosono, *J. Am. Chem. Soc.*, 2006, **128**(31), 10012.
116. Y. Kamihara, T. Watanabe, M. Hirano and H. Hosono, *J. Am. Chem. Soc.*, 2008, **130**(11), 3296.

117. R. Zhi-An, L. Wei, Y. Jie, Y. Wei, S. Xiao-Li, C. Zheng, C. Guang-Can, D. Xiao-Li, S. Li-Ling, Z. Fang and Z. Zhong-Xian, *Chin. Phys. Lett.*, 2008, **25**(6), 2215.
118. T. Hanna, Y. Muraba, S. Matsuishi, N. Igawa, K. Kodama, S.-i. Shamoto and H. Hosono, *Phys. Rev. B: Condens. Matter Mater. Phys.*, 2011, **84**(2), 024521.
119. H. Hosono, K. Tanabe, E. Takayama-Muromachi, H. Kageyama, S. Yamanaka, H. Kumakura, M. Nohara, H. Hiramatsu and S. Fujitsu, *Sci. Technol. Adv. Mater.*, 2015, **16**(3), 033503.
120. D. Johrendt, *J. Mater. Chem.*, 2011, **21**(36), 13726.
121. D. v. Delft and P. Kes, *Phys. Today*, 2010, **63**(9), 38.
122. W. Meissner and R. Ochsenfeld, *Naturwissenschaften*, 1933, **21**(44), 787.
123. J. Bardeen, L. N. Cooper and J. R. Schrieffer, *Phys. Rev.*, 1957, **108**(5), 1175.
124. L. N. Cooper, *Phys. Rev.*, 1956, **104**(4), 1189.
125. M. K. Wu, J. R. Ashburn, C. J. Torng, P. H. Hor, R. L. Meng, L. Gao, Z. J. Huang, Y. Q. Wang and C. W. Chu, *Phys. Rev. Lett.*, 1987, **58**(9), 908.
126. C. C. Tsuei and J. R. Kirtley, *Rev. Mod. Phys.*, 2000, **72**(4), 969.
127. A. Schilling, M. Cantoni, J. D. Guo and H. R. Ott, *Nature*, 1993, **363**(6424), 56.
128. I. I. Mazin, D. J. Singh, M. D. Johannes and M. H. Du, *Phys. Rev. Lett.*, 2008, **101**(5), 057003.
129. G. Wu, H. Chen, Y. L. Xie, Y. J. Yan, T. Wu, R. H. Liu, X. F. Wang, D. F. Fang, J. J. Ying and X. H. Chen, *Phys. Rev. B: Condens. Matter Mater. Phys.*, 2008, **78**(9), 092503.
130. A. S. Sefat, A. Huq, M. A. McGuire, R. Jin, B. C. Sales, D. Mandrus, L. M. D. Cranswick, P. W. Stephens and K. H. Stone, *Phys. Rev. B: Condens. Matter Mater. Phys.*, 2008, **78**(10), 104505.
131. C. H. Lee, A. Iyo, H. Eisaki, H. Kito, M. T. Fernandez-Diaz, R. Kumai, K. Miyazawa, K. Kihou, H. Matsuhata, M. Braden and K. Yamada, *J. Phys. Soc. Jpn.*, 2008, **77**, 44.
132. H. Kotegawa, T. Kawazoe, H. Tou, K. Murata, H. Ogino, K. Kishio and J.-i. Shimoyama, *J. Phys. Soc. Jpn.*, 2009, **78**(12), 123707.
133. A. J. Corkett, D. G. Free, S. J. Cassidy, S. Ramos and S. J. Clarke, *J. Solid State Chem.*, 2014, **216**, 91.
134. Y. Tojo, T. Shibuya, T. Nakamura, K. Shoji, H. Fujioka, M. Matoba, S. Yasui, M. Itoh, S. Iimura, H. Hiramatsu, H. Hosono, S. Hirai, W. Mao, S. Kitao, M. Seto and Y. Kamihara, *J. Phys.: Condens. Matter*, 2019, **31**(11), 115801.
135. S. Sato, H. Ogino, N. Kawaguchi, Y. Katsura, K. Kishio, J. Shimoyama, H. Kotegawa and H. Tou, *Supercond. Sci. Technol.*, 2010, **23**(4), 5.

Subject Index

References to figures are given in *italic* type. References to tables are indicated by **bold** type.